T0202275

15

CRM
SERIES

Centro
di Ricerca
Matematica
Ennio De Giorgi

Geometric Partial Differential Equations
proceedings

edited by
Antonin Chambolle, Matteo Novaga
and Enrico Valdinoci

EDIZIONI
DELLA
NORMALE

© 2013 Scuola Normale Superiore Pisa

ISBN 978-88-7642-472-4
e-ISBN 978-88-7642-473-1

Contents

Preface

This book is the outcome of a conference held in September 2012 at the Centro De Giorgi of the Scuola Normale in Pisa. The aim of the conference was to discuss recent results on nonlinear analysis, and more specifically on geometric partial differential equations.

These equations arise in several models from physics, biology, material science, image processing and applied mathematics in general, and attracted a lot of attention in recent years. The typical question is studying a set or a function with particular properties (isoperimetric problems, shape optimization, free boundary problems, minimal surfaces), or the way a set or a function evolves (front propagation, mean curvature flow, total variation flow, travelling waves, Ginzburg-Landau problems).

A characteristic feature of these problems is the ubiquitous onset of singularities, and a detailed analysis of this phenomenon is often a very difficult task. As classical solutions are generally smooth, the presence of singularities motivates the definition of weak solutions, among which we mention viscosity solutions, varifold solutions and currents. The analysis of these objects requires the development of advanced tools from different branches of mathematics, such as calculus of variations, geometric measure theory, partial differential equations and differential geometry.

In this book, particular attention is paid to the study of special self-similar solutions, such as solitons, homothetic self-shrinkers and travelling waves, which are crucial for understanding both the nature of singularities and the asymptotic behaviour of solutions. Qualitative properties of solutions, such as convexity or monotonicity, are also considered.

Antonin Chambolle
Matteo Novaga
Enrico Valdinoci

Authors' affiliations

N. D. ALIKAKOS – Department of Mathematics, University of Athens, Panepistemiopolis, 15784 Athens, Greece and Institute for Applied and
Computational Mathematics, Foundation of Research and Technology – Hellas, 71110 Heraklion, Crete, Greece
nalikako@math.uoa.gr

S. AMATO – SISSA, Via Bonomea 265, 34136, Trieste, Italia
samato@sissa.it

G. BELLETTINI – Dipartimento di Matematica, Università di Roma Tor Vergata, via della Ricerca Scientifica 1, 00133 Roma, Italia
and
INFN Laboratori Nazionali di Frascati (LNF), via E. Fermi 40, Frascati 00044 Roma, Italia
belletti@mat.uniroma2.it

A. CHAMBOLLE – CMAP, Ecole Polytechnique, CNRS, 91128 Palaiseau Cedex, France
antonin.chambolle@cmap.polytechnique.fr

S. CHOI – Department of Mathematics, University of Arizona, 617 N. Santa Rita Ave., Tucson, AZ 85721-0089, USA
schoi@math.arizona.edu

D. CHRISTODOULOU – Departement Mathematik, ETH, Rämistrasse 101, 8092 Zürich, Switzerland
demetri.christodoulou@math.ethz.ch

L. DUPAIGNE – LAMFA, UMR CNRS 7352, Université Picardie Jules Verne, 33, rue St Leu, 80039 Amiens, France
louis.dupaigne@math.cnrs.fr

A. FARINA – LAMFA, UMR CNRS 7352, Université Picardie Jules Verne 33, rue St Leu, 80039 Amiens, France

MI-HO GIGA – Graduate School of Mathematical Science, University of Tokyo, Komaba 3-8-1, Meguro-ku, Tokyo 153-8914, Japan
mihogiga@ms.u-tokyo.ac.jp

Y. GIGA – Graduate School of Mathematical Science, University of Tokyo, Komaba 3-8-1, Meguro-ku, Tokyo 153-8914, Japan
labgiga@ms.u-tokyo.ac.jp

M. GOLDMAN – Max Planck Institute for Mathematics in the Sciences, Inselstrasse 22, 04103 Leipzig, Germany
goldman@mis.mpg.de

I. C. KIM – Department of Mathematics, UCLA, Los Angeles, CA 90095-1555, USA
ikim@math.ucla.edu

A. LEMENANT – LJLL, Université Paris-Diderot, CNRS, 75205 Paris Cedex 13, France
lemenant@ljll.univ-paris-diderot.fr

M. MORINI – Dipartimento di Matematica, Università di Parma, Parco Area delle Scienze 53/a, 43124 Parma, Italia
massimiliano.morini@unipr.it

A. NAKAYASU – Graduate School of Mathematical Science, University of Tokyo, Komaba 3-8-1, Meguro-ku, Tokyo 153-8914, Japan
ankys@ms.u-tokyo.ac.jp

M. NOVAGA – Dipartimento di Matematica, Università di Pisa, Largo Bruno Pontecorvo 5, 56127 Pisa, Italia
novaga@dm.unipi.it

G. ORLANDI – Dipartimento di Informatica, Università di Verona, Strada le Grazie 15, 37134 Verona, Italia
giandomenico.orlandi@univr.it

M. Paolini – Dipartimento di Matematica, Università Cattolica "Sacro Cuore", Via Trieste 17, 25121 Brescia, Italia
paolini@dmf.unicatt.it

A. Pisante – Dipartimento di Matematica, Sapienza Università di Roma, Piazzale Aldo Moro, 5, 00185 Roma, Italia
pisante@mat.uniroma1.it

G. Pisante – Dipartimento di Matematica e Fisica, Seconda Università degli Studi di Napoli, Viale Lincoln, 5, 81100 Caserta, Italia
giovanni.pisante@unina2.it

M. Ponsiglione – Dipartimento di Matematica, Sapienza Università di Roma, Piazzale Aldo Moro 5, 00185 Roma, Italia
ponsigli@mat.uniroma1.it

B. Sirakov – Pontifícia Universidade Católica do Rio de Janeiro Departamento de Matemática, Rua Marquês de São Vicente, 225, Gávea Rio de Janeiro - RJ, CEP 22451-900, Brasil
bsirakov@mat.puc-rio.br

Y. Sire – LATP, Université Aix-Marseille, CNRS, Technopole de Chateau-Gombert, 9, rue F. Joliot Curie, 13453 Marseille Cedex 13, France
sire@cmi.univ-mrs.fr

A. Stancu – Department of Mathematics and Statistics, Concordia University, Montreal, QC, Canada, H3G 1M8
stancu@mathstat.concordia.ca

On the structure of phase transition maps for three or more coexisting phases

Nicholas D. Alikakos

Abstract. We consider system (1.1) below. After stating certain general facts which do not depend on the structure of W, we focus on the phase transition case. For symmetric W's the main issues have been resolved, and we summarize them here. Next we recall the De Giorgi conjecture in the scalar case, and we point to a Bernstein type theorem that appears appropriate for (1.1). Finally we state a result on the hierarchical structure of the equivariant solutions.

1 Introduction

This paper is partly based on a lecture delivered by the author at the ERC workshop "Geometric Partial Differential Equations" held in Pisa in September 2012. What is presented in the following is an expanded version of that lecture.

Specifically, we consider the system

$$\Delta u - W_u(u) = 0, \text{ for } u : \mathbb{R}^n \to \mathbb{R}^m, \tag{1.1}$$

with $W \in C^2(\mathbb{R}^m; \mathbb{R}_+)$, $W \geq 0$, where $W_u := (\partial W/\partial u_1, \ldots, \partial W/\partial u_n)^\top$, and occasionally with additional hypotheses introduced later on. We refer the reader to Part I in [9] for general information and motivation for system (1.1).

The paper is organized as follows. In Section 2 we present the basics of the problem for general potentials, in Section 3 we study symmetric potentials for the phase transition model and establish the existence of equivariant connection maps, in Sections 4 and 5 we present and prove a related Bernstein-type theorem, and in Section 6 we discuss the hierarchical structure of the equivariant connection maps. Sections 2, 3, and 6 are restricted to statements of results with explanations but without proofs, referring to published papers or preprints for the details. In contrast, in Sections 4 and 5 we give the background and detailed proofs.

The author was partially supported through the project PDEGE – Partial Differential Equations Motivated by Geometric Evolution, co-financed by the European Union – European Social Fund (ESF) and national resources, in the framework of the program Aristeia of the 'Operational Program Education and Lifelong Learning' of the National Strategic Reference Framework (NSRF).

ACKNOWLEDGEMENTS. The author would like to acknowledge the warm hospitality of the Department of Mathematics of Stanford University in the spring semester of 2012, during which part of this paper was written. Special thanks are due to Rafe Mazzeo, George Papanicolaou, Lenya Ryzhik, Rick Schoen, and Brian White. Section 4 is very much influenced from discussions with Rick Schoen and lectures and material provided by Brian White.

2 The basics for general potentials

We recall some known facts for (1.1). The system is the Euler–Lagrange equation for the *free energy functional*

$$J(u; \mathbb{R}^n) := \int_{\mathbb{R}^n} \left(\frac{1}{2}|\nabla u|^2 + W(u) \right) \mathrm{d}x,$$

where $\nabla u = (\partial u_i / \partial u_j)$, for $i = 1, \dots, m$, $j = 1, \dots, n$, and $| \cdot |$ is the Euclidean norm of the matrix.

One important difference of (1.1) with its scalar counterpart (for $m = 1$) is that in that case the structure of bounded entire solutions does not depend very much on W. In contrast, for the system there two distinguished examples with very distinct behavior: The class of *phase transition potentials*, that is, W's with a finite number of global minima (wells) $a_1, \dots, a_N$, with $W(a_i) = 0$, and the class of *Ginzburg–Landau potentials*, for example, the potential $W(u) = \frac{1}{4}(|u|^2 - 1)^2$ (disconnected versus connected zero sets). In the phase transition case and under suitable rescaling the free energy concentrates on minimal hypersurfaces or Plateau complexes (see [14]), while in the Ginzburg–Landau case it concentrates on higher-codimension objects known as vortices, and otherwise the solution converges to a harmonic map (see [18]).

Equation (1.1) can be written as a divergence-free condition, that is,

$$\operatorname{div} T = (\nabla u)^\top (\Delta u - W_u(u)) = 0,$$

for the stress-energy tensor

$$T_{ij}(u, \nabla u) := u_{,i} \cdot u_{,j} - \delta_{ij} \left(\frac{1}{2}|\nabla u|^2 + W(u) \right).[1]$$

In this context it was introduced in [3], but as it turns out it is a particularization of a general formalism well-known to the physicists [40].

[1] The sharp-interface limit $T_\varepsilon \to T_0 = \sigma(\nabla d \otimes \nabla d - \mathrm{Id})$ is the orthogonal projection to the tangent space of the interface S separating the two phases, where $d = \operatorname{dist}(x, S)$ and σ is the associated interface energy.

The divergence-free formulation has certain important consequences. For example, one can derive the monotonicity formula

$$\frac{\mathrm{d}}{\mathrm{d}R}\left(\frac{1}{R^{n-2}}\int_{|x-x_0|<R}\left(\frac{1}{2}|\nabla u|^2+W(u)\right)\mathrm{d}x\right)\geq 0,$$

from which Liouville-type theorems follow (see [3]). For instance,

$$\begin{cases} J(u;\, B_R)=o(R^{n-2}) \text{ as } R\to+\infty, \text{ for } n\geq 3, & \text{implies } u\equiv \text{constant},\\ J(u;\, B_R)=o(\log R) \text{ as } R\to+\infty, \text{ for } n=2, & \text{implies } u\equiv \text{constant}. \end{cases}$$

In particular,

$$\int_{\mathbb{R}^n}\left(\frac{1}{2}|\nabla u|^2+W(u)\right)\mathrm{d}x<+\infty \quad \text{implies} \quad u\equiv \text{constant}, \tag{2.1}$$

for $n\geq 2$. Note that (2.1) is a source of difficulty for constructing solutions to (1.1) via the direct method. It was Farina [27] who first derived the monotonicity formula above and its implication (2.1) in the context of the Ginzburg–Landau system and Modica [41] who derived (2.1) for $m=1$.

In the scalar ODE case ($n=1$, $m=1$), for solutions of (1.1) with limits at infinity, one has the elementary *equipartition relation*

$$\frac{1}{2}|u_x|^2=W(u).$$

In the scalar PDE case ($n\geq 2$, $m=1$), Modica [41] established the estimate

$$\frac{1}{2}|\nabla u|^2\leq W(u), \tag{2.2}$$

(see also [20]). The analog of estimate (2.2) is false for systems in general. All the known counterexamples (see [27, pages 389–390]) involve Ginzburg–Landau potentials. One implication of (2.2) would be the stronger monotonicity formula

$$\frac{\mathrm{d}}{\mathrm{d}R}\left(\frac{1}{R^{n-1}}\int_{|x-x_0|<R}\left(\frac{1}{2}|\nabla u|^2+W(u)\right)\mathrm{d}x\right)\geq 0,$$

already known for the scalar case (see [42]).

Another implication of the divergence-free formulation is a Pohozaev-type identity (see [7])

$$\frac{n-2}{2}\int_\Omega|\nabla u|^2\,\mathrm{d}x+n\int_\Omega W(u)\,\mathrm{d}x+\frac{1}{2}\int_{\partial\Omega}(x-x_0)\cdot\nu\,|\nabla u|^2\,\mathrm{d}S=0,$$

where ν is the outward normal and $x_0 \in \Omega$ arbitrary, for solutions of the system

$$\begin{cases} \Delta u - W_u(u) = 0, & \text{in } \Omega \subset \mathbb{R}^n, \\ u = a, & \text{on } \partial\Omega, \text{ with } W(a) = 0. \end{cases}$$

Gui [36] has developed certain identities for the system, which he calls 'Hamiltonian', and points out their relationship with the classical Pohozaev identity (see, for example, [26]). His identities can also be derived via the stress-energy tensor. Here is a sample: Let $n = 2$, $m = 2$, and let u be a solution to (1.1) satisfying the estimate

$$|u(x_1, x_2) - a_\pm| \leq C\,\mathrm{e}^{-c|x_1|}, \text{ for all } x_2 \in [M, N],$$

with $-\infty \leq M, N \leq +\infty$ and $W(a_\pm) = 0$. Then,

$$\int_{\mathbb{R}} \left(\frac{1}{2} \left(|u_{x_1}(x_1, x_2)|^2 - |u_{x_2}(x_1, x_2)|^2 \right) - W(u(x_1, x_2)) \right) \mathrm{d}x_1 = \text{constant},$$

for all $x_2 \in (M, N)$.

Having asymptotic information on the solution along certain hyperplanes as $|x| \to +\infty$ and also convergence to a minimum of W away from them (see Section 4), one can measure the flux of the stress-energy tensor over large spheres in order to derive balance conditions relating the angles between the hyperplanes, thus deriving rigidity-type results. For the phase transition case and for a triple-well potential, Gui [36] has derived such a result in the planar case $n = 2$, $m = 2$, thus relating the angles of a triple junction to the surface energies. This was extended to the three-dimensional case $n = 3$, $m = 3$, in [5]. Related also is the work of Kowalczyk, Liu, and Pacard [39].

3 Symmetric phase transition potentials Existence of equivariant connection maps

In this section we restrict ourselves to the phase transition case for potentials that respect the symmetries of a finite reflection group G acting on $\mathbb{R}^n$ (see [35]) and we look for equivariant solutions

$$u(gx) = gu(x), \text{ for all } x \in \mathbb{R}^n \text{ and } g \in G.$$

The first results in this direction are due to Bronsard, Gui, and Schatzman [19] for $n = 2$, $m = 2$, and G the group of reflections of the equilateral triangle. Later, the work was extended by Gui and Schatzman [37] to $n = 3$, $m = 3$, and G the group of symmetries of the regular tetrahedron. These two special groups are particularly important as they are related to triple junctions on the plane and to quadruple junctions in three-dimensional space, which are minimal objects (cones) for the related sharp-interface problem.

In work with Fusco [10] we considered the general case of a reflection group and looked for an abstract result. Consider the following very general hypotheses.

Hypothesis 1 (N nondegenerate global minima). *The potential W is of class C^2 and satisfies $W(a_i) = 0$, for $i = 1, \dots, N$, and $W > 0$ on $\mathbb{R}^m \setminus \{a_1, \dots a_N\}$. Furthermore, there holds $v^\top \partial^2 W(u)\, v \geq 2c^2|v|^2$, for $v \in \mathbb{R}^m$ and $i = 1, \dots, N$.*

Hypothesis 2 (Symmetry). *The potential W is invariant under a finite reflection group G acting on $\mathbb{R}^m$ (Coxeter group), that is,*

$$W(gu) = W(u), \text{ for all } g \in G \text{ and } u \in \mathbb{R}^m.$$

Moreover, we assume that there exists $M > 0$ such that $W(su) \geq W(u)$, for $s \geq 1$ and $|u| = M$.

Hypothesis 3 (Location and number of global minima). *Let $F \subset \mathbb{R}^m$ be a fundamental region of G. We assume that the closure $\overline{F}$ contains a single global minimum of W, say a_1, and let G_{a_1} be the subgroup of G that leaves a_1 fixed.*

We set

$$D = \mathrm{Int}\{\cup g\overline{F} \mid g \in G_{a_1}\},$$

and notice that by the invariance of W it follows that the number of minima of W is

$$N = \frac{|G|}{|G_{a_1}|},$$

where here $|\cdot|$ is the order of the group.

We recall from [10] several examples. For $G = \mathcal{H}_2^3$, the group of symmetries of the equilateral triangle on the plane, we can take as F the $\frac{\pi}{3}$ sector. If $a_1 \in F$, then $N = 6$, while if a_1 is on the walls, then $N = 3$. In higher dimensions we have more options since we can place a_1 in the interior of $\overline{F}$, in the interior of a face, on an edge, and so on. For example, if $G = \mathcal{W}^*$, the group of symmetries of the cube in three-dimensional space, then $|G| = 48$. If the cube is situated with its center at the origin and its vertices at the eight points $(\pm 1, \pm 1, \pm 1)$, then we can take as F the simplex generated by $s_1 = e_1 + e_2 + e_3$, $s_2 = e_2 + e_3$, and $s_3 = e_3$, where the e_i's are the standard basis vectors. We have then the following options:

(i) At the origin, $N = 1$.
(ii) On the edge s_3, $N = 6$.

(iii) On the edge s_1, $N = 8$.
(iv) On the edge s_2, $N = 12$.
(v) In the interior of a face, $N = 24$.
(vi) In the interior of the fundamental region, $N = 48$.

We have the following theorem.

Theorem 3.1 ([10, 4, 33]). *Under Hypotheses* 1–3, *there exists a classical entire equivariant solution* $u : \mathbb{R}^n \to \mathbb{R}^m$ *to system* (1.1) *such that*

(i) $|u(x) - a_1| \leq K\mathrm{e}^{-k\,\mathrm{dist}(x,\partial D)}$, *for* $x \in D$ *and for positive constants* k, K,

(ii) $u(\overline{F}) \subset \overline{F}$ *and* $u(D) \subset D$ *(positivity).*[2]

As a consequence of (i), *the solution* u connects *the* $N = |G|/|G_{a_1}|$ *global minima of* W *in the sense that*

$$\lim_{\lambda \to +\infty} u(\lambda g \eta) = g a_1, \text{ for all } g \in G,$$

uniformly for η *in compact subsets of* $D \cap \mathbb{S}^{n-1}$.

Remark 3.2. We need a clarification concerning the dimensions n, m and the group G that is acting by Hypothesis 2 on the target but also on the domain since the solutions are equivariant. If $n \geq m$, then the group G can be embedded in the domain space via a natural homomorphism. For example, consider $n = 3$, $m = 2$, and G the group of symmetries of the equilateral triangle. On the other hand, if $n < m$, the existence of such a homomorphism is more problematic and in general there is no such embedding. For example, consider $n = 2$, $m = 3$, and take as G the group associated to the tetrahedron. For relevant information we refer to Bates, Fusco, and Smyrnelis [15]. Our notation $u(\overline{F})$ and $u(\lambda g\eta)$ tacitly assumes the homomorphism in the case $n \neq m$.

The theorem above was proved in Alikakos and Fusco [10] under an additional hypothesis. Subsequently, the author gave a simplified proof in [4] and, finally, in Fusco [33] the extra hypothesis was removed and the theorem was proved under the hypotheses above.

As it was mentioned in (2.1), there holds

$$J(u; \mathbb{R}^n) = +\infty, \text{ where } J(u, \Omega) = \int_\Omega \left(\frac{1}{2}|\nabla u|^2 + W(u)\right) \mathrm{d}x,$$

[2] Smyrnelis has established that $u(F) \subset F$ for certain groups (personal communication).

for the solution constructed above. However, the solutions are constructed variationally and possess the following minimization property (see [11]), which defines the notion of a *local minimizer* (*cf.* [2]), that is,

$$J(u;\Omega) = \min J(v;\Omega), \text{ such that } v = u \text{ on } \partial\Omega, \tag{3.1}$$

over all domains Ω that are bounded, smooth, and open, but not necessarily symmetric, and over all equivariant positive maps v (*cf.* (ii) in the statement of Theorem 3.1) in $W^{1,2}(\Omega;\mathbb{R}^m)$.

Remarks 3.3. The constructed solution u is a minimizer in the equivariant positive class. For the equilateral triangle group on the plane and the regular tetrahedron group in three-dimensional space, one would expect that the solution constructed is a minimizer in the class of all $W^{1,2}(\Omega;\mathbb{R}^n)$-maps.

It appears that the positivity of u is not an implication of the minimizing property (3.1) for arbitrary G, without extra hypotheses on W.

Looking for equivariant solutions is of course a convenience. However, at this point all the existence results for system (1.1) known to the author involve hypotheses of symmetry (except for $n = 1$).

4 A related Bernstein-type theorem – Background

We recall De Giorgi's conjecture [23] for the scalar equation, for $n \geq 2$, $m = 1$.

Conjecture 4.1 (De Giorgi). For the equation

$$\Delta u - W'(u) = 0,$$

with $W(u) = \frac{1}{4}(u^2 - 1)^2$, under the hypotheses that $u : \mathbb{R}^n \to \mathbb{R}$ is in $C^2(\mathbb{R}^n; [-1,1])$ with $\partial u/\partial x_n > 0$, is it true that the level sets of u are hyperplanes, at least for $n \leq 8$?

A more restricted version of the conjecture involves the additional hypothesis $\lim_{x_n \to \pm\infty} u(x) = \pm 1$.

This conjecture was established by Ghoussoub and Gui [34] for $n = 2$, Ambrosio and Cabré [13] for $n = 3$, and Savin [44] in the restricted form for $4 \leq n \leq 8$. Finally, it was disproved for $n \geq 9$ by del Pino, Kowalczyk, and Wei [24,25]. We refer to the survey paper by Farina and Valdinoci [30], where in addition several extensions to a variety of related equations are given.

Some of the ingredients behind the formulation of this conjecture are

(i) the Bernstein theorem for graphs,
(ii) the relationship between monotonicity and stability,

(iii) the solution of the ODE (the *heteroclinic connection*)

$$\frac{\mathrm{d}^2 U}{\mathrm{d}\eta^2} - W'(U) = 0, \text{ with } \lim_{\eta \to \pm\infty} U(\eta) = \pm 1$$

(unique up to translations),

(iv) the phase transition problem for two phases.

The conclusion in the conjecture is equivalent to showing that

$$u(x) = U\left(\frac{a \cdot x - c}{\sqrt{2}}\right),$$

for some $a \in \mathbb{R}^n$, with $|a| = 1$, and $c \in \mathbb{R}$, that is, $u(x) = U(Px)$, where P is the orthogonal projection to the normal direction of the level sets.

In formulating the analog of the conjecture for systems one should keep in mind that

(i) Tangent planes are special cases of tangent cones. Moreover, minimizing tangent cones have cylindrical structure, that is,

$$C = V \times \tilde{C}, \text{ with } \tilde{C} \text{ minimizing in } V^{\perp},$$

and the cone C is translation invariant 'along V' ($V = \{0\}$ is an option). Also, the Liouville object in this context is the cone at infinity. The Bernstein-type theorem therefore should involve a cone.

(ii) Monotonicity is not related in general to stability for systems.

(iii) The solution $u : \mathbb{R}^n \to \mathbb{R}^{n-k}$ *a posteriori* should be of the form

$$u(x) = \hat{u}(Px),$$

where P is an orthonormal projection on an $(n-k)$-dimensional plane, $\hat{u} : \mathbb{R}^{n-k} \to \mathbb{R}^{n-k}$ is a *connection* map, equivariant as in Theorem 3.1 in Section 3.

(iv) For three or more phases the order parameter should be a vector since otherwise there is no connection between the extreme phases. For example, for coexistence of three phases we need at least a two-dimensional order parameter, thus a partitioning of $\mathbb{R}^2$ in three parts.

(v) For the analog of the restricted conjecture we refer to Section 5 in the present paper.

Our purpose in this section is to present a sample of such Bernstein-type theorems for the simplest nontrivial case, the triple junction in $\mathbb{R}^3$, that corresponds to one of the two singular minimizing cones in $\mathbb{R}^3$ (see [50]). The formulation of such theorems in terms of cones goes back to Fleming

[32] and is subsequently developed in Morgan [43]. However, here we also want to emphasize partitions as the natural setup. For this reason we present in detail White's approach [51] because it improves Almgren's [12] and because of its simplicity, and also for making our treatment as self-contained as possible. Our presentation here is also based on Chan's thesis [21], written under White's supervision.

Finally, we mention the paper of Fazly and Ghoussoub [31], which extends the methods of the scalar equation to systems, as far as this can possibly be done, by assuming certain monotonicities on the components of the solution which amount to

$$\frac{\partial^2 W(u)}{\partial u_i \partial u_j} \leq 0, \ \text{for } i \neq j.$$

For a special system of two equations, we mention the papers of Berestycki, Lin, Wei, and Zhao [16], Berestycki, Terracini, Wang, and Wei [17], Farina [28], and Farina and Soave [29].

We recall here some basic background on partitions and geometric measure theory (see [45,51,52]).

4.1 Minimizing partitions

Consider an open set $U \subset \mathbb{R}^n$ occupied by N immiscible fluids, or phases. Associated to each pair of phases i and j there is a surface energy density e_{ij}, with $e_{ij} > 0$ for $i \neq j$, and $e_{ij} = e_{ji}$, with $e_{ii} = 0$. Hence, if A_i denotes the subset of U occupied by phase i, then U is the disjoint union

$$U = A_1 \cup A_2 \cup A_2 \cup \cdots \cup A_N,$$

and the energy of the partition $A = \{A_i\}_{i=1}^N$ is

$$E(A) = \sum_{0<i<j\leq N} e_{ij}\, \mathbb{M}(\partial A_i \cap \partial A_j),$$

where $\mathbb{M}$ (for *mass*) stands for the measure of the interface. For $n = 3$ it will simply be the area of $\partial A_i \cap \partial A_j$.

If U is unbounded, for example $U = \mathbb{R}^n$ (we say then that A is *complete*), the quantity above in general will be infinity. Thus, for each W open, with $W \Subset U$, we consider the energy

$$E(A; W) = \sum_{0<i<j\leq N} e_{ij}\, \mathbb{M}(I_{ij} \cap W), \ \text{where } I_{ij} := \partial A_i \cap \partial A_j.$$

Definition 4.2. The partition A is a *minimizing* N-partition if given any $W \Subset U$ and any N-partition A' of U with

$$\bigcup_{i=1}^{N}(A_i \vartriangle A_i') \Subset W, \tag{4.1}$$

we have

$$E(A; W) \leq E(A'; W).$$

The symmetric difference $A_i \vartriangle A_i'$ of the sets A_i and A_i' is defined as their union minus their intersection, that is, $A_i \vartriangle A_i' = (A_i \cup A_i') \setminus (A_i \cap A_i')$.

4.2 Flat chains with coefficients in a group

Let G be an abelian group with norm $|\cdot|$, such that $|g| \geq 0$, with $|g| = 0$ if and only if $g = 0$, for all $g \in G$, and

$$|g + h| \leq |g| + |h|, \text{ for all } g, h \in G.$$

Then, $(G, |\cdot|)$ is a metric space and we will assume that it is complete and separable. In our case G will be a finite group.

Fix $\mathbb{R}^n$ and a compact convex set $\mathbb{K}$ in $\mathbb{R}^n$. For each integer $k \geq 0$ consider the abelian group of all formal finite sums of the form $\sum g_i P_i$, where $g_i \in G$ and where P_i is a k-dimensional oriented compact convex polyhedron in $\mathbb{K}$. We form the quotient group obtained by identifying gP with $-g\tilde{P}$, whenever P and $\tilde{P}$ coincide but have opposite orientations. Also, identify gP and $gP_1 + gP_2$, whenever P can be subdivided into P_1 and P_2.

The resulting abelian group $\mathcal{P}_k(\mathbb{K}; G)$ is called the group of *polyhedral k-chains* on $\mathbb{K}$ with coefficients in G. Define the *boundary homomorphism* $\partial : \mathcal{P}_k \to \mathcal{P}_{k-1}$ by

$$\partial\left(\sum g_i P_i\right) := \sum g_i \partial P_i.$$

Note that any polyhedral k-chain T can be written as a linear combination $\sum_i g_i[P_i]$ of nonoverlapping polyhedra, that is, polyhedra with disjoint interiors. Then, the *flat norm* of the chain is defined to be

$$W(T) = \inf_Q \{\mathbb{M}(T - \partial Q) + \mathbb{M}(Q)\},$$

where the infimum is over all polyhedral $(k+1)$-chains Q.

The flat norm makes $\mathcal{P}_k(\mathbb{K}; G)$ into a metric space. The completion of this metric space is denoted by $\mathcal{F}_k(\mathbb{K}; G)$ and its elements are called *flat k-chains* in $\mathbb{K}$ with coefficients in G. By uniform continuity, functionals such as the flat norm and operations such as addition and boundary extend in a unique way from polyhedral chains to flat chains. The mass norm in $\mathcal{P}_k(\mathbb{K}; G)$ extends to a linear semicontinuous functional in $\mathcal{F}_k(\mathbb{K}; G)$.

Suppose that every bounded closed subset of G is compact. A fundamental compactness theorem for flat chains asserts that, given any sequence $T_i \in \mathcal{F}_k(\mathbb{K}; G)$ with $\mathbb{M}(T_i)$ and $\mathbb{M}(\partial T_i)$ uniformly bounded, there is a W-convergent subsequence. More generally, one can define the flat chains in $\mathbb{R}^n$ with compact support, $\mathcal{F}_k(\mathbb{R}^n; G)$, meaning that each element vanishes outside a certain compact convex set. Then, the compactness theorem holds for a sequence $T_i \in \mathcal{F}_k(\mathbb{R}^n; G)$, with $\operatorname{supp} T_i \subset \mathbb{K}$, for $\mathbb{K}$ independent of i. The symbol '$\rightharpoonup$' denotes convergence in the flat norm.

4.3 Flat chains of top dimension

[See [51].] Polyhedral n-chains in $\mathbb{R}^n$ with compact support can be identified with the set of piecewise-constant functions

$$g : \mathbb{R}^n \to G,$$

that vanish outside a compact convex set $\mathbb{K}$. Here, two functions that differ only on a set of measure zero are regarded as the same. 'Piecewise constant' means locally constant except along a finite collection of hyperplanes. The identification is as follows. Any such $T \in \mathcal{F}_k(\mathbb{R}^n; G)$ can be written as

$$T = \sum g_i [P_i],$$

where the P_i's are nonoverlapping and inherit their orientations from $\mathbb{R}^n$. We can associate to T the function

$$g : \mathbb{R}^n \to G, \text{ with } g(x) = \begin{cases} g_i, & \text{if } x \text{ is in the interior of } P_i\,, \\ 0, & \text{if } x \text{ is not in the interior of } P_i. \end{cases}$$

Note that the mass norm of T is equal to the L^1 norm of $g(\cdot)$. Also, since there are no nonzero $(n+1)$-chains in $\mathbb{R}^n$, we see from the definition of W that

$$W(T) = \mathbb{M}(T) = \int_{\mathbb{R}^n} |g(x)|\, dx.$$

Consequently, the W-completion of the polyhedral chains (that is, the flat n-chains) is isomorphic to the L^1-completion of the piecewise-constant functions.

Denoting T by $[\mathbb{K}]_{Lg}$, the isomorphism is

$$L^1(\mathbb{K}; G) \ni g \to [\mathbb{K}]_{Lg} \in \mathcal{F}_n(\mathbb{K}; G),$$

with

$$\mathbb{M}([\mathbb{K}]_{Lg}) = W([\mathbb{K}]_{Lg}) = \int_{\mathbb{R}^n} |g|\,dx.$$

Thus, the flat n-chains T on $\mathbb{R}^n$ with compact support can be identified with the $L^1_{\text{loc}}(\mathbb{R}^n; G)$ functions. The flat chains with $\mathbb{M}(\partial T) < +\infty$ correspond to the sets with finite perimeter (Caccioppoli sets). The BV norm of the function g above gives the perimeter, that is,

$$\|g\|_{BV} = \mathbb{M}(\partial T).$$

The compactness for flat chains with

$$\mathbb{M}(T_n) + \mathbb{M}(\partial T_n) < C$$

is equivalent *in this setup* to the compactness of the embedding

$$BV(\Omega) \Subset L^1(\Omega), \text{ for } \Omega \text{ bounded.}$$

The lower semicontinuity of $\mathbb{M}(\partial T)$ with respect to the W-norm is equivalent to the lower semicontinuity of the BV norm with respect to L^1.

4.4 The group of surface tension coefficients (see [51])

The purpose next is the introduction of an appropriate group G so that for the flat chain $T = \sum g_i P_i$, where $P_i = A_i$, with $A = \{A_i\}$ a partition of U, there holds

$$\mathbb{M}(\partial T \llcorner W) = E(A; W). \tag{4.2}$$

First, assume that

$$e_{ik} \leq e_{ij} + e_{jk}, \text{ for all } i, j, k. \tag{4.3}$$

Let G be the free $\mathbb{Z}_2$-module with N generators $f_1, \ldots, f_N$ (one for each phase). White [51] defines a norm in this group such that

$$|f_i - f_j| = e_{ij},$$

and the $\mathbb{Z}_2$-module identifies

$$f_{i_1} - f_{j_1} = f_{i_1} + f_{j_1}.$$

Utilizing this, it is easy to see in calculating ∂T, and $\mathbb{M}(\partial T)$, that (4.2) holds. In this setup, given a partition of U into N measurable sets $A_1, \ldots, A_N$, and $\mathbb{K}$ as above, we associate the flat n-chain

$$T = \mathbb{K}_{Lg},$$

where

$$g(x) = \begin{cases} f_i, & \text{for } x \in A_i \cap \mathbb{K} \\ 0, & \text{for } x \notin A_i \cap \mathbb{K}. \end{cases}$$

Note that if the A_i's have piecewise-smooth boundaries, then (4.2) holds. More generally, equation (4.2) holds whenever the A_i's are Caccioppoli sets, that is, whenever the flat chains have finite mass.

Conversely, given any flat n-chain T, we can represent T as

$$T = \mathbb{K}_{Lg},$$

where $g \in L^1(U \cap \mathbb{K}; G)$. In this article we take $U = \mathbb{R}^n$. We note that in [51] it is shown that the inequalities (4.3) are no real restriction, in the sense that if they are violated, then one can define new coefficients e^*_{ij} out of the old, so that the infimum of E coincides with the infimum of E^* (defined by replacing e_{ij} with e^*_{ij}). Also, it is noted that (4.3) is necessary for E to be lower semicontinuous with respect to the flat norm. Here, we refer also to Section 4.1 in [1].

4.5 Basics on minimizing chains

We recall some standard facts on minimizing chains and later we point out the relationship with minimizing partitions.

Cones If $x_0 \in \mathbb{R}^n$, where S is a k-dimensional flat chain in $\mathbb{R}^n$, then *the cone over S with vertex at x_0* is the flat chain

$$x_0 S = \text{Cone}(S) = h(I \times S), \tag{4.4}$$

where $h(t, x) = (1 - t)x_0 + tx$, for $0 \le t \le 1$, and $x \in S$. We have

$$S = \partial(x_0 S) + x_0 \partial S,$$

and if $S \subset B_r(x_0)$, where $B_r(x_0)$ is the ball with radius r and center at x_0, then

$$\mathbb{M}(x_0\, S) \le \frac{r}{k+1} \mathbb{M}(S).$$

C_x is a cone with vertex at x if, by definition, it is invariant as a set under the homothetic map

$$y \to x + t(y - x), \text{ for all } t > 0 \text{ and } y \in C_x.$$

If S is a k-flat chain in $\mathbb{R}^n$, then S is *mass minimizing* if $\mathbb{M}(S) \le \mathbb{M}(S')$, for all S' with $\partial S' = \partial S$.

If Γ is a *cycle*, that is, $\partial\Gamma = 0$, then

$$L(\Gamma) := \inf\{\mathbb{M}(X) \mid \partial X = \Gamma\}.$$

Consequently, if S is mass minimizing, then $\mathbb{M}(S) = L(\partial S)$.

The flat chain S is *minimizing* if by definition

$$\mathbb{M}(S \llcorner B_r(x)) = L(\partial(S \llcorner B_r(x)),$$

for all $r > 0$ and center x such that $0 < r < \operatorname{dist}(x, \operatorname{supp}\partial S)$. Mass minimizing is minimizing. We allow the options $\partial S = 0$ and $\mathbb{M}(S) = \infty$.

The *monotonicity formula* holds for k-dimensional minimizing flat chains and states that

$$\Theta(S, x, r) := \frac{\mathbb{M}(S \llcorner B_r(x))}{\omega_k r^k} \tag{4.5}$$

is an increasing function of r, where ω_k is the volume of the k-dimensional unit ball. It follows that for minimizing flat chains S, the limit

$$\Theta(S, x) := \lim_{r \to 0} \Theta(S, x, r)$$

exists, and if

$$\Theta(S, x, r) < B, \text{ for } B \text{ independent of } x, r, \tag{4.6}$$

then the limit

$$\Theta(S) := \lim_{r \to +\infty} \Theta(S, x, r)$$

exists and is independent of x. We note that the condition $\Theta(S, x, r) =$ constant in $r > 0$, for x fixed, implies that S is a cone with vertex at x.

The tangent cone (blow-up) Let S be a minimizing flat chain, $x \notin \operatorname{supp}\partial S$, and let $\{\mu_i\}$ be an increasing sequence of positive numbers, with $\mu_i \to +\infty$. Set

$$S_i = \mathcal{D}_{\mu_i}(S - x), \text{ with } \mathcal{D}_{\mu_i}(S - x) = \{\mu_i(y - x) \mid y \in S\}.$$

Then *along a subsequence* there holds $S_i \rightharpoonup C_x$ (by the compactness theorem), where C_x has the properties

(i) $\partial C_x = 0$,
(ii) C_x is a cone,
(iii) $\Theta(C, 0, r) = \Theta(S, x)$, for all $r > 0$,
(iv) C_x is minimizing.

The cone at infinity (blow-down) If instead in the arrangement above $\{\mu_i\}$ is a decreasing sequence, with $\mu_i \to 0$, and if (4.6) holds, then along a subsequence there holds $S_i \rightharpoonup C_\infty$ (by the compactness theorem), where C_∞ has the properties

(i) $\partial C_\infty = 0$,
(ii) C_∞ is a cone,
(iii) $\Theta(C_\infty, 0, r) = \Theta(C_\infty, 0) =: \Theta(S)$,
(iv) C_∞ is minimizing.

Note that C_x and C_∞ are not necessarily unique.

If N^{k-1} is a smooth $(k-1)$-surface, with $k \leq n-1$ and $N^{k-1} \subset \mathbb{S}^{n-1}$ (the unit sphere in $\mathbb{R}^n$), then the *cone over* N^{k-1} is

$$C(N^{k-1}) = \left\{ x \in \mathbb{R}^n \;\middle|\; \frac{x}{|x|} \in N^{k-1} \right\}.$$

If S is smooth, then the projection

$$\frac{1}{R}(S \cap \mathbb{S}^{n-1}_R),$$

of the set $S \cap \mathbb{S}^{n-1}_R$ on the unit sphere tends to C_∞, as $R \to \infty$, provided that (4.6) holds, that is,

$$C_\infty = \lim_{R \to +\infty} C\left(\frac{1}{R}\left(S \cap \mathbb{S}^{n-1}_R\right)\right),$$

where the limit is in the flat norm, and exists along a sequence

$$R_1 < R_2 < \cdots \to +\infty,$$

where C_∞ is the cone at infinity.

Relationship with partitions [See [12, 21, 46].] The concepts in Paragraph 4.5 have exact analogs for partitions defined as flat chains of top dimension in Paragraphs 4.3 and 4.4 above. Specifically,

(i) the concept of the cone is unchanged,
(ii) the mass minimizing flat chain S is replaced by the minimizing partition T (or A), via the definition in (4.1) above,
(iii) the monotonicity formula holds for minimizing partitions,
(iv) the notion of tangent cone and cone at infinity have exact analogs for minimizing partitions.

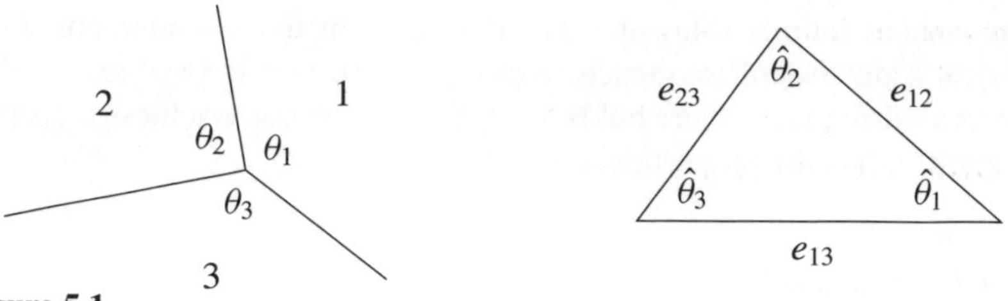

Figure 5.1.

5 A related Bernstein-type theorem – Statements and proofs

We are now ready to state a sample of a Bernstein-type theorem. We begin with $\mathbb{R}^2$.

Theorem 5.1 ($n = 2$). *Let A be a complete minimizing partition in $\mathbb{R}^2$ with $N = 3$ (three phases), with surface tension coefficients satisfying*

$$e_{ik} < e_{ij} + e_{jk}, \text{ for } j \neq i, k \text{ with } i, j, k \in \{1, 2, 3\}. \tag{5.1}$$

Then, ∂A is a triod.

In Figure 5.1 we show a triod with angles θ_1, θ_2, θ_3, and the corresponding triangle with their supplementary angles $\hat{\theta}_i = \pi - \theta_i$. For these angles Young's law holds, that is,

$$\frac{\sin \hat{\theta}_1}{e_{23}} = \frac{\sin \hat{\theta}_2}{e_{13}} = \frac{\sin \hat{\theta}_3}{e_{12}}.$$

We recall that under the condition of the strict triangle inequality for the surface tension coefficients, White has established a general regularity result which applies in particular under (5.1) to A above. His result improves on Almgren's work [12]. Detailed proofs can be found in Chan's thesis [21] (Section 1.6 and pages 10–14). It follows that a priori A consists of triple junctions and line segments, always a finite number in any given open and bounded subset of $\mathbb{R}^2$.

We present the proof of Theorem 5.1 in three steps. The first two are two lemmas that we state next.

Lemma 5.2. *The only minimizing cones are the straight line and the triod.*

Lemma 5.3. *There holds $\mathbb{M}(\partial A \llcorner B_R) \leq CR$.*

Accepting for the time being the lemmas above, we can conclude with the proof of the theorem.

Proof of Theorem 5.1. Let P be one of the junctions and consider the tangent cone C_0 at P, which by Lemma 5.2 is a triod. We have

$$\Theta(C_0, P) = \Theta(A, P) \leq \frac{\mathbb{M}(A \llcorner B_R(P))}{\pi R^2} \leq \lim_{R \to +\infty} \frac{\mathbb{M}(A \llcorner B_R(P))}{\pi R^2} \tag{5.2}$$
$$\leq \Theta(A),$$

where we used the monotonicity formula (4.5) and also the bound provided by Lemma 5.3 above.

Let C_∞ be any of the cones at infinity. Then

$$1 < \Theta(C_0, P) \leq \Theta(A) = \Theta(C_\infty). \tag{5.3}$$

Since C_∞ is minimizing, from Lemma 5.2 we conclude that

$$C_0 = C_\infty, \text{ up to congruence.} \tag{5.4}$$

Returning back to (5.2) and utilizing (5.4), we obtain

$$\frac{\mathbb{M}(A \llcorner B_R(P))}{\pi R^2} = \Theta(A), \text{ for all } R > 0. \tag{5.5}$$

It follows from (5.5) that A is a cone and from the first inequality in (5.3) that it is a singular one, hence a triod by Lemma 5.2. □

Note that the scheme above for proving a Bernstein-type theorem was introduced by Fleming [32], where it was used to give a new proof of the classical Bernstein theorem in $\mathbb{R}^3$. (See page 193 in [22].)

We now give the proofs of Lemmas 5.2 and 5.3.

Proof of Lemma 5.2. We begin by recalling the classical Steiner problem adapted in our weighted setup. Let A, B, C, be three fixed points on the plane. Given a fourth point $\hat{P}$, not necessarily distinct from the other three, consider the weighted sum of the distances of $\hat{P}$ from the vertices of the triangle ABC, that is,

$$e_{12}|\hat{P} - A| + e_{23}|\hat{P} - C| + e_{13}|\hat{P} - B|. \tag{5.6}$$

Suppose that the quantity above is minimized for $P \neq A, B, C$.

We recall Steiner's argument. Let

$$\Gamma = \{Q \in \mathbb{R}^2 \mid F(Q) = F(P)\},$$

where

$$F(Q) = e_{12}|Q - A| + e_{13}|Q - B|.$$

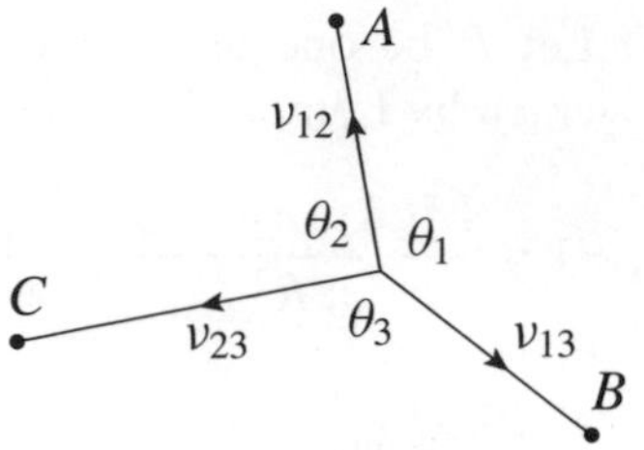

Figure 5.2.

Then, one notes that necessarily the circle with center C and radius $|P - C|$ has to be tangent to the curve Γ at P. Therefore, P in particular solves the problem

$$\min F(Q), \text{ subject to } L(Q) := \langle Q, \nu_{23} \rangle = 0,$$

from which it follows that

$$e_{12}\nu_{12} + e_{13}\nu_{13} = \alpha \nu_{23}, \text{ for some } \alpha \in \mathbb{R}.$$

By replacing A, B with B, C first, and then with C, A, and repeating the argument in each of these cases we obtain respectively

$$e_{13}\nu_{13} + e_{23}\nu_{23} = \beta \nu_{12}, \text{ for some } \beta \in \mathbb{R},$$
$$e_{12}\nu_{12} + e_{23}\nu_{23} = \gamma \nu_{13}, \text{ for some } \gamma \in \mathbb{R}.$$

From these relationships it follows easily that

$$e_{12}\nu_{12} + e_{13}\nu_{13} + e_{23}\nu_{23} = 0,$$

and thus Young's law is established.

We explain under which conditions the quantity in (5.6) is minimized for a $P \neq A, B, C$. It is clear that for the partitioning in Figure 5.2 to be possible, one needs that all the inequalities $\hat{A} < (\pi - \theta_1) + (\pi - \theta_2)$, $\hat{B} < (\pi - \theta_1) + (\pi - \theta_3)$, and $\hat{C} < (\pi - \theta_2) + (\pi - \theta_3)$ should hold, where $\hat{\theta}$ stands for $\pi - \theta$. Thus, in general one needs that there is an arrangement so that all inequalities $\hat{A} < (\pi - \theta_i) + (\pi - \theta_j)$, $\hat{B} < (\pi - \theta_i) + (\pi - \theta_k)$, and $\hat{C} < (\pi - \theta_j) + (\pi - \theta_k)$ hold, where $i, j, k \in \{1, 2, 3\}$, and $\theta_1, \theta_2, \theta_3$ are determined by the triangle in Figure 5.1. For example, if all surface tension coefficients are equal, then $P \neq A, B, C$ if and only if the largest angle of ABC is less than $\frac{2\pi}{3}$.

Next, we show the connectedness of each phase in a minimizing cone C. That is, each phase can appear only once (*cf.* Lemma 1.2 in [21]). We

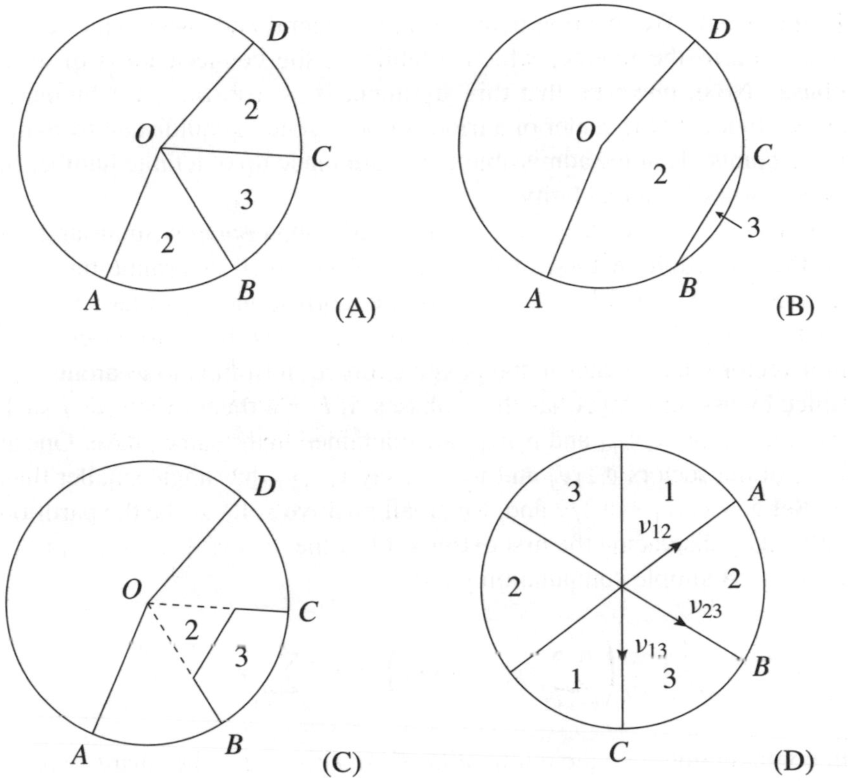

Figure 5.3.

proceed by contradiction. First we exclude the possibility that one phase is separated by another. By regularity, the cone consists of finitely many rays emanating from the origin O. Suppose that we have the arrangement as in Figure 5.3A.

We will exclude the case in Figure 5.3A by comparing with Figure 5.3B. The cone induces a partition A of the unit disk D. Clearly, $\mathbb{M}(\partial A) > \mathbb{M}(\partial A')$, where A' is the partition in Figure 5.3B, and since $A \vartriangle A' \Subset D$, we obtain a contradiction, since C was assumed minimizing. One may object to whether $A \vartriangle A' \Subset D$ is satisfied. For this purpose, we can modify A' as in Figure 5.3C. Thus the only remaining possibility is that the disk is partitioned by a number of triple junctions as in Figure 5.3D.

We will show that only a single triple junction is acceptable. Consider the triangle ABC. By the hypothesis, C is a minimizing cone. Hence O solves the Steiner problem with respect to this triangle. But then, the angles $\theta_1, \theta_2, \theta_3$ have to satisfy Young's law, and in particular have to add up to 2π, which completes the proof of the lemma. □

Remark 5.4. We mention below a more general and also simpler argument due to the referee, which establishes the connectedness of each phase. Note, however, that this argument is no substitute for Steiner's since obviously the center of a triod is not in general equidistant from the three points. That the admissible cones are made up of a finite number of rays follows from regularity.

Let $\nu_1, \ldots, \nu_l$, with $\nu_{l+1} = \nu_1$, ordered counterclockwise around O, be the rays (unit vectors) of the unit disk D which determine the minimizing cone A and let $e_1, \ldots, e_l$ be the corresponding surface tension coefficients. Given $i < j$, we denote by $\nu_i \nu_j \subset D$ the sector swept by a unit vector ν that rotates in the positive direction from ν_i to ν_j around O. Since by assumption A has three phases, if $l \geq 4$ there exist $i \neq j$ such that the sectors $\nu_i \nu_{i+1}$ and $\nu_j \nu_{j+1}$ are contained in the same phase. One at least of the sectors $\nu_{i+1}\nu_j$ and $\nu_{j+1}\nu_i$, say $\nu_{i+1}\nu_j$, has angle smaller than π. Set $\nu = (\nu_{i+1} + \nu_j)/2$ and, for small positive ε, let A_ε be the partition defined by displacing the first extreme O of the rays $\nu_{i+1}, \nu_{i+2}, \ldots, \nu_j$ to $O + \varepsilon\nu$. A simple computation yields

$$\frac{\mathrm{d}}{\mathrm{d}\varepsilon}\bigg|_{\varepsilon=0} \left(\sum_{k=i+1}^{j} e_k |\nu_k + \varepsilon\nu| \right) = \nu \cdot \sum_{k=i+1}^{j} e_k \nu_k < 0,$$

in contradiction with the minimality of A. This establishes that $l = 3$.

To prove Young's law we let A_ε be the partition defined by displacing the first extreme of all the rays to $O + \varepsilon\nu$, where ν is an arbitrary vector. In this case, we obtain

$$\frac{\mathrm{d}}{\mathrm{d}\varepsilon}\bigg|_{\varepsilon=0} \left(\sum_{k=i+1}^{j} e_k |\nu_k + \varepsilon\nu| \right) = \nu \cdot \sum_{k=i+1}^{j} e_k \nu_k.$$

This, the minimality of A, and the arbitrary choice of ν imply that

$$\sum_{k=1}^{N} e_k \nu_k = 0,$$

and since $l = 3$, it follows that $e_1\nu_1 + e_2\nu_2 + e_3\nu_3 = 0$.

Proof of Lemma 5.3. Consider the disk $B(O; R)$. By the regularity of A, the intersection

$$\partial B(O; R) \cap \partial A$$

consists of finitely many points, say k. We now construct a test partition $\tilde{A}$ as follows. First, enlarge the R-disk slightly so that $\mathbb{M}(\partial A \llcorner (B_{R'} \setminus$

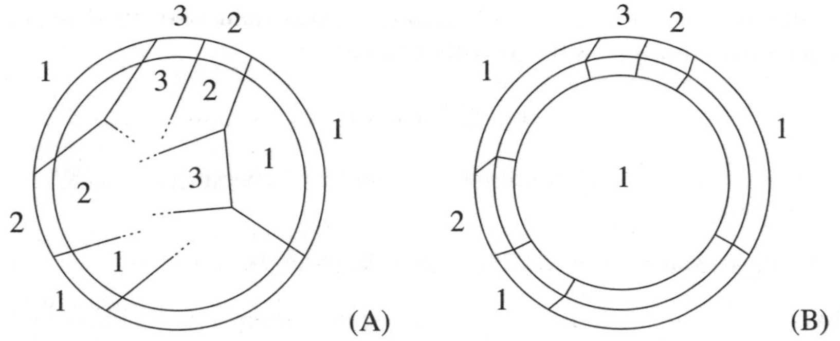

Figure 5.4.

$B_R)) < \varepsilon$, for $R' > R$, with $R' - R \ll 1$. This is possible by the regularity of A. Take $\partial\tilde{A} = \partial A$ inside the ring $R \leq |x| \leq R'$. Next, we introduce an ε^2-layer inside $B(O; R)$, in which we take $\partial\tilde{A}$ to consist of the union of k ε^2-line segments orthogonal to $\partial B(O; R)$, emanating from the k points on $\partial B(O; R)$. Finally, inside $B(O; R - \varepsilon^2)$ take a single phase, say phase 1. (See Figure 5.4B.)

By construction, A and $\tilde{A}$ have the same Dirichlet values in $B(O; R')$. We have

$$\begin{aligned}\mathbb{M}(\partial\tilde{A} \llcorner B_{R'}) &= \mathbb{M}(\partial\tilde{A} \llcorner (B_{R'} \setminus B_R)) + \mathbb{M}(\partial\tilde{A} \llcorner B_R)\\ &\leq \varepsilon + \mathbb{M}(\partial\tilde{A} \llcorner B_R)\\ &\leq \varepsilon + k\varepsilon^2 + 2\pi(R - \varepsilon^2).\end{aligned}$$

Taking now $\varepsilon = \frac{1}{k}$, we obtain $\mathbb{M}(\partial\tilde{A} \llcorner B_{R'}) \leq CR$, and thus, since A is minimizing, the estimate of the lemma follows. □

Remark 5.5. Suppose C is an $(n-1)$-dimensional cone embedded in $\mathbb{R}^n$, $3 \leq n \leq 7$, with zero mean curvature, which is also stable, then C is a hyperplane. This fails for $n \geq 8$. This is a classical result due to Simons (see Appendix B in [45]). The difference with the cones considered here lies in orientation. In the oriented case above there is a globally well-defined unit normal and stability is understood in the class of normal perturbations. The second variation gives the stability inequality

$$\int_C \left\{|\nabla^C z|^2 - |A|^2 z^2\right\} \geq 0,$$

where $|A|$ is the second fundamental form of C. This is not the appropriate formula for the class of unoriented objects we consider here. The perturbations in the unoriented case have also tangential components. Another difference is in the growth estimate

$$\mathbb{M}(B_r \cap \Sigma) \leq Cr^{n-1},$$

which is immediate for minimizing complete hypersurfaces in $\mathbb{R}^n$ (see p. 4 in [22]).

Our purpose next is to establish the following theorem.

Theorem 5.6 **($n = 3$).** *Let A be a complete minimizing partition in $\mathbb{R}^3$ with $N = 3$ (three phases), with surface tension coefficients satisfying*

$$e_{ik} < e_{ij} + e_{jk}, \text{ for } j \neq i, k \text{ with } i, j, k \in \{1, 2, 3\},$$

and with the property that the liquid edges do not intersect. Then,

$$\partial A = \mathbb{R} \times C_{\text{tr}} \quad \text{(cylindrical cone),} \tag{5.7}$$

where C_{tr} is the triod on the plane.

We note that the subset of ∂A whose tangent cones contain exactly three phases is the union of the liquid edges, denoted by $\Sigma_3(A)$. We recall that in the case of equal surface tension coefficients the cone in (5.7) represents one of the two stable types of singularities that soap films can form. This was shown by Taylor [50] who also proved ($e_{ij} = 1$) that if A is a minimizing partition (not necessarily complete), then $\Sigma_3(A)$ consists of a union of $C^{1,\alpha}$ curves. Furthermore, Kinderlehrer, Nirenberg, and Spruck [38] showed that these curves are real analytic.

The basis for Theorem 5.6 above is the following result of Chan [21].

Theorem 5.7 ([21]). *Let A be a complete stable partition in $\mathbb{R}^3$ with $N = 3$ (three phases), where $\Sigma_3(A) = \cup_i \gamma_i$, with γ_i a smooth curve, and suppose that the density at infinity $\Theta(A)$ is finite. Then, $A \setminus \Sigma_3$ is a union of planar pieces.*

A smooth partition is *stable* if the second variation is nonnegative for C^2 compactly supported perturbations Y, that is,

$$\begin{aligned} \frac{\mathrm{d}^2}{\mathrm{d}t^2}\Big|_{t=0} \mathbb{M}(M + tY) &= \int_M (-|B|^2(Y^\perp) - Y^\perp \cdot \Delta_M(Y^\perp)) \\ &\quad - \int_{\partial M} (\operatorname{div}_M Y^\top (Y^\top \cdot \mu) + [Y^\perp, Y^\perp] \cdot \mu) \geq 0, \end{aligned}$$

where M is a regular surface immersed in $\mathbb{R}^3$, the vector field Y is defined in a neighborhood of M, μ is the unit outward conormal on ∂M, $Y^\perp$ and

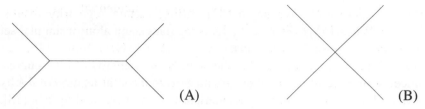

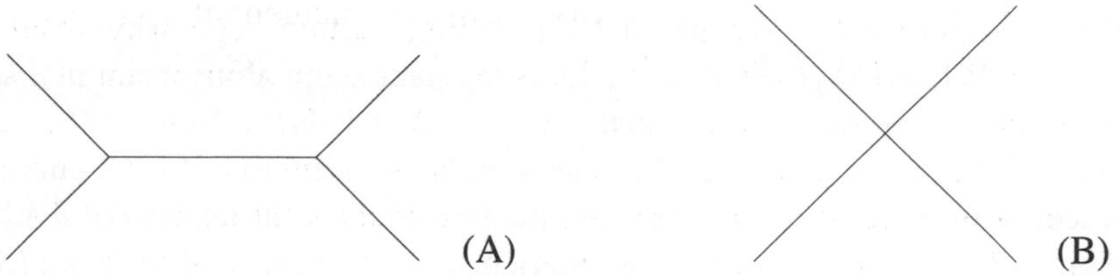

Figure 5.5.

$Y^\top$ are the normal and tangential components of Y respectively, and $|B|$ is the length of the second fundamental form on M (see [47]).

By rearranging the stability inequality above one obtains control of the second fundamental form by terms involving the test variations Y. One of the important ingredients in the proof is the choice of the variations Y at the junctions, where the Young angle conditions are utilized for the matching. The argument finally utilizes a logarithmic cut-off function, that is, the 'logarithmic trick' (see p. 30 in [22]), and renders $|B| = 0$. We note that Chan's result as it stands does not exclude the possibility, for example, of two triple-junctions (see Figure 5.5A).

In Theorem 5.6 we assume that the partition is minimizing, a much stronger condition than stability. The smoothness follows by the regularity results of White mentioned after the statement of Theorem 5.1 above.

The proof of Theorem 5.6 is based on the following lemma.

Lemma 5.8. *There holds* $\mathbb{M}(\partial A \llcorner B_R) \leq CR^2$.

Proof of Theorem 5.6. We employ a dimension reduction argument. Accepting for the moment the estimate in Lemma 5.8, we conclude as follows. By regularity facts (see the proof of Lemma 5.8 below), Chan's theorem above applies and gives that $A \setminus \Sigma_3$ is a union of planar pieces. By the hypothesis that the liquid edges (now straight lines) do not intersect, A has cylindrical structure, $A = A_2 \times \mathbb{R}$, where A_2 is a complete minimizing partition in $\mathbb{R}^2$. Thus, by Theorem 5.1 we are set. □

We note that it is easy to see that the cone at infinity C_∞ for the two triple junction system in Figure 5.5A is as in Figure 5.5B, which is not minimizing.

The result of Theorem 5.6 states in particular that for minimizing 3-partitions, the liquid edge is connected, or equivalently that each phase is connected. This question can be addressed in bounded domains $U \subset \mathbb{R}^n$. In [8], Faliagas and the author establish connectedness for strictly convex U in the class of stable partitions, thus extending to 3-partitions the work of Sternberg and Zumbrun [49].

Proof of Lemma 5.8. We begin by recalling certain regularity results. First, by White [51] (see in [21]), ∂A is regular except along a singular set of at most Hausdorff dimension $n-2=3-2=1$. Next, Simon [46] (see also [21]) applies and gives that the singular set consists of $C^{1,\alpha}$ curves, since for $n=3$ and $N=3$ there are no gaps in the stratification of ∂A by 'spine' dimension. Finally, by Kinderlehrer, Nirenberg, and Spruck [38], these curves are actually real analytic.

The procedure for deriving the estimate is analogous to the two-dimensional case in Lemma 5.3. We consider a ball $B(O;R)$ and look at the intersection $\partial B(O;R) \cap \partial A$, which consists of curves intersecting along a finite number of triple junctions (by our hypothesis that liquid edges do not intersect). Thus on $\mathbb{S}^2(O;R)$ we have a finite network made up of curves and triple junctions. We will be introducing two concentric spheres $\mathbb{S}^2(O;R_1)$ and $\mathbb{S}^2(O;R_2)$, with $R_1 < R < R_2$, and a test partition $\tilde{A}$, which will be obtained by modifying A in the annular region $\mathrm{Ann}(O;R_1,R)$ in a way we explain below, and continued with a single phase in $B(O;R_1)$. Finally, $\tilde{A}$ will coincide with A in $\mathrm{Ann}(O;R,R_2)$. At the end the radii R_1, R_2 will be taken suitably close to R. To begin, we project radially on $\mathbb{S}^2(O;R_1)$ the network of curves and junctions $\partial B(O;R) \cap \partial A$, and also consider the surface swept out by this radial projection, whose mass can be estimated by

$$\mathbb{M}(\partial \tilde{A} \llcorner \mathrm{Ann}(O;R_1,R)) \le L(R)(R-R_1),$$

where $L(R)$ is the total length of $\partial B(O;R) \cap \partial A$, which by regularity is finite. Inside $B(O;R_1)$ we take a single phase, say phase 1. Given $\varepsilon > 0$, by regularity we can take $R_2 - R$ small enough so that

$$\mathbb{M}(\partial A \llcorner \mathrm{Ann}(O;R,R_2)) < \varepsilon.$$

Taking also $L(R)(R-R_1) < \varepsilon$, we obtain

$$\mathbb{M}(\partial \tilde{A} \llcorner B_R) \le \varepsilon + \varepsilon + 4\pi R_1^2 \le 2\varepsilon + 4\pi R^2,$$

and since A and $\tilde{A}$ have the same Dirichlet values in B_R and A is minimizing, the lemma is established. □

6 The hierarchical structure of equivariant connection maps

In this section we study asymptotic properties of the equivariant solutions to system (1.1) produced by Theorem 3.1. To explain the general result we have in mind, we begin with an example. Let $n=3$, $m=3$, and consider a quadruple-well potential W with minima $\{a_1,\ldots,a_4\}$, which

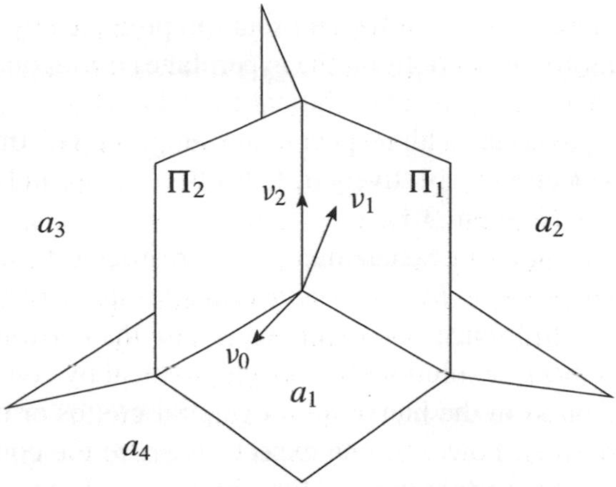

Figure 6.1.

is invariant under the group of symmetries of the regular tetrahedron, that is, $G = \mathcal{T}$. For example (see [9]), consider the potential

$$W(u_1, u_2, u_3) = |u|^4 - \frac{4}{\sqrt{3}}(u_1^2 - u_2^2)u_3 - \frac{2}{3}|u|^2 + \frac{5}{9}.$$

For this choice we have

$$N = \frac{|\mathcal{T}|}{|G_{a_1}|} = \frac{24}{6} = 4, \text{ for the minimum } a_1 = \left(\sqrt{\frac{2}{3}}, 0, \frac{1}{\sqrt{3}}\right),$$

with $D = \{\cup gF \mid g \in G_{a_1}\}$ the simplicial cone generated by

$$\left(0, \sqrt{\frac{2}{3}}, \frac{1}{\sqrt{3}}\right), \quad \left(0, -\sqrt{\frac{2}{3}}, \frac{1}{\sqrt{3}}\right), \quad \left(\sqrt{\frac{2}{3}}, 0, -\frac{1}{\sqrt{3}}\right).$$

Four copies of D partition $\mathbb{R}^3$. Note that the boundary of the partition is made up of six reflection planes and coincides with the tetrahedral minimizing cone in $\mathbb{R}^3$ (see Figure 6.1).

Next, we focus on $\overline{D}$ and its walls Π_1, Π_2. Take $\nu_0 \in \mathrm{Int}(\overline{D})$, $\nu_1 \in \mathrm{Int}(\Pi_1)$, and $\nu_2 \in \mathrm{Int}(\Pi_1 \cap \Pi_2)$, with $|\nu_i| = 1$, for $i = 0, 1, 2$, and consider the limits

$$\lim_{\lambda \to +\infty} u(x + \lambda\nu_0) = u_0(P_0 x), \tag{6.1}$$

$$\lim_{\lambda \to +\infty} u(x + \lambda\nu_1) = u_1(P_1 x), \tag{6.2}$$

$$\lim_{\lambda \to +\infty} u(x + \lambda\nu_2) = u_2(P_2 x), \tag{6.3}$$

for $x \in \mathbb{R}^3$. Here, u is an equivariant solution produced by Theorem 3.1, P_0, P_1, and P_2 are projections on the hyperplanes orthogonal to ν_0, ν_1, ν_2 respectively, and u_0, u_1, u_2 are solutions to (1.1), with $u_i : \mathbb{R}^i \to \mathbb{R}^3$, for $i = 0, 1, 2$, equivariant with respect to the subgroup G_{ν_i} of $\mathcal{T}$ leaving ν_i fixed, and connecting respectively a_1 to itself, a_1 to a_2, and a_1, a_2, and a_3 (in the sense of Theorem 3.1).

The work of Gui and Schatzman [37] is restricted to the tetrahedral group and their solution *by construction* satisfies the list above. We note that one of the differences of Theorem 3.1 and the corresponding theorem in [10] is that our solutions are not hierarchical by construction. It is intractable to build in the hierarchy for general groups or higher dimensions. *A posteriori*, however, one expects to prove the complete hierarchical structure for certain groups, for minimizers. Theorem 3.1 already contains (6.1), with $u_0 \equiv a_1$.

Below we state a general theorem establishing (6.2) in the context of Theorem 3.1. We formulate the theorem in an independent way. Consider the problem

$$\begin{cases} \Delta u - W_u(u) = 0, & \text{in } \Omega \subset \mathbb{R}^n, \\ u = u_0, & \text{on } \partial\Omega, \end{cases} \tag{6.4}$$

for $u : \Omega \to \mathbb{R}^m$, where Ω is a smooth open domain and W is a C^2 potential. We have the following hypotheses on W and Ω.

(H1) *The potential W is symmetric, with*

$$W(u_1, u_2, \ldots, u_m) = W(-u_1, u_2, \ldots, u_m).$$

(H2) *There exists a nondegenerate minimum a_+ of the potential W (cf. Hypothesis* 1 *in Section* 3*) such that $a_1 > 0$ and $0 = W(a_1) \leq W(u)$, for $u \in \mathbb{R}^m$, and such that*

$$W(a_+) < W(u),$$

for $|u - a_+| \leq q_0$, for some $q_0 > 0$, with $u \neq a_+$.

(H3) *There holds*

$$u(Tx) = Tu(x),$$

where T is the reflection with respect to the first coordinate in either x or u, that is, $Tx = (-x_1, x_2, \ldots, x_n)$ and $Tu = (-u_1, u_2, \ldots, u_m)$. Moreover, $u_0(Tx) = Tu_0(x)$.

(H4) *The domain Ω is globally Lipschitz and convex symmetric, in the sense that*

$$(x_1, x_2, \ldots, x_n) \in \Omega \quad \textit{implies} \quad (tx_1, x_2, \ldots, x_n) \in \Omega,$$

for all t, with $|t| \leq 1$.

Finally, we have the following hypotheses on the solution and the connecting orbit.

(H5) *A solution $u : \Omega \to \mathbb{R}^m$ of problem* (6.4) *is a* minimizer *if*

$$J(u; \Omega') = \min J(v; \Omega'), \text{ such that } v = u \text{ on } \partial\Omega',$$

over the class of bounded and smooth domains $\Omega' \subset \Omega$ and over all equivariant maps $v(Tx) = Tv(x)$ in $W^{1,2}(\Omega; \mathbb{R}^m)$. (Cf. (3.1) *in Section* 3.*)*

(H6) *We assume that there is a unique orbit $U : \mathbb{R} \to \mathbb{R}^m$ connecting the minima $a_\pm = (\pm a_1, a_2, \ldots, a_m)$ such that*

$$U'' - W_u(U) = 0, \text{ with } U(\pm\infty) = a_\pm,$$

which is also hyperbolic *in the class of symmetric variations $v(Tx) = Tv(x)$, for $v : \mathbb{R} \to \mathbb{R}^m$, that is, zero is not in the spectrum of the linearized operator*

$$Lv := v'' - W_{uu}(U)v, \text{ for symmetric } v \in W^{1,2}(\mathbb{R}; \mathbb{R}^m).$$

Under the above hypotheses, we have the following theorem.

Theorem 6.1 ([11]). *Assume that hypotheses* (H1)–(H6) *hold. Then, there holds the estimate*

$$|u(x) - U(x_1)| \le K\, \mathrm{e}^{-k\, \mathrm{dist}(x, \partial\Omega)},$$

for $x \in \Omega$ and K, k positive constants.

Finally, we state a theorem under no hypotheses of uniqueness or hyperbolicity of the connecting orbit. Our notation is that of Theorem 3.1 in Section 3 and u is a solution as in that theorem.

Theorem 6.2 ([6]). *Let Π_1 be a wall of D and assume that a_2 is the reflection of a_1 with respect to Π_1. Moreover, assume that the set of orbits connecting a_1 to a_2 is nonempty. Then, there exists a $\nu_1 \in \mathrm{Int}\,\Pi_1$, with $|\nu_1| = 1$, and a sequence $\{\lambda_k\} \to +\infty$ such that*

$$u(x + \lambda_k \nu_1) \to U(P_1 x),$$

where U is a connection between a_1 and a_2, and P_1 is the orthogonal projection to $\Pi_1{}^\perp$.

Remark 6.3. One would expect that a stronger version of the above theorem holds: Given any $\nu_1 \in \mathrm{Int}(\Pi_1)$ and any sequence $\lambda_k \to +\infty$ there exists a subsequence $\{\lambda_k{}'\}$ of $\{\lambda_k\}$ such that $u(x + \lambda_k{}'\nu_1) \to U(P_1 x)$.

References

[1] G. ALBERTI, *Variational methods for phase transitions, an approach via Γ-convergence*, In: L. Ambrosio and N. Dancer, "Calculus of Variations and Partial Differential Equations", G. Buttazzo, A. Marino, and M. K. V. Murthy (eds.) Springer, Berlin, 2000, 95–114.

[2] G. ALBERTI, L. AMBOSIO and X. CABRÉ, *On a long-standing conjecture of E. De Giorgi: Symmetry in 3D for general nonlinearities and a local minimality property*, Acta Appl. Math. **65** (2001), 9–33.

[3] N. D. ALIKAKOS, *Some basic facts on the system* $\Delta u - W_u(u) = 0$, Proc. Amer. Math. Soc. **139** (2011), 153–162.

[4] N. D. ALIKAKOS, *A new proof for the existence of an equivariant entire solution connecting the minima of the potential for the system* $\Delta u - W_u(u) = 0$, Comm. Partial Diff. Eqs. **37** (2012), 2093–2115.

[5] N. D. ALIKAKOS, P. ANTONOPOULOS and A. DAMIALIS, *Plateau angle conditions for the vector-valued Allen–Cahn equation*, preprint.

[6] N. D. ALIKAKOS and P. W. BATES, in preparation.

[7] N. D. ALIKAKOS and A. C. FALIAGAS, *The stress-energy tensor and Pohozaev's identity for systems*, Acta Math. Scientia **32** (2012), 433–439.

[8] N. D. ALIKAKOS and A. C. FALIAGAS, in preparation.

[9] N. D. ALIKAKOS and G. FUSCO, *Entire solutions to nonconvex variational elliptic systems in the presence of a finite symmetry group*, In: "Singularities in Nonlinear Evolution Phenomena and Applications", M. Novaga and G. Orlandi (eds), Publications of the Scuola Normale Superiore, CRM Series, Birkhäuser, 2009, 1–26.

[10] N. D. ALIKAKOS and G. FUSCO, *Entire solutions to equivariant elliptic systems with variational structure*, Arch. Rat. Mech. Anal. **202** (2011), 567–597.

[11] N. D. ALIKAKOS and G. FUSCO, *On the asymptotic behavior of symmetric solutions of the elliptic system* $\Delta u = W_u(u)$ *in unbounded domains*, preprint.

[12] F. J. ALMGREN, JR., "Existence and Regularity Almost Everywhere of Solutions to Elliptic Variational Problems with Constraints", Mem. Amer. Math. Soc. **165**, 1976.

[13] L. AMBROSIO and X. CABRÉ, *Entire solutions of semilinear elliptic equations in* $\mathbb{R}^3$ *and a conjecture of De Giorgi*, J. Amer. Math. Soc. **13** (2000), 725–739.

[14] S. BALDO, *Minimal interface criterion for phase transitions in mixtures of Cahn–Hilliard fluids*, Ann. Inst. Henri Poincaré, Anal. Non Linéaire **7** (1990), 67–90.

[15] P. W. BATES, G. FUSCO and P. SMYRNELIS, *Crystalline and other entire solutions of the vector Allen–Cahn equation*, preprint.

[16] H. BERESTYCKI, T.-C. LIN, J. WEI and C. ZHAO, *On phase-separation models: Asymptotics and qualitative properties*, Arch. Rat. Mech. Anal. **208** (2013), 163–200.

[17] H. BERESTYCKI, S. TERRACINI, K. WANG and J. WEI, *Existence and stability of entire solutions of an elliptic system modeling phase separation*, Adv. Math., to appear.

[18] F. BETHUEL, H. BREZIS and F. HÉLEIN, "Ginzburg–Landau Vortices", Progress in Nonlinear Differential Equations and Their Applications, Vol. 13, Birkhäuser, Basel and Boston, 1994.

[19] L. BRONSARD, C. GUI and M. SCHATZMAN, *A three-layered minimizer in* $\mathbf{R}^2$ *for a variational problem with a symmetric three-well potential*, Comm. Pure Appl. Math. **49** (1996), 677–715.

[20] L. CAFFARELLI, N. GAROFALO and F. SEGALA, *A gradient bound for entire solutions of quasi-linear equations and its consequences*, Comm. Pure Appl. Math. **47** (1994), 1457–1473.

[21] C. C. CHAN, "Structure of the Singular set in Energy-minimizing Partitions and Area-minimizing Surfaces in $\mathbf{R}^N$", Ph.D. thesis, Stanford University, 1995.

[22] T. H. COLDING and W. P. MINICOZZI II, "A Course in Minimal Surfaces", Graduate Studies in Mathematics, Vol. 121, American Mathematical Society, Providence, RI, 2011.

[23] E. DE GIORGI, *Convergence problems for functionals and operators*, In: E. De Giorgi, E. Magenes, and U. Mosco, "Recent methods in non-linear analysis", Proc. Int. Meet., Rome, 1978. Pitagora, Bologna, 1979, 131–188.

[24] M. DEL PINO, M. KOWALCZYK and J. WEI, *On De Giorgi's conjecture in dimension* $N \geq 9$, Ann. Math. (2) **174** (2011), 1485–1569.

[25] M. DEL PINO, M. KOWALCZYK and J. WEI *On De Giorgi's conjecture and beyond*, Proc. Natl. Acad. Sci. USA **109** (2012), 6845–6850.

[26] L. C. EVANS, "Partial Differential Equations", Graduate Studies in Mathematics **19**, second edition, American Mathematical Society, Providence, RI, 2010.

[27] A. FARINA, *Two results on entire solutions of Ginzburg–Landau system in higher dimensions*, J. Funct. Anal. **214** (2004), 386–395.

[28] A. FARINA, *Some symmetry results for entire solutions of an elliptic system arising in phase separation*, Discrete Contin. Dyn. Syst., Ser. A, to appear.

[29] A. FARINA and N. SOAVE, *Monotonicity and 1-dimensional symmetry for solutions of an elliptic system arising in Bose–Einstein condensation*, preprint.

[30] A. FARINA and E. VALDINOCI, *The state of the art for a conjecture of de Giorgi and related problems*, In: "Recent Progress on Reaction-diffusion Systems and Viscosity Solutions", H. Ishii, W.-Y. Lin and Y. Du (eds.), World Scientific, Hackensack, NJ, 2009, 74–96.

[31] M. FAZLY and N. GHOUSSOUB, *De Giorgi type results for elliptic systems*, Calc. Var., in press.

[32] W. H. FLEMING, *On the oriented Plateau problem*, Rend. Circ. Mat. Palermo **11** (1962), 69–90.

[33] G. FUSCO, *Equivariant entire solutions to the elliptic system $\Delta u = W_u(u)$ for general G-invariant potentials*, preprint.

[34] N. GHOUSSOUB and C. GUI, *On a conjecture of De Giorgi and some related problems*, Ann. Math. (2) **157** (2003), 313–334.

[35] L. C. GROVE and C. T. BENSON, "Finite Reflection Groups", Graduate Texts in Mathematics, Vol. 99, second edition. Springer-Verlag, Berlin, 1985.

[36] C. GUI, *Hamiltonian identities for elliptic differential equations*, J. Funct. Anal. **254** (2008), 904–933.

[37] C. GUI and M. SCHATZMAN, *Symmetric quadruple phase transitions*, Indiana Univ. Math. J. **57** (2008), 781–836.

[38] D. KINDERLEHRER, L. NIRENBERG and J. SPRUCK, *Regularity in elliptic free boundary problems I*, J. Anal. Math. **34** (1978), 86–119.

[39] M. KOWALCZYK, Y. LIU and F. PACARD, *The classification of four-end solutions to the Allen–Cahn equation on the plane*, Anal. PDE, to appear.

[40] L. D. LANDAU and E. M. LIFSHITZ, "The Classical Theory of Fields", Course of Theoretical Physics **2**, fourth edition, Butterworth-Heinemann, 1975.

[41] L. MODICA, *A gradient bound and a Liouville theorem for nonlinear Poisson equations*, Comm. Pure Appl. Math. **38** (1985), 679–684.

[42] L. MODICA, *Monotonicity of the energy for entire solutions of semilinear elliptic equations*, In: "Partial Differential Equations and the Calculus of Variations", F. Colombini, A. Marino, L. Modica and S. Spagnolo (eds.), Vol. II, Essays in honor of Ennio De Giorgi,

Progress in Nonlinear Differential Equations and Their Applications, Vol. 2, Birkhäuser, Boston, MA, 1989, 843–850.

[43] F. MORGAN, *Harnack-type mass bounds and Bernstein theorems for area-minimizing flat chains modulo ν*, Comm. Partial Diff. Eqs. **11** (1986), 1257–1283.

[44] O. SAVIN, *Regularity of flat level sets in phase transitions*, Ann. Math. (2) **169** (2009), 41–78.

[45] L. SIMON, "Lectures on Geometric Measure Theory", Proc. Centre Math. Anal. **3**. Australian National University, Canberra, 1983.

[46] L. SIMON, *Cylindrical tangent cones and the singular set of minimal submanifolds*, J. Diff. Geom. **38** (1993), 585–652.

[47] M. SPIVAK, "A Comprehensive Introduction to Differential Geometry", Vol. 4, third edition, Publish or Perish, Houston, TX, 1999.

[48] V. STEFANOPOULOS, *Heteroclinic connections for multiple-well potentials: The anisotropic case*, Proc. R. Soc. Edinb. A **138** (2008), 1313–1330.

[49] P. STERNBERG and K. ZUMBRUN, *A Poincaré inequality with applications to volume-constrained area-minimizing surfaces*, J. Reine Angew. Math. **503** (1998), 63–85.

[50] J. TAYLOR, *The structure of singularities in soap-bubble and soap-bubble-like minimal surfaces*, Ann. Math. (2) **103** (1976), 489–539.

[51] B. WHITE, *Existence of least-energy configurations of immiscible fluids*, J. Geom. Anal. **6** (1996), 151–161.

[52] B. WHITE, "Topics in Geometric Measure Theory", Lecture notes, Stanford University, 2012.

The nonlinear multidomain model: a new formal asymptotic analysis

Stefano Amato, Giovanni Bellettini and Maurizio Paolini

Abstract. We study the asymptotic analysis of a singularly perturbed weakly parabolic system of m- equations of anisotropic reaction-diffusion type. Our main result formally shows that solutions to the system approximate a geometric motion of a hypersurface by anisotropic mean curvature. The anisotropy, supposed to be uniformly convex, is explicit and turns out to be the dual of the star-shaped combination of the m original anisotropies.

1 Introduction

The bidomain model, a simplified version of the FitzHugh-Nagumo system, was originally introduced in electrocardiology as an attempt to describe the electric potentials and current flows inside and outside the cardiac cells, see [1, 14–16] and references therein. In spite of the discrete cellular structure, at a macroscopic level the intra (i) and the extra (e) cellular regions can be thought of as two superimposed and interpenetrating continua, thus coinciding with the domain Ω (the physical region occupied by the heart). Denoting the intra and extra cellular electric potentials respectively with $u_{\mathrm{i}} = u_{\mathrm{i}_\epsilon}$, $u_{\mathrm{e}} = u_{\mathrm{e}_\epsilon} : [0, T] \times \Omega \to \mathbb{R}$, the bidomain model can be formulated using the following weakly parabolic system of two singularly perturbed linearly anisotropic reaction-diffusion equations, of variational nature[1]:

$$\begin{cases} \epsilon \partial_t (u_{\mathrm{i}} - u_{\mathrm{e}}) - \epsilon \operatorname{div}\left(M^{\mathrm{i}}(x)\nabla u_{\mathrm{i}}\right) + \dfrac{1}{\epsilon} f\left(u_{\mathrm{i}} - u_{\mathrm{e}}\right) = 0, \\ \epsilon \partial_t (u_{\mathrm{i}} - u_{\mathrm{e}}) + \epsilon \operatorname{div}\left(M^{\mathrm{e}}(x)\nabla u_{\mathrm{e}}\right) + \dfrac{1}{\epsilon} f\left(u_{\mathrm{i}} - u_{\mathrm{e}}\right) = 0, \end{cases} \tag{1.1}$$

[1] See Remark 3.2 below.

coupled with suitable initial and boundary conditions. Here $\epsilon \in (0,1)$ is a small positive parameter, f is the derivative of a double-well potential with minima at $s_\pm$ (the standard choice is $f(s) = \frac{d}{ds}\left((1-s^2)^2\right)$, so that $s_\pm = \pm 1$), and $M^{\mathrm{i}}(x)$, $M^{\mathrm{e}}(x)$ are two symmetric uniformly positive definite matrices.

The whole process that determines $u_{\mathrm{i}}, u_{\mathrm{e}}$, and in particular the behaviour of the transmembrane potential

$$u = u_\epsilon := u_{\mathrm{i}} - u_{\mathrm{e}},$$

is quite complicated: we refer the reader to the already quoted references for a more accurate description of the physiological phenomenon and its mathematical modelization. For our purposes, here it suffices to recall that the transmembrane potential typically exhibits a thin transition region (of order ϵ) which separates the advancing depolarized region where $u_\epsilon \approx s_+$ from the one where $u_\epsilon \approx s_-$, see [4, 7] and references therein. Remarkably, a not negligible *nonlinear anisotropy* occurs in the limit $\epsilon \to 0^+$, because of the fibered structure of the myocardium. To explain the appearence of the anisotropy, let us introduce the riemannian norms $\phi_{\mathrm{i}}, \phi_{\mathrm{e}}$, defined as

$$(\phi_{\mathrm{i}}(x,\xi^*))^2 = \alpha_{\mathrm{i}}(x,\xi^*) := M^{\mathrm{i}}(x)\xi^* \cdot \xi^*,$$
$$(\phi_{\mathrm{e}}(x,\xi^*))^2 = \alpha_{\mathrm{e}}(x,\xi^*) := M^{\mathrm{e}}(x)\xi^* \cdot \xi^*,$$

where ξ^* denotes a generic covector of the dual $(\mathbb{R}^N)^*$ of $\mathbb{R}^N$, $N \geq 2$, and $\cdot$ is the euclidean scalar product. The squared norms α_{i} and α_{e} describe the microscopic structure of the intra and extra cellular regions[2], and their hessians $\frac{1}{2}\nabla^2_{\xi^*}\alpha_{\mathrm{i}}$, $\frac{1}{2}\nabla^2_{\xi^*}\alpha_{\mathrm{e}}$ (with respect to ξ^*) give M^{i} and M^{e} respectively. Then the anisotropy manifests, for instance, recalling the following *formal* result [4]: as $\epsilon \to 0^+$, the zero level set of u_ϵ approximates a geometric motion of a front, evolving by *Φ^o-anisotropic mean curvature flow*, where Φ^o denotes the dual of Φ, and the anisotropy Φ turns out to be the star-shaped combination (see [7]) of ϕ_{i} and ϕ_{e}, *i.e.* its square satisfies

$$\Phi^2 := \left(\frac{1}{\alpha_{\mathrm{i}}} + \frac{1}{\alpha_{\mathrm{e}}}\right)^{-1}, \tag{1.2}$$

supposing a priori that Φ^2 is smooth and uniformly *convex*. This convergence result is substantiated by a Γ-convergence theorem (at the level

[2] ϕ_{i} and ϕ_{e} depend on the spatial variable x, since the fibers' orientation changes from point to point.

of the corresponding actions) to a geometric functional, the integrand of which is strictly related to (1.2), see [1] and Theorem 3.6 below.

Note that Φ is *not riemannian* anymore (*i.e.*, a nonlinear anisotropy in the language of the present paper), and it may also *fail* to be convex (this latter property can be seen through an explicit example described in [7]). Lackness of an underlying scalar product for Φ suggests that it is natural to depart from the riemannian structure of (1.1) and to consider, more generally, the *nonlinear* bidomain model, described by

$$\begin{cases} \epsilon\partial_t(u_{\rm i} - u_{\rm e}) - \epsilon \operatorname{div}\left(T_{\phi_{\rm i}}(x, \nabla u_{\rm i})\right) + \dfrac{1}{\epsilon} f\left(u_{\rm i} - u_{\rm e}\right) = 0, \\ \epsilon\partial_t(u_{\rm i} - u_{\rm e}) + \epsilon \operatorname{div}\left(T_{\phi_{\rm e}}(x, \nabla u_{\rm e})\right) + \dfrac{1}{\epsilon} f\left(u_{\rm i} - u_{\rm e}\right) = 0, \end{cases} \tag{1.3}$$

where now $\phi_{\rm i}$ and $\phi_{\rm e}$ are two smooth symmetric uniformly convex[3] Finsler metrics [2], and setting as before $\alpha_{\rm i} = \phi_{\rm i}^2$, $\alpha_{\rm e} = \phi_{\rm e}^2$, the maps

$$T_{\phi_{\rm i}} := \frac{1}{2}\nabla_{\xi^*}\alpha_{\rm i}, \qquad T_{\phi_{\rm e}} := \frac{1}{2}\nabla_{\xi^*}\alpha_{\rm e}$$

are the so-called duality maps, taking $(\mathbb{R}^N)^*$ into $\mathbb{R}^N$. An analog of the above mentioned formal convergence result, to the Φ^o-anisotropic mean curvature flow, appears to hold also in this nonlinear setting, still assuming Φ^2 to be uniformly convex, see [7], where a starting analysis of the geometric meaning of the star-shaped combination of two anisotropies is also carried on.

It is interesting to remark that, generalizing system (1.3) to an arbitrary number m of Finsler symmetric metrics $\phi_1, \ldots, \phi_m$, leads to rewrite the problem, that we have called the nonlinear multidomain model, in a slightly different and more natural way[4], as follows: we seek functions $w^r = w^r_\epsilon$ satisfying the weakly parabolic system

$$\begin{cases} \epsilon\partial_t u - \epsilon \operatorname{div}\left(T_{\phi_r}(x, \nabla w^r)\right) + \dfrac{1}{\epsilon} f\left(u\right) = 0, \quad r = 1, \ldots, m, \\ u = \displaystyle\sum_{r=1}^{m} w^r, \end{cases} \tag{1.4}$$

where

$$T_{\phi_r} := \frac{1}{2}\nabla_{\xi^*}\alpha_r \quad \text{and} \quad \alpha_r := \phi_r^2, \qquad r = 1, \ldots, m.$$

[3] Convexity of $\phi_{\rm i}$ and $\phi_{\rm e}$ is of course required in order to ensure well-posedness of (1.3).

[4] When $m = 2$, (1.3) and (1.4) are equivalent, with the positions $u_{\rm i} = w^1$ and $u_{\rm e} = -w^2$.

It is the purpose of the present paper to provide an asymptotic analysis of the zero level set of $u = u_\epsilon$ in (1.4): we will show, in particular, that $\{u_\epsilon(t, \cdot) = 0\}$ converges to the Φ^o-anisotropic mean curvature flow (see (5.54) below), where Φ^2, supposed to be uniformly convex, reads as

$$\Phi^2 := \left(\sum_{r=1}^{m} \frac{1}{\alpha_r} \right)^{-1},$$

thus generalizing the above mentioned convergence result for the linear and nonlinear bidomain models. Our proof, which remains at a *formal level*, is based on a new asymptotic expansion for (1.4), rewritten equivalently as a system of one parabolic equation and $(m-1)$ elliptic equations[5]. The asymptotic expansion is simpler, and at the same time carried on at a higher order of accuracy, with respect to the one exhibited in [7] for the case $m = 2$.

Before passing to describe the content of the paper, two observations are in order. The first one concerns the case in which Φ^2 is known a priori to be uniformly convex: since we are dealing with systems, confirming rigorously the convergence result[6] for the sets $\{u_\epsilon(t, \cdot) = 0\}$ is still an open problem, even in the simplest case (1.1) (see Theorem 3.5 for a precise statement). The second remark concerns the case when Φ is nonconvex[7]: the question arises on what could be in this case the limit behaviour (if any), as $\epsilon \to 0^+$, of solutions to (1.3) (or also to (1.1)). Indeed, for a nonconvex Φ, the corresponding anisotropic mean curvature flow is ill-posed, and consequently highly unstable. The answer to this question seems, at the moment, out of reach, even at a formal level.

Let us now briefly describe the plan of the paper. In Section 2 we recall the definition of star bodies, and we introduce the star-shaped operation for an arbitrary number of star-shaped anisotropies, using the formalism of gauges and radial functions. In Section 3 we recall some known results on the linear and nonlinear bidomain models. The nonlinear multidomain model is introduced in Section 4. Section 5 contains the main result concerning the convergence of $\{u_\epsilon(t, \cdot) = 0\}$ to the Φ^o-anisotropic mean curvature flow.

---

[5] This shows, among other things, the nonlocality of solutions of (1.4).

[6] This, however, could be hopely less hard to prove than a convergence result of the Allen-Cahn's (2×2)-system, to curvature flow of networks, see [11] for a formal result in this direction.

[7] In this case Φ is not the dual of a convex anisotropy.

2 Star-shaped combination of star bodies and of anisotropies

We start with the following definitions. Let V denote either $\mathbb{R}^N$ or its dual $(\mathbb{R}^N)^*$, endowed with the euclidean norm $|\cdot|$. Let $\mathbb{S}^{N-1} := \{v \in V : |v| = 1\}$.

Definition 2.1 (Star-shaped anisotropies). A *star-shaped anisotropy* on V is a continuous function $\phi : V \to [0, +\infty)$, positive out of the origin, and positively one-homogeneous. ϕ is said to be *symmetric* if $\phi(-v) = \phi(v)$ for any $v \in V$.

Definition 2.2 (Convex and linear anisotropies). Let ϕ be a star-shaped anisotropy on V. If ϕ is convex, then it is called a *convex anisotropy*. A convex anisotropy which is the square root of a quadratic form[8] is called a *linear anisotropy*.

Denote with $\mathcal{S}$ the family of *star bodies*:

$$\mathcal{S} := \Big\{ K \subset V \, : \, K \text{ is compact, star-shaped with respect to } 0 \in \operatorname{int}(K), \\ \partial K \text{ is graph over } \mathbb{S}^{N-1} \text{ of a continuous function} \Big\}.$$

Associated with every $K \in \mathcal{S}$, we define the function

$$\phi_K(v) := \inf\{\lambda > 0 \, : \, v \in \lambda K\}, \qquad v \in V,$$

which is sometimes called *gauge* of K, and is a star-shaped anisotropy with $K = \{\phi_K \le 1\}$. A convex set $K \in \mathcal{S}$ is called a *convex body*, see [18,19] and references therein. In this latter case, ϕ_K is usually called *Minkowski functional* of K, see for instance [17],[9] and it is obviously a convex anisotropy.

For $K \in \mathcal{S}$, the function

$$\phi_K^o(v^*) := \sup\{\langle v^*, v\rangle \, : \, v \in K\}, \qquad v^* \in V^*,$$

where $\langle\cdot,\cdot\rangle$ is the duality (identified with the euclidean scalar product $\cdot$) between V and its dual V^*, is called *support function* of K. It is often denoted by h_K and is also called the dual of ϕ_K (or also the anisotropy dual to ϕ_K). The corresponding set $K^o := \{\phi_K^o \le 1\}$ is called the *dual* of K and it is always a convex body, see again [17]; equivalently ϕ_K^o is always a convex anisotropy on V^*.

---

[8] A linear anisotropy is obviously symmetric.

[9] When K is symmetric with respect to the origin, K is said a symmetric convex body, and ϕ_K turns out to be a norm equivalent to the euclidean one.

For $K \in \mathcal{S}$, let $\varrho_K : \mathbb{S}^{N-1} \to (0, +\infty)$ be the *radial function* of K (see for instance [19]) defined as

$$\varrho_K(\nu) := \sup\left\{\lambda \geq 0 \,:\, \lambda\nu \in K\right\}, \qquad \nu \in \mathbb{S}^{N-1}. \tag{2.1}$$

The function ϱ_K is extended (keeping the same symbol) in a one-homogeneous way on the whole of V, *i.e.*, $\varrho_K(v) = |v|\varrho_K(\frac{v}{|v|})$ for any $v \in V \setminus \{0\}$. Notice that

$$\varrho_K(\nu) = \frac{1}{\phi_K(\nu)}, \qquad \nu \in \mathbb{S}^{N-1}, \tag{2.2}$$

and

$$K = \{\lambda\nu \,:\, 0 \leq \lambda \leq \varrho_K(\nu),\ \nu \in \mathbb{S}^{N-1}\}.$$

Remark 2.3. The previous definitions of ϕ_K, ϕ_K^o and ϱ_K can be generalized, by allowing a continuous dependence on the space variable x in some open subset Ω of $\mathbb{R}^N$. In this way we have that $\phi_K = \phi_K(x, v)$, as well as $\varrho_K = \varrho_K(x, v)$, are defined for $(x, v) \in \Omega \times V$ and $\phi_K^o(x, v^*)$ is defined for $(x, v^*) \in \Omega \times V^*$[10]. In the present paper, however, we will be mostly interested in space-independent anisotropies.

Assumption: in this paper we deal only with sets $K \in S$ having smooth boundary. In the case K is a convex body, we will always suppose that K is smooth and uniformly convex, so that K^o is also smooth and uniformly convex[11]. In this case, we say that ϕ_K^2 (or also that ϕ_K) is smooth and uniformly convex.

Also, for simplicity all anisotropies we consider will be assumed to be symmetric.

Now, consider $K_1,\ K_2 \in \mathcal{S}$. We let $\varrho_{K_1} \star \varrho_{K_2} : \mathbb{S}^{N-1} \to (0, +\infty)$ be defined as follows [7]:

$$\varrho_{K_1} \star \varrho_{K_2}(\nu) := \sqrt{\left(\varrho_{K_1}(\nu)\right)^2 + \left(\varrho_{K_2}(\nu)\right)^2}, \qquad \nu \in \mathbb{S}^{N-1}.$$

Again, $\varrho_{K_1} \star \varrho_{K_2}$ is extended (keeping the same symbol) in a one-homogeneous way on the whole of V.

---

[10] A continuous function $\phi : \Omega \times V \to [0, +\infty)$ is called an *inhomogeneous star-shaped anisotropy* on Ω, provided $\phi(x, \cdot)$ is positively one-homogeneous for any $x \in \Omega$, and there exist two constants c, C with $0 < c \leq C < +\infty$ such that $c|v| \leq \phi(x, v) \leq C|v|$ for any $x \in \Omega$ and $v \in V$. If in addition $\phi(x, \cdot)$ is convex for every $x \in \Omega$, then ϕ is called a (inhomogeneous) convex anisotropy (or also a Finsler metric) on Ω. Eventually, if $\phi(x, \cdot)$ is the square root of a quadratic form, then ϕ is a Riemannian metric (an inhomogeneous linear anisotropy).

[11] Hence, our Finsler metrics will be smooth and uniformly convex, in the sense that for any $x \in \Omega$, the function $\phi(x, \cdot)$ is uniformly convex and smooth.

Definition 2.4 (Star-shaped combination of two sets). Given $K_1, K_2 \in \mathcal{S}$, we define the *star-shaped combination*

$$K_1 \star K_2$$

of K_1 and K_2 as the set whose radial function coincides with $\varrho_{K_1} \star \varrho_{K_2}$:

$$\varrho_{K_1 \star K_2} := \varrho_{K_1} \star \varrho_{K_2}.$$

One checks that $K_1 \star K_2 \in \mathcal{S}$, and that the identity element for $\star$ does not belong to $\mathcal{S}$. Moreover

$$K_1 \star K_2 = K_2 \star K_1.$$

It is clear that the set $K_1 \star K_2$ depends on K_1 and K_2 and not only on $K_1 \cup K_2$. However, it cannot be viewed as the union of an enlargement of K_1 with an enlargement of K_2.

The next formula gives the concrete way to compute the star-shaped combination of two sets $K_1, K_2 \in \mathcal{S}$:

$$\partial\,(K_1 \star K_2) := \left\{ \sqrt{\lambda_1^2 + \lambda_2^2}\, \nu \;:\; \nu \in \mathbb{S}^{N-1},\ \lambda_j = \varrho_{K_j}(\nu),\ j = 1, 2 \right\}. \quad (2.3)$$

Remark 2.5. The reason for using star bodies, instead of convex sets, in Definition 2.4 is the following: if K_1 and K_2 are two convex bodies, then $K_1 \star K_2$ is not in general a convex body. An explicit counterexample for $N = 2$ is given in [7], and it involves the two ellipses:

$$K_1 := \left\{(x, y) \in \mathbb{R}^2 : x^2 + \rho y^2 = 1\right\},$$
$$K_2 := \left\{(x, y) \in \mathbb{R}^2 : \rho x^2 + y^2 = 1\right\},$$

defined for $\rho > 0$. Then

(i) $K_1 \star K_2$ is (smooth and) strictly convex, for $\rho \in (\frac{1}{3}, 3)$;
(ii) $K_1 \star K_2$ is (smooth and) convex, for $\rho = \frac{1}{3}$ or $\rho = 3$, with zero boundary curvature at the points of intersection with the lines $\{(x, y) \in \mathbb{R}^2 : x \pm y = 0\}$;
(iii) $K_1 \star K_2$ is (smooth and) not convex, for $\rho < \frac{1}{3}$ or $\rho > 3$.

Observe that for any $K_1, K_2, K_3 \in \mathcal{S}$ we have:

$$(\varrho_{K_1} \star \varrho_{K_2}) \star \varrho_{K_3} = \varrho_{K_1} \star (\varrho_{K_2} \star \varrho_{K_3}),$$

or equivalently:

$$\varrho_{K_1 \star K_2} \star \varrho_{K_3} = \varrho_{K_1} \star \varrho_{K_2 \star K_3}.$$

This observation leads to the following definition.

Definition 2.6 (Star-shaped combination of m sets). Given $m \geq 2$ and $K_1, \ldots, K_m \in \mathcal{S}$, we let

$$\mathop{\star}_{j=1}^{m} \varrho_{K_j}(\nu) := \sqrt{\sum_{j=1}^{m} \left(\varrho_{K_j}(\nu)\right)^2}, \qquad \nu \in \mathbb{S}^{N-1}, \tag{2.4}$$

extended (keeping the same symbol) in a one-homogeneous way on the whole of V, and

$$\mathop{\star}_{j=1}^{m} K_j$$

be the set in $\mathcal{S}$ whose radial function is given by $\mathop{\star}_{j=1}^{m} \varrho_{K_j}$.

Again, note that

$$\partial\left(\mathop{\star}_{j=1}^{m} K_j\right) = \left\{ \sqrt{\sum_{j=1}^{m} \lambda_j^2}\, \nu \ : \ \nu \in \mathbb{S}^{N-1},\ \lambda_j = \varrho_{K_j}(\nu),\ j = 1, \ldots, m \right\}.$$

Problem 2.7. An open problem is to characterize those sets in $\mathcal{S}$ obtained as star-shaped combination of m symmetric convex bodies[12], more precisely to characterize the class

$$\left\{ \mathop{\star}_{j=1}^{m} K_j : K_1, \cdots, K_m \text{ smooth symmetric uniformly convex bodies} \right\}. \tag{2.5}$$

Remark 2.8. From (2.2) and (2.4), it follows the formula

$$\left(\phi_{\mathop{\star}_{j=1}^{m} K_j}(\nu)\right)^2 = \left(\sum_{j=1}^{m} \frac{1}{\left(\phi_{K_j}(\nu)\right)^2}\right)^{-1}, \qquad \nu \in \mathbb{S}^{N-1}, \tag{2.6}$$

extended (keeping the same symbol) in a one-homogeneous way on the whole of V.

Definition 2.9 (Combined anisotropy). The function

$$\phi_{\mathop{\star}_{j=1}^{m} K_j}$$

will be called the *star-shaped combination* of $\phi_{K_1}, \ldots, \phi_{K_m}$, or combined anisotropy for short.

---

[12] In [7] some necessary conditions are given in the case $m = 2$, such as the impossibility of cusps or re-entrant corners in $\partial(K_1 \star K_2)$.

According to (2.6), the star-shaped combination of the star-shaped anisotropies $\phi_1, \ldots, \phi_m : V \to [0, +\infty)$ is defined as:

$$\mathop{\star}_{j=1}^{m} \phi_j := \left(\sum_{j=1}^{m} \frac{1}{\phi_j^2} \right)^{-1/2}. \tag{2.7}$$

2.1 On the hessian of the combined anisotropy

Let $\Phi : (\mathbb{R}^N)^* \to [0, +\infty)$ be the star-shaped combination of the star-shaped anisotropies $\phi_1, \ldots, \phi_m : (\mathbb{R}^N)^* \to [0, +\infty)$. Set for notational convenience

$$\alpha := \Phi^2, \qquad \alpha_j := \phi_j^2, \quad j = 1, \ldots, m.$$

Then formula (2.7) can be rewritten as

$$\alpha = \left(\sum_{j=1}^{m} \frac{1}{\alpha_j} \right)^{-1}. \tag{2.8}$$

The aim of this short section is to find an appropriate representation of the hessian

$$\frac{1}{2} \nabla^2 \alpha$$

of α, which will be useful in Section 5. From formula (2.8) it follows:

$$\nabla \alpha = \alpha^2 \sum_{j=1}^{m} \frac{1}{\alpha_j^2} \nabla \alpha_j. \tag{2.9}$$

Set

$$Q := \frac{1}{2} \alpha^2 \sum_{j=1}^{m} \frac{1}{\alpha_j^2} \nabla^2 \alpha_j, \tag{2.10}$$

and

$$Q_0 := \frac{1}{2} \nabla^2 \alpha - Q.$$

From (2.9), we obtain

$$\begin{aligned} Q_0 &= \alpha^3 \left(\sum_{j=1}^{m} \frac{\nabla \alpha_j}{\alpha_j^2} \right) \otimes \left(\sum_{k=1}^{m} \frac{\nabla \alpha_k}{\alpha_k^2} \right) - \alpha^2 \sum_{k=1}^{m} \frac{\nabla \alpha_k \otimes \nabla \alpha_k}{\alpha_k^3} \\ &= \sum_{k=1}^{m} \left(\frac{\alpha^3}{\alpha_k^4} - \frac{\alpha^2}{\alpha_k^3} \right) \nabla \alpha_k \otimes \nabla \alpha_k + \sum_{j \neq k} \frac{\alpha^3}{\alpha_j^2 \alpha_k^2} \nabla \alpha_j \otimes \nabla \alpha_k \qquad (2.11) \\ &= \alpha^2 \sum_{k=1}^{m} \frac{\alpha - \alpha_k}{\alpha_k^4} \nabla \alpha_k \otimes \nabla \alpha_k + \alpha^3 \sum_{j \neq k} \frac{1}{\alpha_j^2 \alpha_k^2} \nabla \alpha_j \otimes \nabla \alpha_k. \end{aligned}$$

For $m = 2$, formulas (2.10) and (2.11) coincide with those given in [7]. Furthermore, we can observe that, as in the case $m = 2$, we have:

$$Q_0(\xi^*)\xi^* = 0, \qquad \xi^* \in (\mathbb{R}^N)^*. \tag{2.12}$$

This relation will be used in the asymptotics, see Section 5.2.4. In order to show (2.12) we use Euler's formula $\nabla\alpha_j(\xi^*)\xi^* = 2\alpha_j(\xi^*)$. We have

$$\begin{aligned}
\frac{1}{2}Q_0(\xi^*)\xi^* &= \alpha^2(\xi^*)\sum_{k=1}^{m}\frac{\alpha(\xi^*)-\alpha_k(\xi^*)}{(\alpha_k(\xi^*))^4}\,\alpha_k(\xi^*)\nabla\alpha_k(\xi^*)\\
&\quad+\alpha^3(\xi^*)\sum_{j\neq k}\frac{1}{(\alpha_j(\xi^*))^2(\alpha_k(\xi^*))^2}\,\alpha_j(\xi^*)\nabla\alpha_k(\xi^*)\\
&=\sum_{k=1}^{m}\Bigg[\frac{\alpha^2(\xi^*)\Big(\alpha(\xi^*)-\alpha_k(\xi^*)\Big)}{(\alpha_k(\xi^*))^3}\\
&\qquad+\frac{\alpha^3(\xi^*)}{(\alpha_k(\xi^*))^2}\sum_{j\neq k}\frac{1}{\alpha_j(\xi^*)}\Bigg]\nabla\alpha_k(\xi^*),
\end{aligned}$$

and each terms in the summation leads (recalling (2.8) and omitting the symbol ξ^*) to:

$$\frac{\alpha^2}{\alpha_k^2}\left[\frac{\alpha-\alpha_k}{\alpha_k}+\alpha\Big(\frac{1}{\alpha}-\frac{1}{\alpha_k}\Big)\right]=0.$$

Using (2.10) and (2.11) we have therefore obtained a representation for

$$\frac{1}{2}\nabla^2\alpha = Q + Q_0.$$

3 The bidomain model

Before starting our analysis on the multidomain model, we briefly summarize some known results on the bidomain model (1.3), *i.e.*, $m = 2$.

Remark 3.1. System (1.3) is equivalent to the following parabolic/elliptic system:

$$\begin{cases}\epsilon\partial_t u - \epsilon\,\mathrm{div}(T_{\phi_{\mathrm{i}}}(x,\nabla w(x)) + \dfrac{1}{\epsilon}f(u) = 0,\\[2mm] \mathrm{div}\Big(T_{\phi_{\mathrm{i}}}(x,\nabla w) + T_{\phi_{\mathrm{e}}}(x,\nabla w - \nabla u)\Big) = 0,\end{cases} \tag{3.1}$$

obtained by taking the difference of the two equations in (1.3), and setting

$$u = u_{\mathrm{i}} - u_{\mathrm{e}}, \qquad w = u_{\mathrm{i}}.$$

Note that, in the linear case, the elliptic equation can be rewritten as

$$-\operatorname{div}\Big(T_{\phi_e}(x, \nabla u)\Big) + \operatorname{div}\Big((T_{\phi_i} + T_{\phi_e})(x, \nabla w)\Big) = 0.$$

Remark 3.2 (Degenerate variational structure). Let F be the primitive of f vanishing in $s_{\pm}$. System (1.3) is the formal gradient flow of the functionals $\mathcal{F}_\epsilon \colon L^2(\Omega;\, \mathbb{R}^2) \to [0, +\infty]$ defined as:

$$\mathcal{F}_\epsilon(v,\omega) := \begin{cases} \displaystyle\int_\Omega \left\{ \frac{\epsilon}{2}\Big[\alpha_i(\nabla v) + \alpha_e(\nabla \omega)\Big] + \frac{1}{\epsilon} F(v - \omega) \right\} dx \\ \qquad\qquad\qquad\qquad\qquad\qquad \text{if } v, \omega \in H^1(\Omega), \\ +\infty \qquad\qquad\qquad\qquad\qquad\quad \text{otherwise,} \end{cases} \tag{3.2}$$

with respect to the degenerate scalar product

$$b\Big((v,\, \omega),\, (\psi_1,\, \psi_2)\Big) := \int_\Omega (v - \omega)(\psi_1 - \psi_2)\, dx.$$

Thus, system (1.3) can be reformulated as:

$$\epsilon b\Big(\partial_t(u_i,\, u_e),\, (\psi_1,\, \psi_2)\Big) + \delta\mathcal{F}_\epsilon\Big((u_i,\, u_e),\ (\psi_1,\, \psi_2)\Big) = 0,$$
$$(\psi_1,\, \psi_2) \in H^1(\Omega;\, \mathbb{R}^2).$$

The following result is proven in [14, Theorem 2], to which we refer for more details.

Theorem 3.3 (Well-posedness in the linear case). *Let $\Omega \subset \mathbb{R}^N$ be a bounded Lipschitz domain. Suppose that*

$\phi_i, \phi_e : \Omega \times (\mathbb{R}^N)^* \to [0, +\infty)$ *are two convex* linear *anisotropies.*

Let $T > 0$ and $\overline{u} \in L^2(\Omega)$. Then there exists a pair

$$(u_i,\, u_e) \in (L^2(0,\, T;\, H^1(\Omega)))^2,$$

uniquely determined up to a family of additive time-dependent constants, with

$$u := u_i - u_e \in C^0\big([0,\ T];\, L^2(\Omega)\big) \cap L^2\big(0,\, T;\, H^1(\Omega)\big),$$
$$\partial_t u \in L^2_{\mathrm{loc}}\big(0,\, T;\, L^2(\Omega)\big),$$

such that (u_i, u_e) solves system (1.1) *distributionally, with initial condition*

$$u(0, \cdot) = \overline{u} \text{ in } \Omega, \tag{3.3}$$

and zero Neumann boundary condition

$$T_{\phi_i}(x, \nabla u_i(t,x)) \cdot \nu_\Omega(x) = T_{\phi_e}(x, \nabla u_e(t,x)) \cdot \nu_\Omega(x) = 0, \qquad (t,x) \in [0,T] \times \Omega, \tag{3.4}$$

where $\nu_\Omega(x)$ *stands for the inward unit vector normal to* $\partial\Omega$ *at point* $x \in \partial\Omega$.

The initial and boundary conditions (3.3), (3.4) are better understood remembering Remark 3.1.

Problem 3.4. To our best knowledge, a well-posedness result for the nonlinear bidomain model (1.3) (even for ϕ_i, ϕ_e independent of x), coupled with (3.3) and (3.4), is an open problem, and it is under investigation.

The next *formal* result is obtained in [4], using an asymptotic expansion argument, developed up to the second order included.

Theorem 3.5 (Formal convergence in the linear case). *Suppose that*

$$\phi_i, \phi_e : \Omega \times (\mathbb{R}^N)^* \to [0,+\infty) \textit{ are two convex } \text{linear } \textit{anisotropies.}$$

Let u_i, u_e *and* $u = u_\epsilon := u_i - u_e$ *be given by Theorem* 3.3, *with initial condition* $\overline{u} = \overline{u}_\epsilon = u_\epsilon(0,\cdot)$ *well-prepared*[13] *and possibly depending on* ϵ, *in particular so that*

$$\{x \in \Omega : \overline{u}_\epsilon(x) = 0\} = \partial E, \qquad \epsilon \in (0,1),$$

where ∂E *is smooth and compact in* Ω.

Suppose furthermore that the combined anisotropy

$$\Phi = \phi_i \star \phi_e \textit{ is uniformly convex.}$$

Then, for any $t \in [0,T]$ *the sets* $\{u_\epsilon(t,\cdot) = 0\}$ *formally converge*[14], *as* $\epsilon \to 0^+$, *to a hypersurface* $\partial E(t)$ *evolving by anisotropic* Φ^o*-mean curvature for* $T > 0$ *sufficiently small, with* $\partial E(0) = \partial E$.

Theorem 3.5 is related to the following result, obtained in [1].

Theorem 3.6 (Γ-convergence in the linear case). *Suppose that*

$$\phi_i, \phi_e : \Omega \times (\mathbb{R}^N)^* \to [0,+\infty) \textit{ are two convex } \text{linear } \textit{anisotropies.}$$

Then

- *there exists the* $\Gamma\big(L^2(\Omega;\mathbb{R}^2)\big) - \lim_{\epsilon\to 0^+} \mathcal{F}_\epsilon = \mathcal{F}$, *and depends only on* $\mathrm{u} = v - \omega$.

---

13 See [4] for the details.

14 With an expected speed rate of order ϵ, up to logarithmic corrections.

- *$\mathcal{F}$ is finite if and only if* $\mathrm{u} \in BV(\Omega; \{s_\pm\})$. *Moreover*

$$\mathcal{F}(v, \omega) = \int_{S_\mathrm{u}} \sigma(x, \nu_\mathrm{u})\, d\mathcal{H}^{N-1}, \tag{3.5}$$

where S_u is the jump set of u, *$\nu_\mathrm{u}(x)$ is a unit normal to S_u at $x \in S_\mathrm{u}$, and σ is a convex symmetric anisotropy*[15].

In addition (assuming for simplicity that ϕ_i and ϕ_e, and hence σ, are independent of x)

- *$\{\sigma(\cdot) \leq 1\}$ contains the convexified of $\{\phi_\mathrm{i} \star \phi_\mathrm{e} \leq 1\}$,*
- *$\{\sigma(\cdot) \leq 1\}$ is contained in the smallest ellipsoid circumscribing the convexified of $\{\phi_\mathrm{i} \star \phi_\mathrm{e} \leq 1\}$ and tangent to it at the intersection with the coordinate axes. Moreover, the strict inclusion holds whenever the two anisotropies are not proportional.*

The following problem has been pointed out in [1].

Problem 3.7. Is it true that the unit ball of σ coincides with the convexified of $\{\phi_\mathrm{i} \star \phi_\mathrm{e} \leq 1\}$?

Problem 3.8. To our best knowledge, in the nonlinear case a Γ-convergence result similar to the one in Theorem 3.6 is an open problem, which is under investigation.

The following *formal* result, generalizing Theorem 3.5, is obtained in [7], using an asymptotic expansion argument, developed up to the first order.

Theorem 3.9. *Theorem* 3.5 *holds when ϕ_i and ϕ_e are two smooth symmetric uniformly convex anisotropies, namely dropping the linearity assumption on T_{ϕ_i} and T_{ϕ_e}.*

Remark 3.10. Set $w^1 = u_\mathrm{i}$, $w^2 = -u_\mathrm{e}$, so that $u := u_\mathrm{i} - u_\mathrm{e} = w^1 + w^2$ and $u_\mathrm{e} = -w^2$. Let also

$$T_{\phi_1} := T_{\phi_\mathrm{i}}, \qquad T_{\phi_2} := T_{\phi_\mathrm{e}}.$$

Then, observing that $T_{\phi_2}(x, -\xi^*) = -T_{\phi_2}(x, \xi^*)$, we can rewrite system (1.3) as

$$\begin{cases} \epsilon\partial_t u(t, x) - \epsilon \mathrm{div}\Big(T_{\phi_1}\big(x, \nabla w^1(t, x)\big)\Big) + \dfrac{1}{\epsilon} f\big(u(t, x)\big) = 0, \\ \epsilon\partial_t u(t, x) - \epsilon \mathrm{div}\Big(T_{\phi_2}\big(x, \nabla w^2(t, x)\big)\Big) + \dfrac{1}{\epsilon} f\big(u(t, x)\big) = 0. \end{cases} \tag{3.6}$$

---

[15] It is also possible to explicitly characterize $\sigma(x, \cdot)$ as an infimum of an appropriate class of vector-valued functions, see [1] for the details.

Note that (3.6), in turn, is equivalent to the parabolic/elliptic system

$$\begin{cases} \epsilon\partial_t u(t,x) - \epsilon \mathrm{div}\Big(T_{\phi_1}\big(x, \nabla w^1(t,x)\big)\Big) + \dfrac{1}{\epsilon} f\big(u(t,x)\big) = 0, \\ \mathrm{div}\Big(T_{\phi_1}\big(x, \nabla w^1(t,x)\big)\Big) = \mathrm{div}\Big(T_{\phi_2}\big(x, \nabla w^2(t,x)\big)\Big). \end{cases} \tag{3.7}$$

This observation will be the starting point of the asymptotic analysis of Section 5.

4 The nonlinear multidomain model

We come now to the main topic of this paper. First of all, in order to treat an arbitrary number m of components, it seems convenient to rewrite the system in a slightly different way[16] (which is the generalization of (3.6)), showing also more clearly the parabolic character of the problem.

Accordingly, let $m \geq 2$, $\phi_1, \dots, \phi_m : (\mathbb{R}^N)^* \to [0, +\infty)$ be smooth symmetric uniformly convex anisotropies, and consider the degenerate system of parabolic PDE's:

$$\begin{cases} \epsilon\partial_t u - \epsilon \mathrm{div}\big(T_{\phi_r}(\nabla w^r)\big) + \dfrac{1}{\epsilon} f\big(u\big) = 0, \ r = 1, \dots, m, \\ u := \displaystyle\sum_{r=1}^{m} w^r, \end{cases} \qquad \text{in } (0,T)\times\Omega, \tag{4.1}$$

in the unknown $(w^1, \dots, w^m) \in L^2(0,T; H^1(\Omega))^m$, where $T_{\phi_r} := \frac{1}{2}\nabla_{\xi^*}\phi_r^2$ is allowed to be nonlinear, $r = 1, \dots, m$ (no summation on the index r is obviously understood in (4.1)).

Our aim is to formally show that, in the limit $\epsilon \to 0^+$, solutions to (4.1) suitably approximate a Φ^o-anisotropic motion by mean curvature, where Φ is the star-shaped combination of the ϕ_r's, under the assumption that Φ is smooth and uniformly convex. We will assume existence of sufficiently smooth solutions to (4.1) (however, recall that even in the case $m = 2$, this is an open problem, see Problem 3.4).

Remark 4.1 (Simplest possible case). Assume that there exists a smooth symmetric uniformly convex anisotropy ϕ such that

for any $r = 1, \dots, m$ there exists $\lambda_r > 0$ so that $\phi_r = \lambda_r \phi$.

---

[16] This, for $m = 2$, corresponds to write $w^1 = u_{\mathrm{i}}$, $w^2 = -u_{\mathrm{e}}$.

If we put $T_\phi := \frac{1}{2}\nabla\phi^2$, system (4.1) can be rewritten as

$$\begin{cases} \epsilon\partial_t u - \epsilon\lambda_r^2 \operatorname{div}\Big(T_\phi\big(\nabla w^r\big)\Big) + \dfrac{1}{\epsilon} f\big(u\big) = 0, \quad r = 1, \dots, m, \\ u = \displaystyle\sum_{r=1}^{m} w^r. \end{cases} \tag{4.2}$$

Suppose also that ϕ is a linear anisotropy, so that

$$\operatorname{div}\big(T_\phi(\nabla u)\big) = \sum_{r=1}^{m} \operatorname{div}\big(T_\phi(\nabla w^r)\big).$$

Dividing each parabolic equation in (4.1) by λ_r^2, summing over $r = 1, \dots, m$ and dividing by $\sum_{r=1}^{m} \frac{1}{\lambda_r^2}$, we obtain

$$\epsilon\partial_t u - \epsilon\left(\sum_{r=1}^{m} \frac{1}{\lambda_r^2}\right)^{-1} \operatorname{div}\Big(T_\phi\big(\nabla u\big)\Big) + \frac{1}{\epsilon} f(u) = 0. \tag{4.3}$$

Hence, by formula (2.7) it follows that u satisfies the scalar Allen-Cahn's equation [17] where we take as anisotropy the star-shaped combination Φ of the original anisotropies, namely

$$\epsilon\partial_t u - \epsilon\operatorname{div}\left(T_\Phi(\nabla u)\right) + \frac{1}{\epsilon} f(u) = 0 \tag{4.4}$$

where as usual $T_\Phi := \frac{1}{2}\nabla\Phi^2$. Under the previous assumptions, we summarize this more precisely as follows. Let be given suitable functions $\overline{u}$ on $\{0\} \times \Omega$ and $d^1, \dots, d^m$ on $[0, T] \times \partial\Omega$, so that $\overline{u} = \sum_{r=1}^{m} d^r$ on $\{0\} \times \partial\Omega$. If $(w^1, \dots, w^m)$ solve (4.2) with an initial condition $\sum_{r=1}^{m} w^r = \overline{u}$ and m (Dirichlet) boundary conditions $w^r = d^r$ for $r = 1, \dots, m$, then $u := \sum_{r=1}^{m} w^r$ solves (4.4), with initial condition $u = \overline{u}$ and Dirichlet boundary condition $u = \sum_{r=1}^{m} d^r$.

[17] Equation (4.4) has been studied for instance in [6, 8] in connection with anisotropic mean curvature flow. Following also the algorithm of [9, 12], a different perspective was adopted in [10], where the authors approached the problem of ϕ-anisotropic mean curvature flow in $[0, T] \times \mathbb{R}^N$ (taking in the present discussion for simplicity ϕ smooth and uniformly convex) considering, in place of the nonlinear operator $\operatorname{div}(T_\phi(\nabla u))$, its linearized version in Fourier space. In this way, the corresponding solution turns out to be the convolution of the initial condition with the kernel $(t, x) \to \mathcal{F}^{-1}(\sigma_\phi^t(\xi))(x)$, $\mathcal{F}$ being the Fourier transform in the ξ-variable, and $\sigma_\phi^t(\xi) := e^{-4\pi^2 t(\phi^o(\xi))^2}$ for any $\xi \in \mathbb{R}^N$. This linearization based on Fourier transform could be, in principle, applied also to (4.1).

Conversely, we can solve (4.2) with initial condition $u = \overline{u}$ and m (Dirichlet) boundary conditions $w^r = d^r$ for $r = 1, \dots, m$, by first solving the parabolic equation (4.4) (with boundary condition given by $u = \sum_{r=1}^m d^r$) and subsequently solving the first $m - 1$ linear elliptic equations at each time t to recover the unknowns $w^1, \dots, w^{m-1}$, and hence also the last one $w^m := u - \sum_{r=1}^{m-1} w^r$. These elliptic equations are obtained by subtracting the first equation in (4.2) from (4.4) and read as

$$\lambda_r^2 \mathrm{div}\Big(T_\phi(\nabla w^r)\Big) = \mathrm{div}\Big(T_\Phi(\nabla u)\Big), \tag{4.5}$$

with (Dirichlet) boundary condition $w^r = d^r$.

In the special case

$$\lambda_r^2 d^r = \lambda_s^2 d^s, \qquad r, s = 1, \dots, m \tag{4.6}$$

or equivalently

$$\lambda_r^2 \Big(\sum_{s=1}^m \frac{1}{\lambda_s^2} \Big) d^r = \sum_{s=1}^m d^s, \tag{4.7}$$

we can recover the unknowns w^r as

$$w^r := \frac{1}{\lambda_r^2} \Big(\sum_{s=1}^m \frac{1}{\lambda_s^2} \Big)^{-1} u, \qquad r = 1, \dots, m.$$

Remark 4.2. Generalizing the previous cases ($m = 2$), one can transform (4.1) into a parabolic equation and $(m - 1)$ elliptic equations. This suggests a way to assign initial/boundary conditions for (4.1), in the form of one initial condition, and m Neumann or Dirichlet boundary conditions.

5 Formal asymptotics of the multidomain model

In this section we perform a new formal asymptotic expansion for the nonlinear multidomain model, assuming for simplicity

$$f(s) = \frac{d}{ds}\left((1 - s^2)^2\right),$$

in particular $s_\pm = \pm 1$. The computations will be simpler, and at the same time more general[18], than those made in [7]. Due to the strong reaction term, we expect the sum $u_\epsilon := \sum_{r=1}^m w_\epsilon^r$ to assume values near

[18] This will be apparent particularly in the inner expansion of Section 5.2 below.

to ± 1 in most of the domain with a thin, smooth, transition region where it transversally crosses the unstable zero of f. We will denote by $\Omega^{\pm} = \cup_{t=0}^{T}(\{t\} \times \Omega^{\pm}(t))$ the two phases. This motivates the use of matched asymptotics in the outer $\Omega^- \cup \Omega^+$ region (*outer expansion*) and in the transition layer (*inner expansion*).

As a formal consequence (see (5.54) below), the front generated by (4.1) propagates with the same law, up to an error of order $\mathcal{O}(\epsilon)$, as the front generated by a Φ^o-anisotropic mean curvature flow starting from a smooth hypersurface $\partial E \subset \Omega$, where Φ is the star-shaped combination of the m original smooth uniformly convex[19] anisotropies $\phi_1, \dots, \phi_m$.

Remembering Remark 3.10, assuming independence of x of all ϕ_r, we write the system in the convenient form

$$\begin{cases} \epsilon \partial_t u_\epsilon - \epsilon \operatorname{div}\Big(T_{\phi_1}(\nabla w_\epsilon^1)\Big) + \dfrac{1}{\epsilon} f(u_\epsilon) = 0, \\ \operatorname{div}\Big(T_{\phi_r}(\nabla w_\epsilon^r)\Big) = \operatorname{div}\Big(T_{\phi_s}(\nabla w_\epsilon^s)\Big), \qquad 1 \le r,\ s \le m, \\ u_\epsilon = \displaystyle\sum_{r=1}^{m} w_\epsilon^r. \end{cases} \tag{5.1}$$

This system consists of one parabolic equation and $(m-1)$ elliptic equations, to be coupled with an initial condition at $\{t = 0\}$, which in particular is required to satisfy

$$\{u_\epsilon(0, \cdot) = 0\} = \partial E, \qquad \epsilon \in (0, 1), \tag{5.2}$$

and m either Neumann or Dirichlet boundary conditions at $\cup_{t=0}^{T}(\{t\} \times \partial\Omega)$. We restore in this section the notational dependence on ϵ for $u = u_\epsilon$ and all $w^r = w_\epsilon^r$.

5.1 Outer expansion

Given $r = 1, \dots, m$, we expand formally u_ϵ and w_ϵ^r in terms of $\epsilon \in (0, 1)$:

$$u_\epsilon = u_0 + \epsilon u_1 + \epsilon^2 u_2 + \dots, \qquad w_\epsilon^r = w_0^r + \epsilon w_1^r + \epsilon^2 w_2^r + \dots$$

Substituting these expressions into the parabolic equation in (5.1) and using the expansion

$$f(u_\epsilon) = f(u_0) + \epsilon f'(u_0) u_1 + \epsilon^2 \left(\frac{u_1^2 f''(u_0)}{2} + f'(u_0) u_2 \right) + \mathcal{O}(\epsilon^3),$$

[19] Convexity of all ϕ_r is necessary in order to make the multidomain model well-posed.

we get

$$f(u_0) = 0, \qquad u_1 f'(u_0) = 0.$$

Hence, excluding $u_0 = 0$ (the unstable zero of f), we get in $(0, T) \times \Omega$,

$$u_0 \in \{1, -1\}, \tag{5.3}$$

$$u_1 \equiv 0. \tag{5.4}$$

We denote by

$$\Sigma_0(t), \qquad t \in (0, T), \tag{5.5}$$

the jump set of $u_0(t, \cdot)$.

Coming back to the elliptic equations in (5.1), we find

$$\begin{cases} \operatorname{div}\Big(T_{\phi_r}(\nabla w_0^r)\Big) = \operatorname{div}\Big(T_{\phi_s}(\nabla w_0^s)\Big) & 1 \le r,\ s \le m \\ \displaystyle\sum_{r=1}^m w_0^r = u_0 \implies \sum_{r=1}^m \nabla w_0^r = 0, \end{cases} \tag{5.6}$$

where the last implication is a consequence of (5.3).

Note also that

$$u_2 = \frac{1}{f'(u_0)} \operatorname{div}\Big(T_{\phi_r}(\nabla w_0^r)\Big), \qquad r = 1, \ldots, m. \tag{5.7}$$

Remark 5.1. (5.6) is a system of $(m-1)$ *nonlinear* elliptic equations in the $(m-1)$ unknown functions w_0^r (for $r = 2, \ldots, m$), since we can solve the algebraic equation in (5.6) with respect to w_0^1.

Remark 5.2. It is important to notice that the boundary conditions across the limit interface $\Sigma_0(t)$, to be coupled with (5.6), will arise by matching the outer expansion with the inner expansion, see (5.65) and (5.68) (jump conditions and jump of the normal derivative). We assume the elliptic problem expressed by (5.6), (5.65), (5.68) (and augmented with Neumann or Dirichlet boundary conditions on $\partial\Omega$) to be solvable, thus providing w_0^r for every $r = 1, \ldots, m$, and therefore u_2 by (5.7).

If we now perform a Taylor expansion for T_{ϕ_r}, we obtain

$$T_{\phi_r}(\eta^* + \epsilon\zeta^*) = T_{\phi_r}(\eta^*) + \epsilon M^r(\eta^*)\zeta^* + \mathcal{O}(\epsilon^2),$$

where $M^r = \frac{1}{2}\nabla^2\alpha_r$, which can be used in the elliptic equations of (5.1) to get equations for w_1^r for any $r = 1, \ldots, m$, namely:

$$\operatorname{div}\Big(M^r(\nabla w_0^r)\nabla w_1^r\Big) = \operatorname{div}\Big(M^s(\nabla w_0^s)\nabla w_1^s\Big), \qquad 1 \le r,\ s \le m.$$

Moreover, from the relation $\sum_{r=1}^{m} w_\epsilon^r = u_\epsilon$, and recalling from (5.4) that $u_1 = 0$, we obtain

$$\sum_{r=1}^{m} w_1^r = 0. \tag{5.8}$$

By solving this latter algebraic equation with respect (for instance) to w_1^1, and substituting it into the previous equation we obtain a system of $(m-1)$ *linear* elliptic equations in the unknowns w_1^r, for $r = 2, \ldots, m$.

Remark 5.3. The outer expansion has been performed without assuming Φ to be convex.

5.2 Inner expansion

For any $\epsilon \in (0, 1)$ let us consider the set

$$E_\epsilon(t) := \{x \in \Omega \, : \, u_\epsilon(t, x) \geq 0\},$$

the boundary of which will be denoted by

$$\Sigma_\epsilon(t) = \{x \in \Omega \, : \, u_\epsilon(t, x) = 0\}. \tag{5.9}$$

Our aim is to formally identify the geometric evolution law of $\Sigma_\epsilon(t)$ as $\epsilon \to 0^+$.

For $r = 1, \ldots, m$ we seek the shape, in the transition layer, of functions w_ϵ^r satisfying

$$\epsilon^2 \partial_t u_\epsilon - \epsilon^2 \mathrm{div}\Big(T_{\phi_r}(\nabla w_\epsilon^r)\Big) + f(u_\epsilon) = 0, \qquad r = 1, \ldots, m, \tag{5.10}$$

with $u_\epsilon = \sum_{r=1}^{m} w_\epsilon^r$. We put, as usual,

$$\alpha_r := \phi_r^2, \qquad T_{\phi_r} := \frac{1}{2}\nabla\alpha_r, \qquad M^r := \frac{1}{2}\nabla^2\alpha_r, \qquad r = 1, \ldots, m,$$

so that, by Euler's identities for homogeneus functions, we have

$$\alpha_r(\xi^*) = T_{\phi_r}(\xi^*) \cdot \xi^* = M^r(\xi^*)\xi^* \cdot \xi^*, \qquad \xi^* \in (\mathbb{R}^N)^*. \tag{5.11}$$

Remember that the matrix M^r depends on the covector ξ^*, unless ϕ_r is a linear anisotropy (*i.e.*, unless T_{ϕ_r} is linear).

5.2.1 Main assumptions and basic notation We assume in this section that

the star shaped combination Φ^2 is smooth,
symmetric and uniformly convex.

This allows to look at Φ as the dual of a smooth uniformly convex anisotropy φ defined in $\mathbb{R}^N$,

$$\Phi = \varphi^o, \qquad \text{namely} \qquad \varphi = \Phi^o. \tag{5.12}$$

Keeping the simpler symbol φ instead of Φ^o, we can accordingly introduce the φ-anisotropic distance d_φ (*i.e.*, $d_\varphi(x, y) = \varphi(y - x)$), and the φ-signed distance function from $\Sigma_\epsilon(t)$ (positive in the interior of $E_\epsilon(t)$):

$$d_\epsilon^\varphi(t, x) := d_\varphi\big(x, \mathbb{R}^N \setminus E_\epsilon(t)\big) - d_\varphi\big(x, E_\epsilon(t)\big).$$

Following [4], it is convenient to introduce the stretched variable y defined as

$$y = y_\epsilon^\varphi(t, x) := \frac{d_\epsilon^\varphi(t, x)}{\epsilon}.$$

We parametrize $\Sigma_\epsilon(t)$ with a parameter

$$s \in \Sigma, \tag{5.13}$$

Σ a fixed reference $(N-1)$-dimensional smooth manifold, and the function $x(s, t; \epsilon)$ gives the position in Ω of the point s at time t.

We let, for x in a tubular neighbourhood of $\Sigma_\epsilon(t)$,

$$n_\epsilon^\varphi(t, x) := -T_\Phi(\nabla d_\epsilon^\varphi(t, x)) \tag{5.14}$$

be the (outward) Cahn-Hoffman's vector field (remember the notation in (5.12)), for which we suppose the expansion:

$$n_\epsilon^\varphi := n_0^\varphi + \epsilon n_1^\varphi + \dots$$

Points on the evolving manifold $\Sigma_\epsilon(t)$ are assumed to move in the direction of n_ϵ^φ, *i.e.*,

$$\partial_t x(s, t; \epsilon) = V_\epsilon^\varphi \, n_\epsilon^\varphi,$$

where V_ϵ^φ is positive for an expanding set, and where we assume the validity of the following expansion:

$$V_\epsilon^\varphi = V_0^\varphi + \epsilon V_1^\varphi + \epsilon^2 V_2^\varphi + \dots \tag{5.15}$$

The anisotropic projection of a point x on $\Sigma_\epsilon(t)$ is denoted by $s_\epsilon^\varphi(t, x)$, and satisfies

$$\partial_t s_\epsilon^\varphi = 0. \tag{5.16}$$

Hence

$$\partial_t d_\epsilon^\varphi(t,x) = V_\epsilon^\varphi\big(s_\epsilon^\varphi(t,x),\, t\big). \tag{5.17}$$

We also recall (see [6, 8]) that div $\big(T_\Phi(\nabla d_\epsilon^\varphi)\big)$ gives the anisotropic mean curvature of the level hypersurface at that point and can be approximated by the anisotropic mean curvature κ_ϵ^φ of $\Sigma_\epsilon(t)$ (positive when $E_\epsilon(t)$ is uniformly convex) as follows

$$\begin{aligned} \operatorname{div}\big(T_\Phi(\nabla d_\epsilon^\varphi(t,x))\big) = {} & -\kappa_\epsilon^\varphi(s_\epsilon^\varphi(t,x),t) \\ & - \epsilon y_\epsilon^\varphi h_\epsilon^\varphi(s_\epsilon^\varphi(t,x),t) + \mathcal{O}(\epsilon^2 (y_\epsilon^\varphi)^2) \end{aligned} \tag{5.18}$$

for a suitable h_ϵ^φ depending on the local shape of $\Sigma_\epsilon(t)$. We assume the expansions

$$\kappa_\epsilon^\varphi = \kappa_0^\varphi + \epsilon\kappa_1^\varphi + \mathcal{O}(\epsilon^2), \qquad h_\epsilon^\varphi = h_0^\varphi + \mathcal{O}(\epsilon). \tag{5.19}$$

With abuse of notation, for a given ϵ, we let $x(y;\, s,\, t)$ be the point of Ω having signed distance ϵy and projection s on $\Sigma_\epsilon(t)$. We have

$$x(y;\, s,\, t) = x(s,\, t) - \epsilon y n_\epsilon^\varphi + \mathcal{O}(\epsilon^2 y^2). \tag{5.20}$$

For a given ϵ, the triplet $(y;\, s,\, t)$ parametrizes a tubular neighbourhood of $\bigcup_{t\in(0,T)}(\{t\}\times\Sigma_\epsilon(t))$. We look for functions $U_\epsilon(y;\, s,\, t)$ and $W_\epsilon^r(y;\, s,\, t,\, x)$ $(r = 1, \dots, m)$ respectively so that

$$u_\epsilon(t,x) = U_\epsilon\left(\frac{d_\epsilon^\varphi(t,x)}{\epsilon},\, s_\epsilon^\varphi(t,x),\, t\right), \tag{5.21}$$

$$w_\epsilon^r(t,x) = W_\epsilon^r\left(\frac{d_\epsilon^\varphi(t,x)}{\epsilon},\, s_\epsilon^\varphi(t,x),\, t,\, x\right), \qquad r = 1, \dots, m, \tag{5.22}$$

with

$$\sum_{r=1}^m W_\epsilon^r = U_\epsilon. \tag{5.23}$$

Remark 5.4. Formula (5.21) defines uniquely the function U_ϵ, since to every (t, x) there corresponds uniquely the triplet (y, s, t). This observation does not apply to (5.22), in view of the explicit dependence of the functions W_ϵ^r on x.

We shall write

$$W_\epsilon^r = W_0^r + \epsilon W_{1,\epsilon}^r = W_0^r + \epsilon W_1^r + \epsilon^2 W_{2,\epsilon}^r, \qquad r = 1, \dots, m, \tag{5.24}$$

where W_0^r and W_1^r are allowed to depend explicitly on x (and hence on ϵ). We suppose the remainders $W_{1,\epsilon}^r$, $W_{2,\epsilon}^r$ to be bounded as $\epsilon \to 0^+$.

We let also

$$S^r := \frac{1}{2}\nabla^3\alpha_r = \nabla M^r, \qquad r = 1, \dots, m,$$

be the 3-indices, (-1)-homogeneus completely symmetric tensor given by the third derivatives of $\frac{1}{2}\alpha_r$: in components we have

$$S^r_{ijk} := \nabla_k M^r_{ij}, \qquad r = 1, \dots, m,$$

where $\nabla_k = \frac{\partial}{\partial \xi^*_k}$. Finally, for any $k, j = 1, \dots, N$, we introduce the notation

$$M^r_{.k} := \left(M^r_{1k} \dots M^r_{Nk}\right), \qquad S^r_{.jk} := \left(S^r_{1jk} \dots S^r_{Njk}\right), \quad r = 1, \dots, m.$$

Warning: We will adopt the convention of summation on repeated indices, *with the exception of the index* r, for which the explicit symbol $\sum_{r=1}^m$ will be always used. For instance, in formulas (5.28), (5.32), (5.33), (5.34) and (5.83) below, no summation on r is understood.

5.2.2 Preliminary expansions Now we begin to Taylor–expand all terms in (5.10). We have, using the convention of summation on repeated indices,

$$\epsilon^2\partial_t u_\epsilon = \epsilon^2 U_{\epsilon s_\beta}\, \partial_t s^\varphi_{\epsilon\beta} + \epsilon U'_\epsilon\, \partial_t d^\varphi_\epsilon + \epsilon^2 U_{\epsilon t} = \epsilon U'_\epsilon V^\varphi_\epsilon + \epsilon^2 U_{\epsilon t}, \quad (5.25)$$

where we used (5.16) and (5.17).

We write

$$U_\epsilon = U_0 + \epsilon U_{1,\epsilon} = U_0 + \epsilon U_1 + \epsilon^2 U_{2,\epsilon}, \quad (5.26)$$

where we require U_0 and U_1 not to depend on ϵ.

Using Taylor's expansion of the nonlinearity f, we get

$$f(U_\epsilon) = f(U_0) + \epsilon U_{1,\epsilon} f'(U_0) + \frac{1}{2}\epsilon^2 (U_{1,\epsilon})^2 f''(U_0) + \mathcal{O}(\epsilon^3). \quad (5.27)$$

To expand the divergence term, we need some additional work. First of all, by Taylor–expanding the operator T_{ϕ_r}, we get

$$T_{\phi_r}(\eta^* + \epsilon\zeta^*) = T_{\phi_r}(\eta^*) + \epsilon M^r(\eta^*)\zeta^* + \frac{1}{2}\epsilon^2 S^r_{.jk}(\eta^*)\zeta^*_j\zeta^*_k + \mathcal{O}(\epsilon^3),$$

so that, for any $r = 1, \dots, m$,

$$
\begin{aligned}
&\epsilon^2 T_{\phi_r}(\nabla w_\epsilon^r) \\
&= T_{\phi_r}\left(\epsilon W_\epsilon^{r\prime} \nabla d_\epsilon^\varphi + \epsilon^2 W_{\epsilon s_\beta}^r \nabla s_{\epsilon\beta}^\varphi + \epsilon^2 \nabla W_\epsilon^r\right) \\
&= \epsilon W_\epsilon^{r\prime}\, T_{\phi_r}(\nabla d_\epsilon^\varphi) + \epsilon^2 W_{\epsilon s_\beta}^r M^r(\nabla d_\epsilon^\varphi)\nabla s_{\epsilon\beta}^\varphi + \epsilon^2 M^r(\nabla d_\epsilon^\varphi)\nabla W_\epsilon^r \\
&\quad + \frac{1}{2W_\epsilon^{r\prime}}\epsilon^3 S_{\cdot jk}^r(\nabla d_\epsilon^\varphi)\left[W_{\epsilon s_\beta}^r \partial_{x_j} s_{\epsilon\beta}^\varphi + \partial_{x_j} W_\epsilon^r\right]\left[W_{\epsilon s_\beta}^r \partial_{x_k} s_{\epsilon\beta}^\varphi + \partial_{x_k} W_\epsilon^r\right] \\
&\quad + \mathcal{O}(\epsilon^4).
\end{aligned}
\tag{5.28}
$$

Remark 5.5. Since we still have to apply the divergence operator (which produces an extra ϵ^{-1} factor), we need to go through the ϵ^3 term in (5.28). We also observe that the term $\mathcal{O}(\epsilon^4)$ in (5.28) is actually a term of order $\mathcal{O}\left(\epsilon^4 \frac{1}{W_\epsilon^{r\prime 2}}\right)$ which, a posteriori, turns out to be of order $\mathcal{O}(\epsilon^4)$: indeed, from (5.44) below it follows that $W_\epsilon^{r\prime}$ is nonvanishing in the transition layer.

We now recall that by Euler's identities for homogeneous functions we have

$$
T_{\phi_r}(\xi^*) = \nabla_i T_{\phi_r}(\xi^*)\xi_i^*, \qquad \xi^* \in (\mathbb{R}^N)^*, \tag{5.29}
$$

which implies

$$
T_{\phi_r}(\nabla d_\epsilon^\varphi) \cdot \nabla s_{\epsilon\beta}^\varphi = M^r(\nabla d_\epsilon^\varphi)\nabla s_{\epsilon\beta}^\varphi \cdot \nabla d_\epsilon^\varphi, \qquad r = 1, \dots, m. \tag{5.30}
$$

Differentiating (5.29) with respect to ξ_k^* and using the notation $\nabla_{ik}^2 = \frac{\partial^2}{\partial\xi_k^*\partial\xi_i^*}$, we also have

$$
\nabla_{ik}^2 T_{\phi_r}(\xi^*)\xi_i^* = S_{\cdot ik}^r \xi_i^* = 0 \in \mathbb{R}^N, \qquad \xi^* \in (\mathbb{R}^N)^*, \quad k = 1, \dots, N,
$$

which implies

$$
S_{ijk}^r(\nabla d_\epsilon^\varphi)\nabla_i d_\epsilon^\varphi = 0, \qquad j, k = 1, \dots, N, \quad r = 1, \dots, m. \tag{5.31}
$$

For any $r = 1, \dots, m$, we compute, using (5.30),

$$
\begin{aligned}
&\epsilon^2 \mathrm{div}(M^r(\nabla d_\epsilon^\varphi)\nabla W_\epsilon^r) \\
&= \epsilon^2 \partial_{x_i}\left(M_{ij}^r(\nabla d_\epsilon^\varphi) W_{\epsilon x_j}^r\right) \\
&= \epsilon T_{\phi_r}(\nabla d_\epsilon^\varphi) \cdot \nabla W_\epsilon^{r\prime} \\
&\quad + \epsilon^2 W_{\epsilon x_j}^r \mathrm{div}\left(M_{\cdot j}^r(\nabla d_\epsilon^\varphi)\right) + \epsilon^2 M^r(\nabla d_\epsilon^\varphi)\nabla s_{\epsilon\beta}^\varphi \cdot \nabla W_{\epsilon s_\beta}^r \\
&\quad + \epsilon^2 W_{\epsilon x_i x_j}^r M_{ij}^r(\nabla d_\epsilon^\varphi).
\end{aligned}
\tag{5.32}
$$

By differentiating (5.28) we obtain, using also (5.11),

$$
\begin{aligned}
&\epsilon^2 \mathrm{div}\left(T_{\phi_r}(\nabla w_\epsilon^r)\right)\\
&= \alpha_r(\nabla d_\epsilon^\varphi) W_\epsilon^{r''} + 2\epsilon W_{\epsilon s_\beta}^{r'} \, T_{\phi_r}(\nabla d_\epsilon^\varphi) \cdot \nabla s_{\epsilon\beta}^\varphi \\
&\quad + 2\epsilon \, T_{\phi_r}(\nabla d_\epsilon^\varphi) \cdot \nabla W_\epsilon^{r'} + \epsilon W_\epsilon^{r'} \, \mathrm{div}(T_{\phi_r}(\nabla d_\epsilon^\varphi)) \\
&\quad + \epsilon^2 W_{\epsilon s_\beta s_\delta}^r \, M^r(\nabla d_\epsilon^\varphi)\nabla s_{\epsilon\beta}^\varphi \cdot \nabla s_{\epsilon\delta}^\varphi + \epsilon^2 M^r(\nabla d_\epsilon^\varphi)\nabla s_{\epsilon\beta}^\varphi \cdot \nabla W_{\epsilon s_\beta}^r \\
&\quad + \epsilon^2 W_{\epsilon s_\beta}^r \mathrm{div}\left(M^r(\nabla d_\epsilon^\varphi)\nabla s_{\epsilon\beta}^\varphi\right) \\
&\quad + \epsilon^2 W_{\epsilon x_j}^r \mathrm{div}\left(M_{\cdot j}^r(\nabla d_\epsilon^\varphi)\right) + \epsilon^2 \, M^r(\nabla d_\epsilon^\varphi)\nabla s_{\epsilon\beta}^\varphi \cdot \nabla W_{\epsilon s_\beta}^r \\
&\quad + \epsilon^2 W_{\epsilon x_i x_j}^r M_{ij}^r(\nabla d_\epsilon^\varphi) \\
&\quad + \mathcal{O}(\epsilon^3),
\end{aligned}
\tag{5.33}
$$

where we notice that no contribution of order larger than $\mathcal{O}(\epsilon^3)$ can come from the $\mathcal{O}(\epsilon^3)$ term in (5.28) — because they can only be produced *via* differentiation with respect to y, which in turn gives rise to a scalar product between $\nabla d_\epsilon^\varphi$ and the tensor $S^r(\nabla d_\epsilon^\varphi)$ (which in the end vanishes, due to Euler's identities (5.31)).

Hence, in terms of U_ϵ and W_ϵ^r, the expansion of the r-th parabolic equation in (5.10), for $r = 1, \ldots, m$, reads as, using also (5.25),

$$
\begin{aligned}
0 = \; & -\alpha_r(\nabla d_\epsilon^\varphi) W_\epsilon^{r''} + f(U_\epsilon) \\
& + \epsilon\Big(V_\epsilon^\varphi U_\epsilon' - 2W_{\epsilon s_\beta}^{r'} T_{\phi_r}(\nabla d_\epsilon^\varphi) \cdot \nabla s_{\epsilon\beta}^\varphi - 2T_{\phi_r}(\nabla d_\epsilon^\varphi) \cdot \nabla W_\epsilon^{r'} \\
& \qquad - W_\epsilon^{r'} \mathrm{div}\left(T_{\phi_r}(\nabla d_\epsilon^\varphi)\right)\Big) \\
& + \epsilon^2 \Big(U_{\epsilon t} - W_{\epsilon s_\beta s_\delta}^r M^r(\nabla d_\epsilon^\varphi)\nabla s_{\epsilon\beta}^\varphi \cdot \nabla s_{\epsilon\delta}^\varphi \\
& \qquad - 2M^r(\nabla d_\epsilon^\varphi)\nabla s_{\epsilon\beta}^\varphi \cdot \nabla W_{\epsilon s_\beta}^r \\
& \qquad - W_{\epsilon s_\beta}^r \mathrm{div}\Big(M^r(\nabla d_\epsilon^\varphi)\nabla s_{\epsilon\beta}^\varphi\Big) - W_{\epsilon x_j}^r \mathrm{div}\left(M_{\cdot j}^r(\nabla d_\epsilon^\varphi)\right) \\
& \qquad - W_{\epsilon x_i x_j}^r M_{ij}^r(\nabla d_\epsilon^\varphi)\Big) \\
& + \mathcal{O}(\epsilon^3).
\end{aligned}
\tag{5.34}
$$

5.2.3 Order 0 Recall [8] that $\nabla d_\epsilon^\varphi$ satisfies the anisotropic eikonal equation

$$
(\Phi(\nabla d_\epsilon^\varphi))^2 = 1 \tag{5.35}
$$

in the evolving transition layer.

Assuming the formal expansion

$$d_\epsilon^\varphi = d_0^\varphi + \epsilon d_1^\varphi + \epsilon^2 d_2^\varphi + \mathcal{O}(\epsilon^3), \tag{5.36}$$

where $d_0^\varphi(t,\cdot)$ is the φ-signed distance from $\Sigma_0(t)$ (positive in the interior of $\{u_0(t,\cdot)=1\}$), equation (5.35) leads to

$$\begin{aligned} 1 = {} & \Phi^2(\nabla d_0^\varphi) + 2\epsilon T_\Phi(\nabla d_0^\varphi)\cdot \nabla d_1^\varphi \\ & + \epsilon^2\Big(2T_\Phi(\nabla d_0^\varphi)\cdot \nabla d_2^\varphi + \nabla T_\Phi(\nabla d_0^\varphi)\nabla d_1^\varphi \cdot \nabla d_1^\varphi\Big) + \mathcal{O}(\epsilon^3), \end{aligned} \tag{5.37}$$

which in particular entails:

$$\Phi^2(\nabla d_0^\varphi) = 1, \tag{5.38}$$

$$T_\Phi(\nabla d_0^\varphi)\cdot \nabla d_1^\varphi = 0, \tag{5.39}$$

$$2T_\Phi(\nabla d_0^\varphi)\cdot \nabla d_2^\varphi + \nabla T_\Phi(\nabla d_0^\varphi)\nabla d_1^\varphi \cdot \nabla d_1^\varphi = 0$$

(the latter equation will not be used in what follows).

Using formula (2.7), equation (5.38) reads as

$$\sum_{r=1}^{m} \frac{1}{\alpha_r(\nabla d_0^\varphi(t,x))} = 1, \tag{5.40}$$

again for all x in a suitable tubular neighbourhood of $\Sigma_\epsilon(t)$.

Remark 5.6 (Weights). The quantities

$$\frac{1}{\alpha_r(\nabla d_0^\varphi)}, \qquad r = 1,\dots,m$$

can be used as "weights" to obtain a weighted mean of equations (5.34). This observation will be crucial in the sequel.

Collecting all terms of order zero in ϵ from each parabolic equation (5.34), dividing by $\alpha_r(\nabla d_0^\varphi)$, summing $r = 1,\dots,m$ and using (5.40), we obtain

$$-U_0'' + f(U_0) = 0, \tag{5.41}$$

where we used expansions (5.24), (5.26), (5.27), (5.36) for $U_\epsilon, W_\epsilon^r, f(U_\epsilon), d_\epsilon^\varphi$, and we have employed (5.23).

The only admissible solution of (5.41) (see for instance [4, 5]) is the standard standing wave

$$U_0(y,s,t) = \gamma(y), \qquad y \in \mathbb{R}, \tag{5.42}$$

where $\gamma(y) = \mathrm{tgh}(cy)$ (here c is a constant only depending on f); in particular, U_0 does not depend on (s, t).

Now we can recover each of the m functions W_0^r, $r = 1, \ldots, m$, by substituting $f(U_0) = U_0''$ into (5.34):

$$\alpha_r(\nabla d_0^\varphi) W_0^{r\prime\prime} = U_0'' = \gamma''.$$

Hence

$$W_0^{r\prime\prime} = \frac{1}{\alpha_r(\nabla d_0^\varphi)} U_0'' = \frac{1}{\alpha_r(\nabla d_0^\varphi)} \gamma'', \qquad r = 1, \ldots, m. \tag{5.43}$$

We also get by integration[20]

$$W_0^{r\prime} = \frac{1}{\alpha_r(\nabla d_0^\varphi)} U_0' = \frac{1}{\alpha_r(\nabla d_0^\varphi)} \gamma', \qquad r = 1, \ldots, m. \tag{5.44}$$

Remark 5.7. The functions $W_0^{r\prime}$ depend explicitly on x (and on t) through the coefficient $\frac{1}{\alpha_r(\nabla d_0^\varphi)}$. They are, on the other hand, independent of s.

5.2.4 Order 1 Let us consider the terms of order ϵ in equations (5.34). To this aim, we use the representation of $\frac{1}{2}\nabla^2\alpha = Q + Q_0$ given in section 2.1 for $\alpha = \Phi^2$, namely

$$Q = \alpha^2 \sum_{r=1}^m \frac{1}{\alpha_r^2} M^r,$$

where

$$Q_0(\xi^*)\xi^* = 0, \qquad \xi^* \in (\mathbb{R}^N)^*. \tag{5.45}$$

Remember that by Euler's identities for homogeneous functions we have

$$T_\Phi(\xi^*) = \frac{1}{2}\nabla^2\alpha(\xi^*)\xi^* = \left(Q(\xi^*) + Q_0(\xi^*)\right)\xi^*, \qquad \xi \in (\mathbb{R}^N)^*.$$

Hence, using (5.45),

$$\begin{aligned} T_\Phi(\nabla d_0^\varphi) &= \Big(Q(\nabla d_0^\varphi) + Q_0(\nabla d_0^\varphi)\Big)\nabla d_0^\varphi \\ &= Q(\nabla d_0^\varphi)\nabla d_0^\varphi \\ &= (\alpha(\nabla d_0^\varphi))^2 \sum_{r=1}^m \frac{1}{(\alpha_r(\nabla d_0^\varphi))^2} M^r(\nabla d_0^\varphi)\nabla d_0^\varphi \\ &= \sum_{r=1}^m \frac{1}{(\alpha_r(\nabla d_0^\varphi))^2} M^r(\nabla d_0^\varphi)\nabla d_0^\varphi, \end{aligned} \tag{5.46}$$

where the last equality follows from (5.35).

[20] See Section 5.2.5 below, and in particular equation (5.63).

Therefore

$$\operatorname{div}\left(T_\Phi(\nabla d_0^\varphi)\right) = \sum_{r=1}^{m} \operatorname{div}\left(\frac{1}{(\alpha_r(\nabla d_0^\varphi))^2} T_{\phi_r}(\nabla d_0^\varphi)\right). \tag{5.47}$$

For each $r = 1, \ldots, m$, we now collect all terms of order one in (5.34).

Remembering once more that $U_0 = \gamma$ and $W_0^{r\prime}$ do not depend explicitly on s and t so that in particular $W_{0s_\beta}^{r\prime} = 0$, we obtain

$$\begin{aligned} &-\alpha_r(\nabla d_0^\varphi) W_1^{r\prime\prime} - 2W_0^{r\prime\prime} T_{\phi_r}(\nabla d_0^\varphi) \cdot \nabla d_1^\varphi + f'(\gamma)U_1 \\ &+ \gamma' V_0^\varphi - 2T_{\phi_r}(\nabla d_0^\varphi) \cdot \nabla W_0^{r\prime} - W_0^{r\prime} \operatorname{div}\left(T_{\phi_r}(\nabla d_0^\varphi)\right) = 0, \end{aligned} \tag{5.48}$$

where we have taken into account that the term

$$-\alpha_r(\nabla d_0^\varphi) W_1^{r\prime\prime} - 2W_0^{r\prime\prime} T_{\phi_r}(\nabla d_0^\varphi) \cdot \nabla d_1^\varphi + f'(\gamma)U_1$$

arises from the expansion at the order ϵ of the first line on the right hand side of (5.34).

Using formula (5.44), equation (5.48) can be rewritten as

$$\begin{aligned} &-\alpha_r(\nabla d_0^\varphi) W_1^{r\prime\prime} - 2\gamma'' \frac{T_{\phi_r}(\nabla d_0^\varphi) \cdot \nabla d_1^\varphi}{\alpha_r(\nabla d_0^\varphi)} + f'(\gamma)U_1 \\ &+ \gamma' V_0^\varphi - \alpha_r(\nabla d_0^\varphi)\gamma' \left[\frac{1}{(\alpha_r(\nabla d_0^\varphi))^2} \operatorname{div}\left(T_{\phi_r}(\nabla d_0^\varphi)\right) \right. \\ &\qquad \left. + \frac{2}{\alpha_r(\nabla d_0^\varphi)} T_{\phi_r}(\nabla d_0^\varphi) \cdot \nabla \frac{1}{\alpha_r(\nabla d_0^\varphi)} \right] = 0. \end{aligned} \tag{5.49}$$

Since $\nabla \frac{1}{\alpha_r^2} = \frac{2}{\alpha_r} \nabla \frac{1}{\alpha_r}$, the expression in square brackets is simply

$$\operatorname{div}\left(\frac{1}{(\alpha_r(\nabla d_0^\varphi))^2} T_{\phi_r}(\nabla d_0^\varphi)\right), \qquad r = 1, \ldots, m. \tag{5.50}$$

Recalling (5.47), the sum over $r = 1, \ldots, m$ of the latter divergences gives $\operatorname{div}\left(T_\Phi(\nabla d_0^\varphi)\right)$. The weighted sum of equations (5.49) finally produces

$$-\mathcal{L}(U_1) = \gamma' \left[V_0^\varphi - \operatorname{div}\left(T_\Phi(\nabla d_0^\varphi)\right) \right],$$

where

$$\mathcal{L}(g) := -g'' + f'(\gamma) g,$$

and we make use of (5.39).

Recall now that from (5.18) and the expansions of κ_ϵ^φ it follows

$$\operatorname{div}\left(T_\Phi(\nabla d_\epsilon^\varphi)\right) = -\kappa_0^\varphi - \epsilon\kappa_1^\varphi - \epsilon y h_0^\varphi + \mathcal{O}(\epsilon^2 y^2), \tag{5.51}$$

in particular

$$\operatorname{div}\left(T_\Phi(\nabla d_0^\varphi)\right) = -\kappa_0^\varphi.$$

We then obtain

$$-\mathcal{L}(U_1) = \gamma'\Big[V_0^\varphi + \kappa_0^\varphi\Big]. \tag{5.52}$$

We recall now from [3–5] that for equation $-\mathcal{L}(g) = v$ to be solvable, we must enforce the orthogonality condition

$$\int_{\mathbb{R}} \gamma' v \, dy = 0. \tag{5.53}$$

This and (5.52) imply the remarkable fact

$$V_0^\varphi = -\kappa_0^\varphi, \tag{5.54}$$

so that

$$U_1 = 0. \tag{5.55}$$

Remark 5.8 (Convergence to anisotropic mean curvature flow). Note carefully that (5.54) justifies the convergence of solutions of system (1.4) to Φ^o-anisotropic mean curvature flow.

Substituting (5.54) and (5.55) in (5.49), dividing by $\alpha_r(\nabla d_0^\varphi)$ and recalling that the square bracket in (5.49) equals (5.50), we end up with the equation for W_1^r, for any $r = 1, \ldots, m$:

$$\begin{aligned} W_1^{r\prime\prime} = {} & \frac{1}{\alpha_r(\nabla d_0^\varphi)}\gamma' V_0^\varphi - \gamma' \operatorname{div}\left(\frac{1}{(\alpha_r(\nabla d_0^\varphi))^2} T_{\phi_r}(\nabla d_0^\varphi)\right) \\ & - 2\gamma'' \frac{T_{\phi_r}(\nabla d_0^\varphi)\cdot\nabla d_1^\varphi}{(\alpha_r(\nabla d_0^\varphi))^2} \\ = {} & \frac{1}{\alpha_r(\nabla d_0^\varphi)}\gamma' \operatorname{div}\left(T_\Phi(\nabla d_0^\varphi)\right) - \gamma' \operatorname{div}\left(\frac{1}{(\alpha_r(\nabla d_0^\varphi))^2} T_{\phi_r}(\nabla d_0^\varphi)\right) \\ & - 2\gamma'' \frac{T_{\phi_r}(\nabla d_0^\varphi)\cdot\nabla d_1^\varphi}{(\alpha_r(\nabla d_0^\varphi))^2}, \end{aligned} \tag{5.56}$$

since, from (5.51) and (5.54),

$$\operatorname{div}\left(T_\Phi(\nabla d_0^\varphi)\right) = V_0^\varphi.$$

As a consequence, recalling (5.40), (5.47) and (5.39), we have

$$\sum_{r=1}^m W_1^{r\prime\prime} = U_1'' = 0, \tag{5.57}$$

where the last equality follows from (5.55).

Equation (5.56) can be written as[21]

$$W_1^{r''} = \gamma' \Big[\operatorname{div} \Big(\frac{1}{\alpha_r(\nabla d_0^\varphi)} T_\Phi(\nabla d_0^\varphi) \Big) - \operatorname{div} \Big(\frac{1}{(\alpha_r(\nabla d_0^\varphi))^2} T_{\phi_r}(\nabla d_0^\varphi) \Big) - T_\Phi(\nabla d_0^\varphi) \cdot \nabla \frac{1}{\alpha_r(\nabla d_0^\varphi)} \Big] - 2\gamma'' \frac{T_{\phi_r}(\nabla d_0^\varphi) \cdot \nabla d_1^\varphi}{(\alpha_r(\nabla d_0^\varphi))^2}. \tag{5.58}$$

From (5.57) it follows that $\sum_{r=1}^m W_1^r$ minus a linear function vanishes, namely

$$\sum_{r=1}^m W_1^r - U_1 = C_1 y + C_0.$$

We now claim that $C_0 = C_1 = 0$, and hence

$$\sum_{r=1}^m W_1^r = U_1 (= 0). \tag{5.59}$$

The constant C_0 turns out to be zero for the following argument: as a consequence of (5.23) and (5.9),

$$0 = U_\epsilon(0, t, x) = \sum_{r=1}^m W_\epsilon^r(0, t, x), \qquad \epsilon \in (0, 1),$$

which implies

$$\sum_{r=1}^m W_i^r(0, t, x) = 0, \qquad i \geq 0$$

and hence $C_0 = 0$.

For what concerns the constant C_1, we have, using (5.72) below and (5.39),

$$C_1 = \sum_{r=1}^m W_1^{r'} = \sum_{r=1}^m \Big\{ (\gamma - 1)\Theta^r - 2\gamma' \frac{T_{\phi_r}(\nabla d_0^\varphi) \cdot \nabla d_1^\varphi}{(\alpha_r(\nabla d_0^\varphi))^2} + w_0^{r'} \Big\}$$
$$= \sum_{r=1}^m \Big\{ (\gamma - 1)\Theta^r + w_0^{r'} \Big\}.$$

[21] Although written in a somewhat different form, this result coincides with that of [4], where d_ϵ^φ has not been expanded (hence d_ϵ^φ appears in place of d_0^φ in (5.58), and accordingly the last addendum is not present).

On the other hand, from (5.70) below, it follows $\sum_{r=1}^{m} w_0^{r\prime} = 0$, so that $C_1 = (\gamma - 1)\sum_{r=1}^{m} \Theta^r$. In order to conclude the proof of claim (5.59) it is enough to observe that $\sum_{r=1}^{m} \Theta^r = 0$, as a consequence of the expression of Θ^r in (5.72), and of (5.40) and (5.47), and so $C_1 = 0$.

5.2.5 Matching procedure We are now in a position to recover the first term w_0^r of the outer expansion of w_ϵ^r, by adding to (5.6) a jump condition for w_0^r and a condition for $n_0^\varphi \cdot \nabla w_0^r$ across the interface $\Sigma_0(t)$, defined as the boundary of the external phase $\{u_0(t,\cdot) = 1\}$ (see (5.5)). We set

$$\Sigma_\epsilon(t) = \{x + \epsilon\sigma_1(s,t)n_0^\varphi + \mathcal{O}(\epsilon^2) : x \in \Sigma_0(t)\}, \tag{5.60}$$

for a suitable $\sigma_1 \colon \Sigma \times \mathbb{R} \to \mathbb{R}$, where Σ is the reference manifold in (5.13).

We will make use of the change of variables (5.20), and we will match the two expansions in the region of common validity $|y| \to +\infty$ and x approaching $\Sigma_\epsilon(t)$:

$$w_\epsilon^r\left(t, x(s,t) - \epsilon y n_\epsilon^\varphi + \mathcal{O}(\epsilon^2 y^2)\right) \approx W_\epsilon^r(y; s, t, x(s,t) - \epsilon y n_\epsilon^\varphi + \mathcal{O}(\epsilon^2 y^2)).$$

By expanding the left and right hand sides, understanding that w_ϵ^r is computed at points $x(s,t) \in \Sigma_\epsilon(t)$, we get

$$w_\epsilon^r - \epsilon y n_\epsilon^\varphi \cdot \nabla w_\epsilon^r + \mathcal{O}(\epsilon^2 y^2) \approx W_\epsilon^r - \epsilon y n_\epsilon^\varphi \cdot \nabla W_\epsilon^r + \mathcal{O}(\epsilon^2 y^2), \quad r = 1, \dots, m.$$

Expanding w_ϵ^r, W_ϵ^r in powers of ϵ, and matching the first two orders, we get in particular

$$\lim_{y \to \pm\infty} W_0^r(y, s(t,x), t, x) = w_0^r(t,x), \tag{5.61}$$

and

$$\begin{aligned}\lim_{y \to \pm\infty} \Big\{ & W_1^r(y, s(t,x), t, x) - w_1^r(t,x) \\ & - y\Big(n_0^\varphi \cdot \nabla W_0^r(y, s(t,x), t, x) - n_0^\varphi \cdot \nabla w_0^r(t,x)\Big)\Big\} = 0,\end{aligned} \tag{5.62}$$

where w_0^r and w_1^r are evaluated at each side of the interface according to when y goes to plus or minus infinity.

Equality (5.61) in particular suggests

$$\lim_{y \to \pm\infty} W_0^{r\prime}(y, s(t,x), t, x) = 0, \qquad r = 1, \dots, m, \tag{5.63}$$

and the jump $[\![w_0^r]\!]$ of w_0^r across the interface is given by

$$[\![w_0^r]\!](s(t,x),t) = \int_{\mathbb{R}} W_0^{r\prime}(y,s(t,x),t,x)\,dy, \qquad r = 1,\dots,m. \tag{5.64}$$

From (5.44) we get

$$[\![w_0^r]\!] = \frac{c_0}{\alpha_r(\nabla d_0^\varphi)}, \qquad r = 1,\dots,m, \tag{5.65}$$

where

$$c_0 := \int_{\mathbb{R}} \gamma'\,dy \in (0,+\infty).$$

To obtain the equation involving the conormal derivative, we formally differentiate equation (5.62):

$$\begin{aligned}&\lim_{y\to\pm\infty}\left\{W_1^{r\prime}(y,s(t,x),t,x) - n_0^\varphi\cdot\nabla W_0^r(y,s(t,x),t,x)\right\}\\&= -n_0^\varphi\cdot\nabla w_0^r(t,x),\end{aligned} \tag{5.66}$$

where we used also the fact that

$$\lim_{y\to\pm\infty} y\, n_0^\varphi\cdot\nabla W_0^{r\prime}(y,s(t,x),t,x) = 0,$$

since $\nabla W_0^{r\prime} = \gamma'\nabla\frac{1}{\alpha_r(\nabla d_0^\varphi)}$ by (5.44) and γ' decays exponentially to 0 as $y\to\pm\infty$. For the same reason, $W_1^{r\prime}$ is also bounded, thus

$$\begin{aligned}&-[\![n_0^\varphi\cdot\nabla w_0^r]\!](t,x)\\&= \int_{\mathbb{R}}\left(W_1^{r\prime\prime}(y,s(t,x),t,x) - n_0^\varphi\cdot\nabla W_0^{r\prime}(y,s(t,x),t,x)\right)dy.\end{aligned} \tag{5.67}$$

Coupling (5.67) with (5.56) and (5.44), and recalling from (5.14) that $n_0^\varphi = -T_\Phi(\nabla d_0^\varphi)$ we end up with

$$-[\![n_0^\varphi\cdot\nabla w_0^r]\!] = c_0\,\mathrm{div}\left[\frac{1}{\alpha_r(\nabla d_0^\varphi)}T_\Phi(\nabla d_0^\varphi) - \frac{1}{\alpha_r^2(\nabla d_0^\varphi)}T_{\phi_r}(\nabla d_0^\varphi)\right], \qquad r = 1,\dots,m. \tag{5.68}$$

The two jump conditions on w_0 across $\Sigma_0(t)$, together with the *far field* equation (5.6) and appropriate boundary conditions at $\partial\Omega$ allow to retrieve a unique solution w_0.

If we integrate (5.44), and use the matching condition for w_0^r in (5.61), we get for W_0^r the expression

$$W_0^r = \frac{1}{\alpha_r(\nabla d_0^\varphi)}(\gamma - 1) + w_0^{r+}(s,t), \qquad r = 1, \dots, m, \tag{5.69}$$

where w_0^{r+} is the trace on $\Sigma_0(t)$ of w_0^r from the external phase $\{u_0(t,\cdot) = 1\}$. In particular

$$\sum_{r=1}^{m} w_0^{r+} = 1. \tag{5.70}$$

Thus

$$\begin{aligned} &W^r_{0s_\beta} = w^{r+}_{0s_\beta}, \qquad \nabla W_0^r = (\gamma - 1)\nabla \frac{1}{\alpha_r}, \qquad \nabla W^r_{0s_\beta} = 0, \\ &W^r_{0s_\beta s_\delta} = w^{r+}_{0s_\beta s_\delta}, \qquad W^r_{0x_ix_j} = (\gamma - 1)\partial_{x_ix_j}\frac{1}{\alpha_r}. \end{aligned} \tag{5.71}$$

In a similar fashion we can integrate (5.56), and use the matching condition (5.66), to get, for any $r = 1, \dots, m$,

$$\begin{aligned} W_1^{r\prime} &= (\gamma - 1)\left\{\frac{1}{\alpha_r(\nabla d_0^\varphi)}\operatorname{div}\left(T_\Phi(\nabla d_0^\varphi)\right) - \operatorname{div}\left(\frac{1}{\alpha_r^2(\nabla d_0^\varphi)}T_{\phi_r}(\nabla d_0^\varphi)\right)\right\} \\ &\quad - 2\gamma' \frac{T_{\phi_r}(\nabla d_0^\varphi)\cdot \nabla d_1^\varphi}{(\alpha_r(\nabla d_0^\varphi))^2} + w_0^{r\prime}(s,t) \\ &= (\gamma - 1)\Theta^r(t,x) - 2\gamma' \frac{T_{\phi_r}(\nabla d_0^\varphi)\cdot \nabla d_1^\varphi}{(\alpha_r(\nabla d_0^\varphi))^2} + w_0^{r\prime}(s,t), \end{aligned} \tag{5.72}$$

where

$$w_0^{r\prime} := T_\Phi(\nabla d_0^\varphi)\cdot \nabla w_0^{r+},$$

and Θ^r is a shorthand for the expression in braces. Observe that only the last term explicitly depends on s, while the other terms depend on y (by means of γ) and on x (by means of Θ^r). Thus

$$W^{r\prime}_{1s_\beta} = w^{r\prime}_{0s_\beta} \tag{5.73}$$

$$\nabla W_1^{r\prime} = (\gamma - 1)\nabla\Theta^r - 2\gamma'\nabla\left(\frac{T_{\phi_r}(\nabla d_0^\varphi)\cdot \nabla d_1^\varphi}{(\alpha_r(\nabla d_0^\varphi))^2}\right). \tag{5.74}$$

Remark 5.9. Note that the jump of the *conormal* derivative $[\![n_0^\varphi \cdot \nabla w_0^r]\!]$ vanishes in the special case of equal anisotropic ratio which, in our context, consists of choosing, for every $r = 1, \dots, m$, $\alpha_r := \lambda_r \overline{\alpha}$ with some given smooth symmetric uniformly convex squared anisotropy $\overline{\alpha}$ and positive λ_r (indeed, in this case eikonal equation (5.40) leads to $\overline{\alpha}(\nabla d_0^\varphi) = \sum_{r=1}^m \lambda_r^{-1}$).

Remark 5.10. Given $r = 1, \dots, m$, the function $W_1^r(\cdot, t, x)$ is expected to have linear growth at infinity (independent of ϵ)[22]; observe, however, that $\sum_{r=1}^m W_1^r(\cdot, t, x) = 0$, see (5.59).

5.2.6 Order 2 We end our asymptotic analysis considering the $\mathcal{O}(\epsilon^2)$ terms in equation (5.34), which represents an improvement with respect to [7] (in which expansions are performed only up to the order $\mathcal{O}(\epsilon)$ and $m = 2$). Recall that $U_0' = \gamma'$ depends only on y and that $U_1 = 0$. Then the terms of order $\mathcal{O}(\epsilon^2)$ arising from the first line on the right hand side of (5.34) are:

$$\begin{aligned} &- \alpha_r(\nabla d_0^\varphi) W_2^{r\prime\prime} - 2W_1^{r\prime\prime} T_{\phi_r}(\nabla d_0^\varphi) \cdot \nabla d_1^\varphi \\ &- \Big[2T_{\phi_r}(\nabla d_0^\varphi) \cdot \nabla d_2^\varphi + M^r(\nabla d_0^\varphi)\nabla d_1^\varphi \cdot \nabla d_1^\varphi\Big] W_0^{r\prime\prime} + f'(\gamma)U_2. \end{aligned} \tag{5.75}$$

The terms of order $\mathcal{O}(\epsilon)$ arising from the terms in the round parentheses in the second line of (5.34) are:

$$\begin{aligned} &\gamma' V_1^\varphi - 2W_{1s_\beta}^{r\prime} T_{\phi_r}(\nabla d_0^\varphi) \cdot \nabla s_{0\beta}^\varphi - 2T_{\phi_r}(\nabla d_0^\varphi) \cdot \nabla W_1^{r\prime} \\ &- 2M^r(\nabla d_0^\varphi)\nabla d_1^\varphi \cdot \nabla W_0^{r\prime} \\ &- W_1^{r\prime} \mathrm{div}\big(T_{\phi_r}(\nabla d_0^\varphi)\big) - W_0^{r\prime} \mathrm{div}\big(M^r(\nabla d_0^\varphi)\nabla d_1^\varphi\big). \end{aligned} \tag{5.76}$$

Note that, using (5.73), if we set

$$\mathrm{A}^r := -2W_{1s_\beta}^{r\prime} T_{\phi_r}(\nabla d_0^\varphi) \cdot \nabla s_{0\beta}^\varphi, \tag{5.77}$$

then A^r is independent of y, hence

$$\int_{\mathbb{R}} \gamma' \mathrm{A}^r \, dy = \mathrm{A}^r \int_{\mathbb{R}} \gamma' \, dy = c_0 \mathrm{A}^r. \tag{5.78}$$

Remark 5.11. The term A^r is independent of d_1^φ.

---

[22] Differently with respect to $W_0^r(\cdot, t, x)$, which is expected to be bounded at infinity.

The terms of order $\mathcal{O}(1)$ arising from the terms in the round parentheses in the third and fourth lines of (5.34) are:

$$\begin{aligned} &- W^r_{0s_\beta s_\delta} M^r(\nabla d_0^\varphi)\nabla s_{0\beta}^\varphi \cdot \nabla s_{0\delta}^\varphi - 2M^r(\nabla d_0^\varphi)\nabla s_{0\beta}^\varphi \cdot \nabla W^r_{0s_\beta} \\ &- W^r_{0s_\beta}\mathrm{div}\Big(M^r(\nabla d_0^\varphi)\nabla s_{0\beta}^\varphi\Big) - W^r_{0x_j}\mathrm{div}\left(M^r_{\cdot j}(\nabla d_0^\varphi)\right) \\ &- W^r_{0x_i x_j} M^r_{ij}(\nabla d_0^\varphi) =: \mathrm{B}^r. \end{aligned} \tag{5.79}$$

Observe that, from (5.69) and (5.71), it follows that the y-dependence of B^r is through γ only in the term $W^r_{0x_j}$.

Remark 5.12. The term B^r is independent of d_1^φ.

Collecting together (5.75), (5.76) and (5.79) we get

$$\begin{aligned} &- \alpha_r(\nabla d_0^\varphi) W_2^{r\prime\prime} - 2W_1^{r\prime\prime} T_{\phi_r}(\nabla d_0^\varphi)\cdot\nabla d_1^\varphi \\ &- \Big[2T_{\phi_r}(\nabla d_0^\varphi)\cdot\nabla d_2^\varphi + M^r(\nabla d_0^\varphi)\nabla d_1^\varphi \cdot d_1^\varphi\Big] W_0^{r\prime\prime} + f'(\gamma)U_2 \\ &\gamma' V_1^\varphi - 2W^{r\prime}_{1s_\beta} T_{\phi_r}(\nabla d_0^\varphi)\cdot\nabla s_{0\beta}^\varphi - 2T_{\phi_r}(\nabla d_0^\varphi)\cdot\nabla W_1^{r\prime} \\ &- 2M^r(\nabla d_0^\varphi)\nabla d_1^\varphi \cdot \nabla W_0^{r\prime} - W_1^{r\prime}\mathrm{div}\big(T_{\phi_r}(\nabla d_0^\varphi)\big) \\ &- W_0^{r\prime}\mathrm{div}\big(M^r(\nabla d_0^\varphi)\nabla d_1^\varphi\big) + \mathrm{B}^r. \end{aligned} \tag{5.80}$$

Before continuing, let us write (5.72) in the form

$$W_1^{r\prime} = -2\gamma' \frac{T_{\phi_r}(\nabla d_0^\varphi)\cdot\nabla d_1^\varphi}{(\alpha_r(\nabla d_0^\varphi))^2} + \widetilde{\mathrm{C}}^r, \tag{5.81}$$

where

$$\widetilde{\mathrm{C}}^r := (\gamma - 1)\Theta^r + w_0^{r\prime},$$

so that $\widetilde{\mathrm{C}}^r$ depends on y only through the term $\gamma(y)\Theta^r(t,x)$, and therefore

$$\int_{\mathbb{R}} \gamma' \,\widetilde{\mathrm{C}}^r\, dy = c_0\left(-\Theta^r + w_0^{r\prime}\right). \tag{5.82}$$

Remark 5.13. The term $\widetilde{\mathrm{C}}^r$ is independent of d_1^φ.

Substituting (5.43), (5.44), (5.56), (5.81) into (5.80), and reordering terms we get, for any $r = 1, \dots, m$,

$$
\begin{aligned}
0 = &-\alpha_r(\nabla d_0^\varphi) W_2^{r\prime\prime} + U_2 f'(\gamma) + \gamma' V_1^\varphi \\
&+ \gamma' \left\{ 2T_{\phi_r}(\nabla d_0^\varphi)\cdot\nabla d_1^\varphi \left[\frac{\kappa_0^\varphi}{\alpha_r(\nabla d_0^\varphi)} + \operatorname{div}\left(\frac{T_{\phi_r}(\nabla d_0^\varphi)}{(\alpha_r(\nabla d_0^\varphi))^2} \right) \right] \right\} \\
&+ 4\gamma' T_{\phi_r}(\nabla d_0^\varphi)\cdot\nabla\left(\frac{T_{\phi_r}(\nabla d_0^\varphi)\cdot\nabla d_1^\varphi}{(\alpha_r(\nabla d_0^\varphi))^2} \right) \\
&- 2\gamma' M^r(\nabla d_0)\nabla d_1^\varphi\cdot\nabla\left(\frac{1}{\alpha_r(\nabla d_0^\varphi)} \right) \\
&- \gamma' \frac{1}{\alpha_r(\nabla d_0^\varphi)} \operatorname{div}\big(M^r(\nabla d_0^\varphi)\nabla d_1^\varphi\big) \\
&+ 2\gamma' \frac{T_{\phi_r}(\nabla d_0^\varphi)\cdot\nabla d_1^\varphi}{(\alpha_r(\nabla d_0^\varphi))^2} \operatorname{div}\left(T_{\phi_r}(\nabla d_0^\varphi)\right) \\
&+ \mathrm{A}^r + \mathrm{B}^r + \mathrm{C}^r + \gamma''\mathrm{D}^r,
\end{aligned}
\tag{5.83}
$$

where $\mathrm{C}_r := -2T_{\phi_r}(\nabla d_0^\varphi)\cdot\nabla\widetilde{\mathrm{C}}^r - \widetilde{\mathrm{C}}^r \operatorname{div}(T_{\phi_r}(\nabla d_0^\varphi))$, and

$$
\begin{aligned}
\mathrm{D}^r := &\left(\frac{2T_{\phi_r}(\nabla d_0^\varphi)\cdot\nabla d_1^\varphi}{\alpha_r(\nabla d_0^\varphi)} \right)^2 \\
&- \frac{\left[2T_{\phi_r}(\nabla d_0^\varphi)\cdot\nabla d_2^\varphi + M^r(\nabla d_0^\varphi)\nabla d_1^\varphi\cdot\nabla d_1^\varphi \right]}{\alpha_r(\nabla d_0^\varphi)}.
\end{aligned}
$$

Remark 5.14. Note that D^r depends on d_1^φ, however

$$
\int_{\mathbb{R}} \gamma'\gamma''\,\mathrm{D}^r\,dy = \mathrm{D}^r \int_{\mathbb{R}} \gamma'\gamma''\,dy = 0. \tag{5.84}
$$

Let us now focus the attention to (5.83), where for the moment we neglect the first line and the term $\mathrm{A}^r + \mathrm{B}^r + \mathrm{C}^r + \gamma''\mathrm{D}^r$ and omit the factor γ': dividing by $\alpha_r(\nabla d_0^\varphi)$ we have

$$
\begin{aligned}
&\frac{2T_{\phi_r}(\nabla d_0^\varphi)\cdot\nabla d_1^\varphi}{(\alpha_r(\nabla d_0^\varphi))^2}\kappa_0^\varphi + \frac{2T_{\phi_r}(\nabla d_0^\varphi)\cdot\nabla d_1^\varphi}{\alpha_r(\nabla d_0^\varphi)} \operatorname{div}\left(\frac{T_{\phi_r}(\nabla d_0^\varphi)}{(\alpha_r(\nabla d_0^\varphi))^2} \right) \\
&+4\frac{T_{\phi_r}(\nabla d_0^\varphi)}{\alpha_r(\nabla d_0^\varphi)}\cdot\nabla\left(\frac{T_{\phi_r}(\nabla d_0^\varphi)\cdot\nabla d_1^\varphi}{(\alpha_r(\nabla d_0^\varphi)^2} \right) + 2\frac{T_{\phi_r}(\nabla d_0^\varphi)\cdot\nabla d_1^\varphi}{(\alpha_r(\nabla d_0^\varphi))^3} \operatorname{div}\left(T_{\phi_r}(\nabla d_0^\varphi)\right) \\
&- \frac{2}{\alpha_r(\nabla d_0^\varphi)} M^r(\nabla d_0)\nabla d_1^\varphi\cdot\nabla\left(\frac{1}{\alpha_r(\nabla d_0^\varphi)} \right) \\
&- \frac{1}{(\alpha_r(\nabla d_0^\varphi))^2} \operatorname{div}\big(M^r(\nabla d_0^\varphi)\nabla d_1^\varphi\big)
\end{aligned}
\tag{5.85}
$$

Observe now that the first term in (5.85) will disappear when summing up on $r = 1, \ldots, m$, thanks again to (5.46) and (5.38). Moreover, the two terms in last line of (5.85) can be put together giving $\mathrm{div}\left(\frac{M^r(\nabla d_0^\varphi)\nabla d_1^\varphi}{(\alpha_r(\nabla d_0^\varphi))^2}\right)$ so that, summing up on r, we get:

$$\underbrace{2\sum_{r=1}^{m}\frac{T_{\phi_r}(\nabla d_0^\varphi)\cdot\nabla d_1^\varphi}{\alpha_r(\nabla d_0^\varphi)}\mathrm{div}\left(\frac{T_{\phi_r}(\nabla d_0^\varphi)}{(\alpha_r(\nabla d_0^\varphi))^2}\right)}_{:=E}$$
$$+\underbrace{4\sum_{r=1}^{m}\frac{T_{\phi_r}(\nabla d_0^\varphi)}{\alpha_r(\nabla d_0^\varphi)}\cdot\nabla\left(\frac{T_{\phi_r}(\nabla d_0^\varphi)\cdot\nabla d_1^\varphi}{\alpha_r^2(\nabla d_0^\varphi)}\right)}_{:=F}$$
$$+\underbrace{2\sum_{r=1}^{m}\frac{T_{\phi_r}(\nabla d_0^\varphi)\cdot\nabla d_1^\varphi}{(\alpha_r(\nabla d_0^\varphi))^3}\mathrm{div}T_{\phi_r}(\nabla d_0^\varphi)}_{:=G}\underbrace{-\sum_{r=1}^{m}\mathrm{div}\left(\frac{M^r(\nabla d_0^\varphi)\nabla d_1^\varphi}{(\alpha_r(\nabla d_0^\varphi))^2}\right)}_{:=H}. \tag{5.86}$$

Recall now that $-\kappa_1^\varphi - yh_0^\varphi = \mathrm{div}\big(\nabla T_\Phi(\nabla d_0^\varphi)\nabla d_1^\varphi\big)$. Using formulas (2.10), (2.11), and the relations $\nabla\alpha_r = 2T_{\phi_r}$, $\Phi^2(\nabla d_0^\varphi) = 1$, we get

$$-\kappa_1^\varphi - yh_0^\varphi = \sum_{r=1}^{m}\mathrm{div}\left(\frac{M^r(\nabla d_0^\varphi)\nabla d_1^\varphi}{(\alpha_r(\nabla d_0^\varphi))^2}\right)$$
$$+\sum_{r=1}^{m}\mathrm{div}\left(\frac{1-\alpha_r(\nabla d_0^\varphi)}{(\alpha_r(\nabla d_0^\varphi))^4}\nabla\alpha_r(\nabla d_0^\varphi)\otimes\nabla\alpha_r(\nabla d_0^\varphi)\nabla d_1^\varphi\right)$$
$$+\sum_{j\neq r}\mathrm{div}\left(\frac{1}{(\alpha_r(\nabla d_0^\varphi))^2(\alpha_j(\nabla d_0^\varphi))^2}\nabla\alpha_r(\nabla d_0^\varphi)\otimes\nabla\alpha_j(\nabla d_0^\varphi)\nabla d_1^\varphi\right).$$

Adding and subtracting the term $4\sum_{r=1}^{m}\mathrm{div}\left(\frac{1}{(\alpha_r(\nabla d_0^\varphi))^4}T_{\phi_r}(\nabla d_0^\varphi)\otimes T_{\phi_r}(\nabla d_0^\varphi)\nabla d_1^\varphi\right)$ it follows

$$-\kappa_1^\varphi - yh_0^\varphi = \sum_{r=1}^{m}\mathrm{div}\left(\frac{M^r(\nabla d_0^\varphi)\nabla d_1^\varphi}{(\alpha_r(\nabla d_0^\varphi))^2}\right)$$
$$+4\sum_{r=1}^{m}\mathrm{div}\left(\frac{1-\alpha_r(\nabla d_0^\varphi)}{(\alpha_r(\nabla d_0^\varphi))^4}(T_{\phi_r}(\nabla d_0^\varphi)\cdot\nabla d_1^\varphi)T_{\phi_r}(\nabla d_0^\varphi)\right)$$
$$+4\sum_{j,r}\mathrm{div}\left(\frac{1}{(\alpha_r(\nabla d_0^\varphi))^2(\alpha_j(\nabla d_0^\varphi))^2}T_{\phi_r}(\nabla d_0^\varphi)\otimes T_{\phi_j}(\nabla d_0^\varphi)\nabla d_1^\varphi\right)$$
$$-4\sum_{r=1}^{m}\mathrm{div}\left(\frac{1}{(\alpha_r(\nabla d_0^\varphi))^4}T_{\phi_r}(\nabla d_0^\varphi)\otimes T_{\phi_r}(\nabla d_0^\varphi)\nabla d_1^\varphi\right).$$

Fixing one of the two indices r, j, for instance r, and summing over the other one $j = 1, \dots, m$, we get

$$\sum_{j,r} \operatorname{div}\left(\frac{1}{(\alpha_r(\nabla d_0^\varphi))^2(\alpha_j(\nabla d_0^\varphi))^2} T_{\phi_r}(\nabla d_0^\varphi) \otimes T_{\phi_j}(\nabla d_0^\varphi)\nabla d_1^\varphi\right) = 0,$$

thanks again to eikonal equation (5.39). We deduce

$$\begin{aligned}
-\kappa_1^\varphi - y h_0^\varphi &= \sum_{r=1}^m \operatorname{div}\left(\frac{M^r(\nabla d_0^\varphi)\nabla d_1^\varphi}{(\alpha_r(\nabla d_0^\varphi))^2}\right) \\
&\quad + 4\sum_{r=1}^m \operatorname{div}\left(\frac{1-\alpha_r(\nabla d_0^\varphi)}{(\alpha_r(\nabla d_0^\varphi))^4}(T_{\phi_r}(\nabla d_0^\varphi)\cdot \nabla d_1^\varphi) T_{\phi_r}(\nabla d_0^\varphi)\right) \\
&\quad - 4\sum_{r=1}^m \operatorname{div}\left(\frac{1}{(\alpha_r(\nabla d_0^\varphi))^4} T_{\phi_r}(\nabla d_0^\varphi) \otimes T_{\phi_r}(\nabla d_0^\varphi)\nabla d_1^\varphi\right) \\
&= \sum_{r=1}^m \operatorname{div}\left(\frac{M^r(\nabla d_0^\varphi)\nabla d_1^\varphi}{(\alpha_r(\nabla d_0^\varphi))^2}\right) \\
&\quad - 4\sum_{r=1}^m \operatorname{div}\left(\frac{1}{(\alpha_r(\nabla d_0^\varphi))^3}(T_{\phi_r}(\nabla d_0^\varphi)\cdot \nabla d_1^\varphi) T_{\phi_r}(\nabla d_0^\varphi)\right) \\
&:= I,
\end{aligned}$$

where we used

$$T_{\phi_r}(\nabla d_0^\varphi) \otimes T_{\phi_r}(\nabla d_0^\varphi)\nabla d_1^\varphi = (T_{\phi_r}(\nabla d_0^\varphi)\cdot \nabla d_1^\varphi) T_{\phi_r}(\nabla d_0^\varphi).$$

We claim now that $\kappa_1^\varphi + y h_0^\varphi$ is equal to (5.86) — namely:

$$E + F + G + H + I = 0. \tag{5.87}$$

We first observe that the first term appearing in I cancels with H, so that it is enough to show

$$E + F + G = 4\sum_{r=1}^m \operatorname{div}\left(\frac{1}{(\alpha_r(\nabla d_0^\varphi))^2}(T_{\phi_r}(\nabla d_0^\varphi)\cdot \nabla d_1^\varphi) T_{\phi_r}(\nabla d_0^\varphi)\right),$$

i.e.,

$$
\begin{aligned}
&2\sum_{r=1}^{m} \frac{T_{\phi_r}(\nabla d_0^{\varphi}) \cdot \nabla d_1^{\varphi}}{\alpha_r(\nabla d_0^{\varphi})} \mathrm{div}\left(\frac{T_{\phi_r}(\nabla d_0^{\varphi})}{(\alpha_r(\nabla d_0^{\varphi}))^2}\right)\\
&+4\sum_{r=1}^{m} \frac{T_{\phi_r}(\nabla d_0^{\varphi})}{\alpha_r(\nabla d_0^{\varphi})} \cdot \nabla\left(\frac{T_{\phi_r}(\nabla d_0^{\varphi}) \cdot \nabla d_1^{\varphi}}{(\alpha_r(\nabla d_0^{\varphi}))^2}\right)\\
&+2\sum_{r=1}^{m} \frac{T_{\phi_r}(\nabla d_0^{\varphi}) \cdot \nabla d_1^{\varphi}}{(\alpha_r(\nabla d_0^{\varphi}))^3} \mathrm{div} T_{\phi_r}(\nabla d_0^{\varphi})\\
&=4\sum_{r=1}^{m} \mathrm{div}\left(\frac{1}{(\alpha_r(\nabla d_0^{\varphi}))^3}(T_{\phi_r}(\nabla d_0^{\varphi}) \cdot \nabla d_1^{\varphi}) T_{\phi_r}(\nabla d_0^{\varphi})\right).
\end{aligned}
\tag{5.88}
$$

The right hand side of (5.88) can be rewritten as

$$
\begin{aligned}
&4\sum_{r=1}^{m} \mathrm{div}\left(\frac{1}{(\alpha_r(\nabla d_0^{\varphi}))^3}(T_{\phi_r}(\nabla d_0^{\varphi}) \cdot \nabla d_1^{\varphi}) T_{\phi_r}(\nabla d_0^{\varphi})\right)\\
&=4\sum_{r=1}^{m} \frac{T_{\phi_r}(\nabla d_0^{\varphi}) \cdot \nabla d_1^{\varphi}}{(\alpha_r(\nabla d_0^{\varphi}))^2} \mathrm{div}\left(\frac{T_{\phi_r}(\nabla d_0^{\varphi})}{\alpha_r(\nabla d_0^{\varphi})}\right)\\
&\quad+4\sum_{r=1}^{m} \frac{T_{\phi_r}(\nabla d_0^{\varphi})}{\alpha_r(\nabla d_0^{\varphi})} \cdot \nabla\left(\frac{T_{\phi_r}(\nabla d_0^{\varphi}) \cdot \nabla d_1^{\varphi}}{\alpha_r^2(\nabla d_0^{\varphi})}\right),
\end{aligned}
$$

so that its last addendum cancels with F. Thus, in order to show (5.87) it remains to prove that

$$
\begin{aligned}
&2\sum_{r=1}^{m} \frac{T_{\phi_r}(\nabla d_0^{\varphi}) \cdot \nabla d_1^{\varphi}}{\alpha_r(\nabla d_0^{\varphi})} \mathrm{div}\left(\frac{T_{\phi_r}(\nabla d_0^{\varphi})}{(\alpha_r(\nabla d_0^{\varphi}))^2}\right)\\
&+2\sum_{r=1}^{m} \frac{T_{\phi_r}(\nabla d_0^{\varphi}) \cdot \nabla d_1^{\varphi}}{(\alpha_r(\nabla d_0^{\varphi}))^3} \mathrm{div} T_{\phi_r}(\nabla d_0^{\varphi})\\
&=4\sum_{r=1}^{m} \frac{T_{\phi_r}(\nabla d_0^{\varphi}) \cdot \nabla d_1^{\varphi}}{(\alpha_r(\nabla d_0^{\varphi}))^2} \mathrm{div}\left(\frac{T_{\phi_r}(\nabla d_0^{\varphi})}{\alpha_r(\nabla d_0^{\varphi})}\right),
\end{aligned}
$$

or equivalently

$$
\begin{aligned}
\sum_{r=1}^{m} \frac{T_{\phi_r}(\nabla d_0^{\varphi}) \cdot \nabla d_1^{\varphi}}{\alpha_r(\nabla d_0^{\varphi})} \Bigg\{ &\mathrm{div}\left(\frac{T_{\phi_r}(\nabla d_0^{\varphi})}{(\alpha_r(\nabla d_0^{\varphi}))^2}\right) + \frac{\mathrm{div}\left(T_{\phi_r}(\nabla d_0^{\varphi})\right)}{(\alpha_r(\nabla d_0^{\varphi}))^2}\\
&- \frac{2}{\alpha_r(\nabla d_0^{\varphi})} \mathrm{div}\left(\frac{T_{\phi_r}(\nabla d_0^{\varphi})}{\alpha_r(\nabla d_0^{\varphi})}\right)\Bigg\} = 0.
\end{aligned}
\tag{5.89}
$$

Using the identity

$$\operatorname{div}\left(\frac{T_{\phi_r}(\nabla d_0^{\varphi})}{(\alpha_r(\nabla d_0^{\varphi}))^2}\right) = \frac{1}{\alpha_r(\nabla d_0^{\varphi})}\operatorname{div}\left(\frac{T_{\phi_r}(\nabla d_0^{\varphi})}{\alpha_r(\nabla d_0^{\varphi})}\right) + \frac{T_{\phi_r}(\nabla d_0^{\varphi})}{\alpha_r(\nabla d_0^{\varphi})}\cdot\nabla\left(\frac{1}{\alpha_r(\nabla d_0^{\varphi})}\right),$$

it follows that, for any $r = 1, \dots, m$, the quantity in braces in (5.89) becomes

$$\frac{T_{\phi_r}(\nabla d_0^{\varphi})}{\alpha_r(\nabla d_0^{\varphi})}\cdot\nabla\left(\frac{1}{\alpha_r(\nabla d_0^{\varphi})}\right) + \frac{\operatorname{div}\left(T_{\phi_r}(\nabla d_0^{\varphi})\right)}{(\alpha_r(\nabla d_0^{\varphi}))^2} - \frac{1}{\alpha_r(\nabla d_0^{\varphi})}\operatorname{div}\left(\frac{T_{\phi_r}(\nabla d_0^{\varphi})}{\alpha_r(\nabla d_0^{\varphi})}\right), \tag{5.90}$$

which is identically zero. This concludes the proof of our claim (5.87).

From (5.83), summing over $r = 1, \dots, m$ and using (5.40) we deduce

$$0 = -U_2'' + U_2 f'(\gamma) + \gamma'(V_1^{\psi} + \kappa_1^{\psi}) + y\gamma' h_0^{\psi} + \sum_{r=1}^{m} \frac{1}{\alpha_r(\nabla d_0^{\varphi})}\left[\mathrm{A}^r + \mathrm{B}^r + \mathrm{C}^r + \gamma''\mathrm{D}^r\right].$$

Note that we have used $U_2 = \sum_{r=1}^{m} W_2^r$: in general it may happen that $U_2 - \sum_{r=1}^{m} W_2^r = \mathcal{O}(\epsilon)$, but we have the freedom[23] to redefine the functions W_2^r up to discrepancies of order $\mathcal{O}(\epsilon)$, and put the subsequent errors in the terms U_3 and W_3^r, which we are not interested in.

Recalling (5.78), (5.82) and (5.84), and observing also that

$$\int_{\mathbb{R}} y\gamma'\gamma'\,dy = 0,$$

(so that the orthogonality condition (5.53) leads to drop out the terms with h_0^{φ}), we end up with the following integrability condition:

$$0 = c_1(V_1^{\varphi} + \kappa_1^{\varphi}) + c_0 G,$$

where

$$c_1 = \int_{\mathbb{R}} (\gamma')^2\,dy$$

[23] This is because enforcing the relation between (t, x) and (y, s, t, x) introduces a dependence on ϵ.

and

$$G = \sum_{r=1}^{m} \frac{1}{\alpha(\nabla d_0^\varphi)} \int_{\mathbb{R}} \gamma' \left(\mathrm{A}^r + \mathrm{B}^r + \mathrm{C}^r\right) dy.$$

The term G is presumably nonzero, which shows that, in general, V_1^φ is nonzero. This is a difference with respect to the formal asymptotic analysis of the anisotropic Allen-Cahn's equation [3–5], and suggests an $\mathcal{O}(\epsilon)$-error estimate between the geometric front and $\Sigma_\epsilon(t)$ (while, in the Allen-Cahn's equation, the estimate can be improved to the order $\mathcal{O}(\epsilon^2)$).

Remark 5.15 (Approximate evolution law and forcing term). The integrability condition for function U_2 relates V_1^φ and κ_1^φ and together with the integrability condition for U_1 leads to the approximate evolution law

$$V_\epsilon = -\kappa_\epsilon^\varphi - \epsilon \frac{c_0}{c_1} G + \mathcal{O}(\epsilon^2)$$

for Σ_ϵ. By dropping the $\mathcal{O}(\epsilon^2)$ term we obtain a new approximation Σ_1 of Σ_ϵ which we assume to have an $\mathcal{O}(\epsilon^2)$ error. This allows in turn to recover the $\mathcal{O}(\epsilon)$ term for the signed distance d_1^φ by taking the difference between the signed distance from $\Sigma_1(t)$ and the signed distance from $\Sigma_0(t)$ and dividing by ϵ. Now we can recover the functions W_1^r (which indeed depend on ∇d_1^φ) and solve the differential equation for U_2 (which also depends on ∇d_1^φ) to get U_2. This argument works provided G does not depend on d_1^φ, since it is also through G that the function U_2 is determined. We see from Remarks 5.11, 5.12, 5.13 and the properties of D^r, that the function G is indeed independent of d_1^φ.

Problem 5.16. Investigate on the existence and regularity of solutions to the elliptic equation (5.6), coupled with (5.65), (5.68), leading to the function w_0^r for any $r = 1, \ldots, m$.

References

[1] L. AMBROSIO, P. COLLI FRANZONE and G. SAVARÉ, *On the asymptotic behaviour of anisotropic energies arising in the cardiac bidomain model*, Interface Free Bound, **2** (2000), 213–266.

[2] D. BAO, S.-S. CHERN and Z. SHEN, "An Introduction to Riemann-Finsler Geometry", Graduate Texts in Mathematics, Vol. 200, Springer-Verlag, 2000.

[3] G. BELLETTINI, "Lecture Notes on Mean Curvature Flow: Barriers and Singular Perturbations", Edizioni Sc. Norm. Sup. Pisa, to appear.

[4] G. BELLETTINI, P. COLLI FRANZONE and M. PAOLINI, *Convergence of front propagation for anisotropic bistable reaction–diffusion equations*, Asymptotic Anal. **15** (1997), 325–358.

[5] G. BELLETTINI and M. PAOLINI, *Quasi–optimal error estimates for the mean curvature flow with a forcing term*, Differential Integral Equations **8** (1995), 735–752.

[6] G. BELLETTINI and M. PAOLINI, *Anisotropic motion by mean curvature in the context of Finsler geometry*, Hokkaido Math. J. **25** (1996), 537–566.

[7] G. BELLETTINI, M. PAOLINI and F. PASQUARELLI, *Non convex mean curvature flow as a formal singular limit of the non linear bidomain model*, Advances in Differential Equations, to appear.

[8] G. BELLETTINI, M. PAOLINI and S. VENTURINI, *Some results on surface measures in Calculus of Variations*, Ann. Mat. Pura Appl. **170** (1996), 329–359.

[9] J. BENCE, B. MERRIMAN and S. OSHER, *Diffusion generated motion by mean curvature*, Computational Crystal Growers Workshop, J. Taylor (ed.), Selected Lectures in Math., Amer. Math. Soc. (1992), 73–83.

[10] E. BONNETIER, E. BRETIN and A. CHAMBOLLE, *Consistency result for a non monotone scheme for anisotropic mean curvature flow*, Interfaces Free Bound. **14** (2012), 1–35.

[11] L. BRONSARD and F. REITICH, *On three-phase boundary motion and the singular limit of a vector-valued Ginzburg-Landau equation* Arch. Ration. Mech. Anal. **124** (1993), 355–379.

[12] A. CHAMBOLLE and M. NOVAGA, *Convergence of an algorithm for the anisotropic and crystalline mean curvature flow*, SIAM J. Math. Anal. **37** (2006), 1878–1987.

[13] P. COLLI FRANZONE, M. PENNACCHIO and G. SAVARÉ, *Multiscale modeling for the bioelectric activity of the heart*, SIAM J. Math. Anal. **37** (2005), 1333–1370.

[14] P. COLLI FRANZONE and G. SAVARÉ, *Degenerate evolution systems modeling the cardiac electric field ad micro and macroscopic level*, Evolution equations, semigroups and functional analysis (Milano 2000), Vol. 50 Progress Nonlin. Diff. Equations Appl. Birkhäuser, Basel (2002), 49–78.

[15] J. COROMILAS, A. L. DILLON and S. M. WIT, *Anisotropy reentry as a cause of ventricular tachyarrhytmias*, Cardiac Electrophysiology: From Cell to Bedside, ch.49, W. B. Saunders Co., Philadelphia (1994), 511–526.

[16] R. M. MIURA, *Accurate computation of the stable solitary wave for the FitzHugh-Nagumo equations*, J. Math. Biol. **13** (1981/82), 247–269.

[17] R. T. ROCKAFELLAR, "Convex Analysis", Princeton University Press, Princeton, 1972.

[18] R. SCHNEIDER, "Convex Bodies: the Brunn-Minkowski Theory", Encyclopedia of Mathematics and its Applications, Vol. 44, Cambridge Univ. Press, 1993.

[19] A. C. THOMPSON, "Minkowski Geometry", Encyclopedia of Mathematics and its Applications, Vol. 63, Cambridge Univ. Press, 1996.

Existence and qualitative properties of isoperimetric sets in periodic media

Antonin Chambolle, Michael Goldman and Matteo Novaga

Abstract. We review and extend here some recent results on the existence of minimal surfaces and isoperimetric sets in non homogeneous and anisotropic periodic media. We also describe the qualitative properties of the homogenized surface tension, also known as stable norm (or minimal action) in Weak KAM theory. In particular we investigate its strict convexity and differentiability properties.

1 Introduction

In Euclidean spaces, it is well known that hyperplanes are local minimizers of the perimeter and that balls are the (unique) solutions to the isoperimetric problem *i.e.* they have the least perimeter among all the sets having a given volume. The situation of course changes for interfacial energies which are no longer homogeneous nor isotropic but it is still natural to investigate the existence of local minimizers which are plane-like and of compact isoperimetric sets in this context. More precisely, for an open set $\Omega \subseteq \mathbb{R}^d$ and a set of finite perimeter E (see [23]), we will consider interfacial energies of the form

$$\mathcal{E}(E,\Omega) := \int_{\partial^* E\cap\Omega} F(x,\nu^E)\,d\mathcal{H}^{d-1}$$

where $\mathcal{H}^{d-1}$ is the $(d-1)$-dimensional Hausdorff measure, ν^E is the internal normal to E, $\partial^* E$ is the reduced boundary of E, and $F(x,p)$ is continuous and periodic in x, convex and one-homogeneous in p with

$$c_0|p| \le F(x,p) \le \frac{1}{c_0}|p| \qquad \forall (x,p)\in\mathbb{R}^d\times\mathbb{R}^d \tag{1.1}$$

The second author thanks E. Spadaro and E. Cinti for interesting discussions about Almgren's Lemma and Remark 4.5.

for some $c_0 > 0$. When $\Omega = \mathbb{R}^d$, we will simply denote by $\mathcal{E}$, the functional $\mathcal{E}(\cdot, \mathbb{R}^d)$. In the following we will denote by $\mathbb{T}$ the d dimensional torus and let $Q := [0, 1)^d$.

In a first part, we review the fundamental result of Caffarelli and De La Llave [12] concerning the existence of plane-like minimizers of $\mathcal{E}$ and we will define a homogenized energy $\phi(p)$ (usually called the stable norm or the minimal action functional), which represents the average energy of a plane-like minimizer in the direction p. The qualitative properties of the minimal action are studied in the second section. The following result was proven in [13] (see also [4,25]).

- If p is "totally irrational" (meaning that there exists no $q \in \mathbb{Z}^d$ such that $q \cdot p = 0$) then $\nabla\phi(p)$ exists.
- The same occurs for any p such that the plane-like minimizers satisfying the strong Birkhoff property give rise to a foliation of the space.
- If there is a gap in this lamination and if $(q_1, \dots, q_k) \in \mathbb{Z}^d$ is a maximal family of independent vectors such that $q_i \cdot p = 0$, then $\partial\phi(p)$ is a convex set of dimension k, and ϕ is differentiable in the directions which are orthogonal to $\{q_1, \dots, q_k\}$. In particular if p is not totally irrational then ϕ is not differentiable at p.
- ϕ^2 is strictly convex.

In the last section, we extend some results of [24] concerning the existence of compact minimizers of the isoperimetric problem

$$\min_{|E|=v} \mathcal{E}(E) \tag{1.2}$$

for every given volume $v > 0$ and show that these minimizers, once rescaled, converge to the Wulff shape associated to the stable norm ϕ.

Let us conclude this introduction by pointing out that, using a deep result of Bourgain and Brezis [8], see also [13, 15, 19], all the results presented here directly extend to functionals of the form

$$\int_{\partial^* E\cap\Omega} F(x, \nu^E) d\mathcal{H}^{d-1} + \int_{E\cap\Omega} g(x)\, dx \tag{1.3}$$

where $g \in L^d(\mathbb{T})$ is a periodic function with zero mean satisfying some smallness assumption (for the results of Section 3 to hold, one needs also that g is Lipschitz continuous).

Notice also that when considering the perimeter *i.e.* when $F(x, p) = |p|$, smooth minimizers of (1.3) satisfy the prescribed mean curvature equation

$$\kappa_E = -g$$

where κ_E is the mean curvature of the set E. The existence of plane-like minimizers of (1.3) can then be rephrased in term of existence of plane-like sets with prescribed mean curvature. On the other hand, in [24], the isoperimetric problem (1.2) was introduced in order to study existence of compact sets with prescribed mean curvature, leading to the proof of the following theorem.

Theorem 1.1. *Let $d \leq 7$ and g be a periodic $C^{0,\alpha}$ function on $\mathbb{R}^d$ with zero average and satisfying a suitable smallness assumption. Then for every $\varepsilon > 0$ there exists $\varepsilon' \in [0, \varepsilon]$ such that there exists a compact solution of*

$$\kappa_E = g + \varepsilon'.$$

2 Plane-like minimizers

In [12], Caffarelli and De La Llave proved the existence of plane-like minimizers of $\mathcal{E}$.

Theorem 2.1. *There exists $M > 0$ depending only on c_0 such that for every $p \in \mathbb{R}^d \setminus \{0\}$ and $a \in \mathbb{R}$, there exists a local minimizer (also called Class A Minimizer) E of $\mathcal{E}$ such that*

$$\left\{x \cdot \frac{p}{|p|} > a + M\right\} \subseteq E \subseteq \left\{x \cdot \frac{p}{|p|} > a - M\right\}. \tag{2.1}$$

Moreover ∂E is connected. A set satisfying the condition (2.1) *is called plane-like.*

Definition 2.2. Let $p \in \mathbb{R}^d \setminus \{0\}$ and let E be a plane-like minimizer of $\mathcal{E}$ in the direction p. We set

$$\phi(p) := |p| \lim_{R\to\infty} \frac{1}{\omega_{d-1} R^{d-1}} \mathcal{E}(E, B_R),$$

where ω_{d-1} is the volume of the unit ball in $\mathbb{R}^{d-1}$.

Caffarelli and De La Llave proved that this limit exists and does not depend on E. In [15], the first author and Thouroude related this definition to the cell formula:

$$\phi(p) = \min\left\{\int_{\mathbb{T}} F(x, p + Dv(x)) \ : \ v \in BV(\mathbb{T})\right\}. \tag{2.2}$$

It is obvious from (2.2) that ϕ is a convex, one-homogeneous function. However, since the problem defining ϕ is not strictly convex, in general the minimizer of (2.2) is not unique. Nevertheless, this uniqueness generically holds (see [13, Theorem 4.23, Theorem B.1]). This is an instance of the so-called Mañé's conjecture. It has been shown in [15] that the minimizers of (2.2) give an easy way to construct plane-like minimizers:

Proposition 2.3. *Let v_p be a minimizer of* (2.2) *then for every $s \in \mathbb{R}$, the set $\{v_p(x) + p \cdot x > s\}$ is a plane-like minimizer of $\mathcal{E}$ in the direction p.*

For $\varepsilon > 0$ and $E \subseteq \mathbb{R}^d$ of finite perimeter, let

$$\mathcal{E}_\varepsilon(E) := \varepsilon^{(d-1)}\mathcal{E}\left(\varepsilon^{-1}E\right) = \int_{\partial^* E} F\left(x/\varepsilon, \nu^E\right)\, d\mathcal{H}^{d-1}.$$

It was shown in [15] (see also [10]) that the convergence of the average energy of plane-like minimizers to the stable norm can also be reinterpreted in term of Γ-convergence [18].

Theorem 2.4. *When $\varepsilon \to 0$, the functionals $\mathcal{E}_\varepsilon$ Γ-converge, with respect to the L^1-convergence of the characteristic functions, to the anisotropic functional*

$$\mathcal{E}_0(E) = \int_{\partial^* E} \phi(\nu^E)\, d\mathcal{H}^{d-1} \qquad E \subseteq \mathbb{R}^d \text{ of finite perimeter.}$$

3 Strict convexity and differentiability properties of the stable norm

In this section we are going to study the differentiability and strict convexity of the stable norm ϕ. It is a geometric analog of the minimal action functional of KAM theory whose differentiability has first been studied by Aubry and Mather [3, 27] for geodesics on the two dimensional torus. The results of Aubry and Mather have then been extended by Moser [31], in the framework of non-parametric integrands, and more recently by Senn [32]. In this context, the study of the set of non-self-intersecting minimizers, which correspond to our plane-like minimizers satisfying the Birkhoff property has been performed by Moser and Bangert [7, 30], whereas the proof of the strict convexity of the minimal action has been recently shown by Senn [33]. Another related problem is the homogenization of periodic Riemannian metrics (geodesics are objects of dimension one whereas in our problem the hypersurfaces are of codimension one). We refer to [9, 11] for more information on this problem.

We define the polar function of F by

$$F^\circ(x, z) := \sup_{\{F(x,p)\leq 1\}} z \cdot p$$

so that $(F^\circ)^\circ = F$. If we denote by $F^*(x,z)$ the convex conjugate of F with respect to the second variable then $\{F^*(x,z)=0\} = \{F^\circ \le 1\}$. We will make the following additional hypotheses on F:

- F is $C^{2,\alpha}(\mathbb{R}^d \times (\mathbb{R}^d \setminus \{0\}))$,
- F is elliptic (that is $F(x,p) - C|p|$ is a convex function of p for some $C > 0$).

With these hypothesis we have [1, 16] [13, Proposition 3.4]

Proposition 3.1. *For any plane-like Class A Minimizer E, the reduced boundary $\partial^* E$ is of class $C^{2,\alpha}$ and $\mathcal{H}^{d-3}(\partial E \setminus \partial^* E) = 0$. Let $E_1 \subseteq E_2$ be two Class A Minimizers with connected boundary, then $\mathcal{H}^{d-3}(\partial E_1 \cap \partial E_2) = 0$.*

In the following we let

$$X := \{z \in L^\infty(\mathbb{T}, \mathbb{R}^d) \, : \, \operatorname{div} z = 0 \, , \; F^\circ(x, z(x)) \le 1 \; a.e\}.$$

Using arguments of convex duality, it is possible to characterize the stable norm as a support function [13, Proposition 2.13].

Proposition 3.2. *There holds*

$$\phi(p) = \sup_{z \in X} \left(\int_{\mathbb{T}} z \right) \cdot p \, .$$

Hence, the subgradient of ϕ at $p \in \mathbb{R}^d$ is given by

$$\partial\phi(p) \; = \; \left\{ \int_{\mathbb{T}} z \, : \, z \in X \, , \, \left(\int_{\mathbb{T}} z \right) \cdot p = \phi(p) \right\} . \qquad (3.1)$$

Since

$$\phi \text{ is differentiable at } p \quad \Longleftrightarrow \quad \partial\phi(p) \text{ is a singleton,}$$

Proposition 3.2 tells us that checking the differentiability of ϕ at a given point p is equivalent to checking whether for any vectorfields $z_1, z_2 \in X$,

$$\left(\int_{\mathbb{T}} z_1 \right) \cdot p = \left(\int_{\mathbb{T}} z_2 \right) \cdot p = \phi(p) \qquad \Longrightarrow \qquad \int_{\mathbb{T}} z_1 = \int_{\mathbb{T}} z_2 .$$

We now introduce the notion of calibration.

Definition 3.3. We say that a vector field $z \in X$ is a periodic calibration of a set E of locally finite perimeter if, we have

$$[z, \nu^E] = F(x, \nu^E) \qquad \mathcal{H}^{d-1} \llcorner \partial^* E - a.e.$$

When no confusion can be made, by calibration we mean a periodic calibration.

In the previous definition, $[z, \nu^E]$ has to be understood in the sense of Anzellotti but is roughly speaking $z(x) \cdot \nu^E(x)$ when it makes sense (see [2, 14]). By the differentiability of $F(x, \cdot)$, this implies that on a calibrated set, the value of z is imposed since (see [14] for a more precise statement)

$$z(x) = \nabla_p F(x, \nu^E) \tag{3.2}$$

Using some arguments of convex analysis and the coarea formula, it is possible to prove the following relation between calibrations and minimizers of (2.2).

Proposition 3.4. *Let $z \in X$ a vector field such that $\left(\int_{\mathbb{T}} z\right) \in \partial\phi(p)$. Then, for any minimizer v_p of* (2.2)*,*

$$[z, Dv_p + p] = F(x, Dv_p + p) \qquad |Dv_p + p| - a.e. \tag{3.3}$$

and for every $s \in \mathbb{R}$, z calibrates the set $E_s := \{v_p + p \cdot x > s\}$. We say that such a vector field z is a calibration in the direction p.

Equation (3.3) is the Euler-Lagrange equation associated to (2.2). Notice that thanks to (3.2), the value of any calibration in the direction p is fixed on ∂E_s. Hence, it is reasonable to expect that if these sets fill a big portion of the space, the average on the torus of any calibration will be fixed which would imply the differentiability of ϕ. One of the important consequences of calibration is that it implies an ordering of the plane-like minimizers.

Proposition 3.5. *Suppose that $z \in X$ calibrates two plane-like minimizers E_1 and E_2 with connected boundaries. Then, either $E_1 \subseteq E_2$, or $E_2 \subseteq E_1$. As a consequence $\mathcal{H}^{d-3}(\partial E_1 \cap \partial E_2) = 0$.*

Using the cell formula we can already prove the strict convexity of ϕ^2.

Theorem 3.6. *The function ϕ^2 is strictly convex.*

Proof. Let p_1, p_2, with $p_1 \neq p_2$, and let $p = p_1 + p_2$. We want to show that, if $\phi(p) = \phi(p_1) + \phi(p_2)$, then p_1 is proportional to p_2, which gives the thesis.

Indeed, we have

$$\begin{aligned}\phi(p) &= \int_{\mathbb{T}} [z_p, p + Dv_p] \\ &= \int_{\mathbb{T}} F(x, p + Dv_p) \\ &\leq \int_{\mathbb{T}} F(x, p + Dv_{p_1} + Dv_{p_2}) \\ &\leq \int_{\mathbb{T}} F(x, p_1 + Dv_{p_1}) + F(x, p_2 + Dv_{p_2}) \\ &= \phi(p_1) + \phi(p_2)\,.\end{aligned}$$

Since $\phi(p) = \phi(p_1)+\phi(p_2)$, it follows that $v_{p_1}+v_{p_2}$ is also a minimizer of (2.2) and, in particular, z_p satisfies

$$[z_p, p_1+Dv_{p_1}]+[z_p, p_2+Dv_{p_2}] \;=\; F(x, p_1+Dv_{p_1})+F(x, p_2+Dv_{p_2})$$

$(|p_1 + Dv_{p_1}| + |p_2 + Dv_{p_2}|)$-a.e., so that

$$[z_p, p_i + Dv_{p_i}] \;=\; F(x, p_i + Dv_{p_i}) \qquad i \in \{1, 2\}\,.$$

This means that z_p is a calibration for the plane-like minimizers

$$\{v_p + p \cdot x \geq s\}\,,\ \{v_{p_1} + p_1 \cdot x \geq s\}\ \text{and}\ \{v_{p_2} + p_2 \cdot x \geq s\}$$

for all $s \in \mathbb{R}$. By Proposition 3.5, it follows that they are included in one another which is possible only if p_1 is proportional to p_2. □

We can also show that ϕ is differentiable in the totally irrational directions.

Proposition 3.7. *Assume p is totally irrational. Then for any two calibrations z, z' in the direction p, $\int_{\mathbb{T}} z\,dx = \int_{\mathbb{T}} z'\,dx$. As a consequence, $\partial\phi(p)$ is a singleton and ϕ is differentiable at p.*

Proof. The fundamental idea is that since p is totally irrational, even if the levelsets $\{v_p + p \cdot x > s\}$ do not fill the whole space, the remaining holes must have finite volume and therefore do not count in the average.

Consider z, z' two calibrations and a solution v_p of (2.2).

Let us show that, for any s,

$$\int_{\{v_p+p\cdot x=s\}} (z(x) - z'(x))\,dx \;=\; 0. \tag{3.4}$$

Let $C_s := \{x : v_p(x) + p\cdot x = s\}$ then $\partial C_s = \partial\{v_p + p\cdot x > s\} \cup \partial\{v_p + p\cdot x \geq s\}$. Moreover, all the C_s have zero Lebesgue measure except for a countable number of values. Consider such a value s. Since z and z' calibrate C_s which is a plane-like minimizer we have $[z, \nu^{C_s}] = [z', \nu^{C_s}]$ on $\partial^* C_s$.

Then, we observe that the sets $C_s^q = Q \cap (C_s - q)$, $q \in \mathbb{Z}^d$, are all disjoint since p is totally irrational and since all the C_s^q are calibrated by z, so that their measures sum up to less than 1.

Let e_i be a vector of the canonical basis of $\mathbb{R}^d$ then by the divergence Theorem (where the integration by parts can by justified thanks to $|C_s| \leq 1$) we compute

$$\int_{C_s} (z(x) - z'(x)) \cdot e_i \, dx = -\int_{C_s} x_i \operatorname{div}(z(x) - z'(x))\, dx = 0\,, \quad (3.5)$$

which gives our claim.

In particular, we obtain that $\int_{\mathbb{R}^d} (z - z')\, dx = \sum_s \int_{C_s} (z - z')\, dx = 0$ hence $\int_Q (z - z')\, dx = 0$. □

When p is not totally irrational, we have to consider a slightly bigger class of plane-like minimizers than those obtained as $\{v_p + p\cdot x > s\}$ for v_p a minimizer of (2.2) and $s \in \mathbb{R}$. Indeed, we must consider all the plane-like minimizers which are maximally periodic.

Definition 3.8. Following [6,25,32] we give the following definitions:

- we say that $E \subseteq \mathbb{R}^d$ satisfies the Birkhoff property if, for any $q \in \mathbb{Z}^d$, either $E \subseteq E + q$ or $E + q \subseteq E$;
- we say that E satisfies the strong Birkhoff property in the direction $p \in \mathbb{Z}^d$ if $E \subseteq E + q$ when $p \cdot q \leq 0$ and $E + q \subseteq E$ when $p\cdot q \geq 0$.

We will let $\mathcal{CA}(p)$ be the set of all the plane-like minimizers in the direction p which satisfy the strong Birkhoff property

Notice that the sets $\{v_p + p\cdot x > s\}$ have the strong Birkhoff property. It can be shown that the sets of this form correspond exactly to the recurrent plane-like minimizers which are those which can be approximated by below or by above by entire translations of themselves (see [13, Proposition 4.18]). For sets satisfying the Birkhoff property there holds [13, Lem. 4.13, Proposition 4.14, Prop 4.15].

Proposition 3.9. *Let E be a Class A minimizer with the Birkhoff property: then it is a plane-like minimizer (with a constant M just depending on c_0 and d), calibrated and ∂E is connected.*

For sets satisfying the Strong Birkhoff property, it can be further proven [13, Theorem 4.19].

Theorem 3.10. *Let z be a calibration in the direction p, then z calibrates every plane-like minimizer with the strong Birkhoff property.*

As a consequence of Proposition 3.5 and Theorem 3.10, for every $p \in \mathbb{R}^d \setminus \{0\}$, the plane-like minimizers of $\mathcal{CA}(p)$ form a lamination of $\mathbb{R}^d$ (possibly with gaps). In light of (3.2), we see that

Proposition 3.11. *If there is no gap in the lamination by plane-like minimizers of $\mathcal{CA}(p)$ then ϕ is differentiable at the point p.*

We are thus just left to prove that if p is not totally irrational (meaning that there exists $q \in \mathbb{Z}^d$ such that $p \cdot q = 0$) and if there is a gap G (whose boundary $\partial E^+ \cup \partial E^-$ is made of two plane-like minimizers of $\mathcal{CA}(p)$) in the lamination then ϕ is not differentiable at the point p. To simplify the notations and the argument, let us consider the case $p = 2$ and $(p, q) = (e_1, e_2)$, the canonical basis of $\mathbb{R}^2$. Let $E_n := \{v_{e_1+\frac{1}{n}e_2} + (e_1+\frac{1}{n}e_2) \cdot x > 0\}$ be plane-like minimizers in the direction $e_1 + \frac{1}{n}e_2$ which intersects G, then up to translations (in the direction e_2), we can assume that there is a subsequence which converges to a set H_+ which also intersects the gap. It can be shown that H_+ is an heteroclinic solution meaning that it is included inside G, satisfies the Birkhoff property (but not the strong one), and that $H_+ \pm ke_2 \to E^\pm$ when $k \in \mathbb{N}$ goes to infinity (see [13, Proposition 4.27]). Moreover H_+ is calibrated by $z_+ := \lim z_n$ where z_n is any calibration in the direction $e_1 + \frac{1}{n}e_2$ (notice that z_+ is then a calibration in the direction e_1). Consider similarly H_- (respectively z_-), an heteroclinic solution in the direction $-e_2$ (respectively a calibration of H_-) then we aim at proving that

$$\int_{G \cap Q} [z_+ - z_-, e_2] > 0$$

which would imply the non differentiability of ϕ at e_1 (in the direction e_2).

Proposition 3.12. *For $t \in [0, 1)$, let $S_t := \{x \cdot e_2 = t\}$ (and $S = S_0$) then almost every $s, t \in \mathbb{R}$, we have*

$$\int_{S_t \cap G} [z_+ - z_-, e_2] = \int_{S_s \cap G} [z_+ - z_-, e_2]. \tag{3.6}$$

In particular,

$$\int_{Q \cap G} [z_+ - z_-, e_2] = \int_{S \cap G} [z_+ - z_-, e_2]. \tag{3.7}$$

Proof. Fix $s < t \in \mathbb{R}$, let $S_s^t := \{x \in Q : s < x \cdot e_2 < t\}$ then

$$0 = \int_{S_s^t \cap G} \operatorname{div}(z_+ - z_-) = \int_{S_t \cap G} [z_+ - z_-, e_2] - \int_{S_s \cap G} [z_+ - z_-, e_2].$$

□

Proposition 3.13. *Let ν be the inward normal to H_q, then*

$$\int_{S \cap G} [z_+ - z_-, e_2] = \int_{\partial^* H_+} [z_+ - z_-, \nu]. \tag{3.8}$$

Proof. We first introduce some additional notation (see Figure 3.1): let

$$\Sigma^+ := \partial^* H_+ \cap \{x \cdot e_2 > 0\} \qquad G^+ := G \cap \{x \cdot e_2 > 0\} \cap H_+^c$$

and

$$S^+ := S \cap G \cap H_+^c.$$

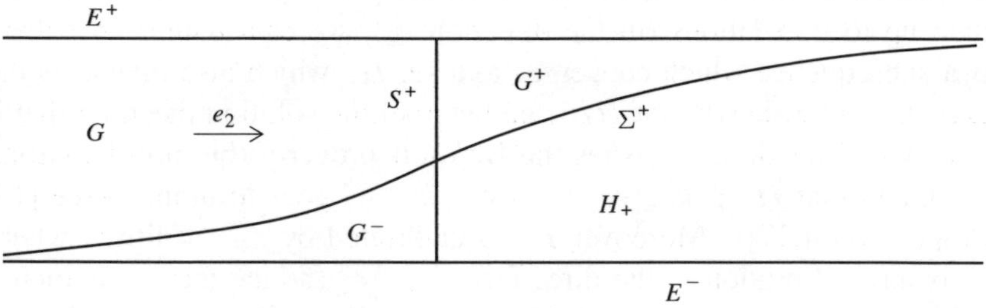

Figure 3.1. Heteroclinic solution in the direction e_2.

Then,

$$0 = \int_{G^+} \operatorname{div}(z_+ - z_-) = \int_{S^+} [z_+ - z_-, e_2] - \int_{\Sigma^+} [z_+ - z_-, \nu]$$

Similarly we define Σ^- and S^- and get

$$\int_{\Sigma^-} [z_+ - z_-, \nu] = \int_{S^-} [z_+ - z_-, e_2].$$

Summing these two equalities we find (3.8). □

We can now conclude. Indeed, since z_+ calibrates H_+ and since $z_- \in X$, on $\partial^* H_+$, there holds $[z_+ - z_-, e_2] = F(x, \nu) - [z_-, \nu] \geq 0$ and thus if

$$\int_{\partial^* H_+} [z_+ - z_-, \nu] = 0$$

then $F(x,\nu) = [z_-, \nu]$ on ∂H_+ and thus z_- calibrates also H_- which would lead to a contradiction since it implies that H_+ and H_- cannot cross. Hence,

$$\int_{G\cap Q} [z_+ - z_-, e_2] = \int_{\partial^* H_+} [z_+ - z_-, \nu] > 0.$$

In conclusion we have (see [13] for a complete proof)

Proposition 3.14. *If there is a gap in the lamination by plane-like minimizers of $\mathcal{CA}(p)$ and if $(q_1, \dots, q_k) \in \mathbb{Z}^d$ is a maximal family of independent vectors such that $q_i \cdot p = 0$, then $\partial\phi(p)$ is a convex set of dimension k, and ϕ is differentiable in the directions which are orthogonal to $\{q_1, \dots, q_k\}$. In particular if p is not totally irrational then ϕ is not differentiable at p.*

4 Existence and asymptotic behavior of isoperimetric sets

In this section we extend some results of [24] on the existence of compact minimizers of the isoperimetric problem (1.2). In addition to (1.1), we will make the hypothesis that F is uniformly Lipschitz continuous in x, *i.e.* there exists $C > 0$ such that

$$|F(x,p) - F(y,p)| \le C|x-y| \qquad \forall (x,y,p) \in \mathbb{R}^d \times \mathbb{R}^d \times \mathbb{S}^{d-1}.$$

Our strategy will differ from the one of [24, Theorem 2.6]. It will instead closely follow [24, Theorem 4.9]. The idea is to use the Direct Method of the calculus of variations together with a kind of concentration compactness argument to deal with the invariance by translations of the problem. Notice that a similar strategy has been used to prove existence of minimal clusters (see [26, Theorem 29.1]). We first recall Almgren's Lemma (see [26, Lem. II.6.18], [28]).

Lemma 4.1. *If E is a set of finite perimeter and A is an open set of $\mathbb{R}^d$ such that $\mathcal{H}^{d-1}(\partial^* E \cap A) > 0$ then there exists $\sigma_0 > 0$ and $C > 0$ such that for every $\sigma \in (-\sigma_0, \sigma_0)$ there exists a set F such that*

- $F \Delta E \Subset A$.
- $|F| = |E| + \sigma$.
- $|\mathcal{E}(F, A) - \mathcal{E}(E, A)| \le C|\sigma|$.

We now prove that any minimizer (if it exists) has to be compact.

Proposition 4.2. *For every $v > 0$, every minimizer E of* (1.2) *has bounded diameter.*

Proof. The proof follows the classical method to prove density estimates for minimizers of isoperimetric problems (see [23]). Fix $v > 0$ and let E be a minimizer of (1.2). Let then $f(r) := |E \backslash B_r|$. Let us assume that the diameter of E is not finite then $f(r) > 0$ for every $r > 0$. Let us assume that $\mathcal{H}^{d-1}(\partial^* E \cap B_1) > 0$ then let σ_0 and C be given by Almgren's Lemma and fix $R > 1$ such that $f(R) \leq \sigma_0$ then for every $r > R$ there exists F such that

- $F \Delta E \Subset B_1$.
- $|F| = |E| + f(r)$
- $|\mathcal{E}(E, B_r) - \mathcal{E}(F, B_r)| \leq C f(r)$.

Letting $G := F \cap B_r$ we have $|G| = |E|$ thus by minimality of E, we find

$$\begin{aligned} \mathcal{E}(E) \leq \mathcal{E}(G) &\leq \mathcal{E}(F, B_r) + \mathcal{H}^{d-1}(\partial B_r \cap F) \\ &\leq \mathcal{E}(E, B_r) + C f(r) + \mathcal{H}^{d-1}(\partial B_r \cap E) \end{aligned}$$

and thus

$$c_0 \mathcal{H}^{d-1}(\partial^* E \backslash B_r) \leq C f(r) + \mathcal{H}^{d-1}(\partial B_r \cap E) - f'(r)$$

then using the isoperimetric inequality we get

$$c_1 f(r)^{\frac{d-1}{d}} \leq C f(r) - f'(r)$$

If now $R_1 > R$ is such that $f(r) \leq \frac{c_1}{2C}$, we get

$$C f(r)^{\frac{d-1}{d}} \leq -f'(r)$$

and thus $\left(f^{1/d}\right)' \leq -C$ which leads to a contradiction. □

Remark 4.3. Adapting the proof of [17] to the anisotropic case, it should be possible to prove the boundedness of minimizers under the weaker assumption that $F(\cdot, p)$ is continuous (using the so-called $\varepsilon - \varepsilon^{\beta}$ property).

We can now prove the existence of compact minimizers for every volume $v > 0$.

Theorem 4.4. *For every $v > 0$ there exists a compact minimizer of* (1.2).

Proof. To simplify the notations, let us assume that $v = 1$. Let E_k be a minimizing sequence meaning that $|E_k| = 1$ and $\mathcal{E}(E_k) \to \inf_{|E|=1} \mathcal{E}(E)$. For every $k \in \mathbb{N}$, let $\{Q_{i,k}\}_{i \in \mathbb{N}}$ be a partition of $\mathbb{R}^d$ into disjoint cubes of equal volume larger than 2, such that the sets $E_k \cap Q_{i,k}$ are of decreasing measure, and let $x_{i,k} = |E_k \cap Q_{i,k}|$. By the isoperimetric inequality [23], there exist $0 < c < C$ such that

$$\begin{aligned} c\sum_i x_{i,k}^{\frac{d-1}{d}} &= c\sum_i \min\left(|E_k \cap Q_{i,k}|, |Q_{i,k}\setminus E_k|\right)^{\frac{d-1}{d}} \\ &\le \sum_i P(E_k, Q_{i,k}) \\ &\le \sum_i c_0\mathcal{E}(E_k, Q_{i,k}) \\ &\le c_0\mathcal{E}(E_k) \le C \end{aligned}$$

hence

$$\sum_{i=1}^{\infty} x_{i,k} = 1 \qquad \text{and} \qquad \sum_{i=1}^{\infty} x_{i,k}^{\frac{d-1}{d}} \le \frac{C}{c}.$$

Since $x_{i,k}$ is nonincreasing with respect to i, it follows that (cf [24, Lem. 4.2]) for any N

$$\sum_{i=N}^{\infty} x_{i,k} \le \frac{C}{c}\frac{1}{N^{1/d}}. \tag{4.1}$$

Up to extracting a subsequence, we can suppose that $x_{i,k} \to \alpha_i \in [0, 1]$ as $k \to +\infty$ for every $i \in \mathbb{N}$, so that by (4.1) we have

$$\sum_i \alpha_i = 1. \tag{4.2}$$

Let $z_{i,k} \in Q_{i,k}$. Up to extracting a further subsequence, we can suppose that $d(z_{i,k}, z_{j,k}) \to c_{ij} \in [0, +\infty]$, and

$$\left(E_k - z_{i,k}\right) \to E_i \quad \text{in the } L^1_{\text{loc}}\text{-convergence}$$

for every $i \in \mathbb{N}$. And it is not very difficult to check that E_i are minimizers of (1.2) under the volume constraint $v_i := |E_i|$. Notice that by Proposition 4.2, each E_i is bounded.

We say that $i \sim j$ if $c_{ij} < +\infty$ and we denote by $[i]$ the equivalence class of i. Notice that E_i equals E_j up to a translation, if $i \sim j$. We want to prove that

$$\sum_{[i]} v_i = 1, \tag{4.3}$$

where the sum is taken over all equivalence classes. For all $R > 0$ let $Q_R = [-R/2, R/2]^d$ be the cube of sidelength R. Then for every $i \in \mathbb{N}$,

$$|E_i| \ge |E_i \cap Q_R| = \lim_{k\to+\infty} \left|\left(E_k - z_{i,k}\right) \cap Q_R\right|.$$

If j is such that $j \sim i$ and $c_{ij} \le \frac{R}{2}$, possibly increasing R we have $Q_{j,k} - z_{i,k} \subset Q_R$ for all $k \in \mathbb{N}$, so that

$$\lim_{k\to+\infty} \left|(E_k - z_{i,k}) \cap Q_R\right| \ge \lim_{k\to+\infty} \sum_{c_{ij}\le\frac{R}{2}} |E_k \cap Q_{j,k}| = \sum_{c_{ij}\le\frac{R}{2}} \alpha_j.$$

Letting $R \to +\infty$ we then have

$$|E_i| \ge \sum_{i\sim j} \alpha_j$$

hence, recalling (4.2),

$$\sum_{[i]} |E_i| \ge 1,$$

thus proving (4.3) (since the other inequality is clear).

Let us now show that

$$\sum_{[i]} \mathcal{E}(E_i) \le \inf_{|E|=1} \mathcal{E}(E). \tag{4.4}$$

Choosing a representative in each equivalence class $[i]$ and reindexing, from now on we shall assume that $c_{ij} = +\infty$ for all $i \ne j$. Let $I \in \mathbb{N}$ be fixed. Then for every $R > 0$ there exists $K \in \mathbb{N}$ such that for every $k \ge K$ and i, j less than I, we have

$$d(z_{i,k}, z_{j,k}) > R.$$

For $k \ge K$ we thus have

$$\begin{aligned}
\mathcal{E}(E_k) &\ge \sum_{i=1}^{I} \int_{\partial E_k \cap (B_R + z_{i,k})} F(x, \nu^{E_k})\, d\mathcal{H}^{d-1} \\
&= \sum_{i=1}^{I} \int_{\partial (E_k - z_{i,k}) \cap B_R} F(x, \nu^{E_k})\, d\mathcal{H}^{d-1} \\
&= \sum_{i=1}^{I} \mathcal{E}(E_k - z_{i,k}, B_R)
\end{aligned}$$

From this, and the lower-semicontinuity of $\mathcal{E}$, we get

$$\inf_{|E|=1} \mathcal{E}(E) \ge \sum_{i=1}^{I} \liminf_{k\to\infty} \mathcal{E}(E_k - z_{i,k}, B_R) \ge \sum_{i=1}^{I} \mathcal{E}(E_i, B_R).$$

Letting $R \to \infty$ and then $I \to \infty$ (if the number of equivalence classes is finite then just take I equal to this number), we find (4.4). Let finally

$d_i := \text{diam}(E_i)$ and $F := \bigcup_i (E_i + 2d_i e_1)$ where e_1 is a unit vector then $|F| = 1$ and

$$\mathcal{E}(F) = \sum_i \mathcal{E}(E_i) \leq \inf_{|E|=1} \mathcal{E}(E)$$

and thus F is a minimizer of (1.2) (notice that by Proposition 4.2, we must have $E_i = \emptyset$ for i large enough).

□

Remark 4.5. Another proof, in the spirit of [24, Theorem 2.6] would consist in proving first existence of compact minimizers of the relaxed problems

$$\min_{E \subset \mathbb{R}^d} \mathcal{E}(E) + \mu||E| - v| \tag{4.5}$$

for $\mu > 0$ using uniform density estimates (see [24, Proposition 2.1, Proposition 2.3]) and then showing that for μ large enough, the minimizers of (4.5) have volume exactly equal to v. In this more general situation with respect to the one studied in [24], instead of relying on the Euler-Lagrange equation as in [24, Theorem 2.6] (which works only in a smooth setting *i.e.* for F elliptic and smooth and for low dimension d) one could argue by contradiction and follow the lines of [20]. Notice also that contrary to [24, Theorem 2.6], this strategy (just as the one adopted here in the proof of Theorem 4.4) would not give quantitative bounds on the diameter.

Remark 4.6. The isoperimetric problem (1.2) is very similar to the isoperimetric problem on manifold with densities which has recently attracted a lot of attention and where similar issues of existence of compact minimizers appear (see [17,28,29]). Notice however that in these works, the media is usually considered as isotropic, meaning that $F(x, p) = f(x)|p|$ with some hypothesis on the behavior at infinity (or with some radial symmetry) of f which is not compatible with periodicity.

Remark 4.7. Using Almgren's Lemma, it is not difficult to see that minimizers of the isoperimetric problem (1.2) are quasi-minimizers of $\mathcal{E}$ (of course without volume constraint anymore) and as such, they enjoy the same regularity properties (see [26, Example 2.13], [16]). In particular, under the hypothesis of Section 3, they are $C^{2,\alpha}$ out of a singular set of $(d-3)$-Hausdorff measure equal to zero.

Let $W = \{\phi^\circ \leq 1\}$ be the Wulff shape associated to ϕ. It is then the (unique) solution to the isoperimetric problem associated to ϕ (see [22])

$$\min_{|E|=|W|} \int_{\partial^* E} \phi(\nu^E) d\mathcal{H}^{d-1}.$$

For $v > 0$, let E_v be a compact minimizer of (1.2). Let $\varepsilon := \left(\frac{|W|}{v}\right)^{1/d}$ and $E_\varepsilon := \varepsilon E_v$ then E_ε is a minimizer of

$$\min_{|E|=|W|} \int_{\partial^* E} F(x/\varepsilon, \nu^E) d\mathcal{H}^{d-1}$$

then using Theorem 2.4 and following the same proof as in Theorem 4.4 (see [24, Theorem 4.9]), we get:

Theorem 4.8. *There exist a sequence of vectors $z_\varepsilon \in \mathbb{R}^d$ such that $E_\varepsilon + z_\varepsilon \to W$ when $\varepsilon \to 0$.*

The asymptotic shape for small volume has been investigated in a very precise way in [21].

References

[1] F.J. ALMGREN, R. SCHOEN and L. SIMON, *Regularity and singularity estimates on hypersurfaces minimizing elliptic variational integrals*, Acta MaTheorem **139** (1977), 217–265.

[2] G. ANZELLOTTI, *Pairings between measures and bounded functions and compensated compactness*, Annali di Matematica Pura ed Applicata **135** (1983), 293–318.

[3] S. AUBRY, *The Devil's staircase Transformation in incommensurate Lattices*, Lecture Notes in MaTheorem **925** (1982), 221–245.

[4] F. AUER and V. BANGERT, *Differentiability of the stable norm in codimension one*, Amer. J. MaTheorem **128** (2006), 215-238.

[5] V. BANGERT, *The existence of gaps in minimal foliations*, Aequationes MaTheorem **34** (1987), 153–166.

[6] V. BANGERT, *A uniqueness theorem for $\mathbb{Z}^n$-periodic variational problems*, Comment. MaTheorem Helvetici **62** (1987), 511–531.

[7] V. BANGERT, *On minimal laminations of the torus*, Ann. IHP Analyse non linéaire **2** (1989), 95–138.

[8] J. BOURGAIN and H. BREZIS, *On the Equation* div $Y = f$ *and Application to Control of Phases*, Jour. of Amer. MaTheorem Soc. **16** (2002), 393–426.

[9] A. BRAIDES, G. BUTTAZZO and I. FRAGALÀ, *Riemannian approximation of Finsler metrics*, Asymptotic Anal. **31** (2002), 177–187.

[10] A. BRAIDES and V. CHIADÒ PIAT, *A derivation formula for convex integral functionals defined on $BV(\Omega)$*, J. Convex Anal. **2** (1995), 69–85.

[11] D. BURAGO, S. IVANOV and B. KLEINER, *On the structure of the stable norm of periodic metrics*, MaTheorem Res. Lett. **4** (1997), 791–808.

[12] L. CAFFARELLI and R. DE LA LLAVE, *Planelike Minimizers in Periodic Media*, Comm. Pure Appl. MaTheorem **54** (2001), 1403–1441.

[13] A. CHAMBOLLE, M. GOLDMAN and M. NOVAGA, *Plane-like minimizers and differentiability of the stable norm*, published J. Geom. Anal., to appear.

[14] A. CHAMBOLLE, M. GOLDMAN and M. NOVAGA, *Fine properties of the subdifferential for a class of one-homogeneous functionals* , preprint.

[15] A. CHAMBOLLE and G. THOUROUDE, *Homogenization of interfacial energies and construction of plane-like minimizers in periodic media through a cell problem*, Netw. Heterog. Media **4** (2009), 127–152.

[16] F. DUZAAR and K. STEFFEN, *Optimal Interior and Boundary Regularity for Almost Minimizers to Elliptic Variational Integrals*, J. Reine angew. MaTheorem **546** (2002), 73–138.

[17] E. CINTI and A. PRATELLI, *The $\varepsilon - \varepsilon^{\beta}$ property, the boundedness of isoperimetric sets in $\mathbb{R}^N$ with density, and some applications*, preprint.

[18] G. DAL MASO, "An Introduction to Γ-convergence", Progress in Nonlinear Differential Equations and their Applications, Vol. 8, Birkhäuser, 1993.

[19] T. DE PAUW and W. PFEFFER, *Distributions for which* $\operatorname{div} v = F$ *has a continuous solution*, Comm. Pure Appl. MaTheorem 61, 2, 2008.

[20] L. ESPOSITO and N. FUSCO, *A remark on a free interface problem with volume constraint*, J. Convex Anal. **18** (2011), 417–426.

[21] A. FIGALLI and F. MAGGI, *On the equilibrium shapes of liquid drops and crystals*, Arch. Ration. Mech. Anal. **201** (2011).

[22] I. FONSECA and S. MÜLLER, *A uniqueness proof for the Wulff theorem*, Proc. Roy. Soc. Edinburgh Sect. A **119**, (1991).

[23] E. GIUSTI, "Minimal Surfaces and Functions of Bounded Variation", Monographs in Mathematics, Vol. 80, Birkhäuser, 1984.

[24] M. GOLDMAN and M. NOVAGA, *Volume-constrained minimizers for the prescribed curvature problem in periodic media*, Calc. Var. and PDE **44** (2012), Issue 3.

[25] H. JUNGINGER-GESTRICH, "Minimizing Hypersurfaces and Differentiability Properties of the Stable Norm", PhD Thesis, Freiburg, 2007.

[26] F. MAGGI, "Sets of Finite Perimeter and Geometric Variational Problems: an Introduction to Geometric Measure Theory", Cambridge Studies in Advanced Mathematics, Vol. 135, Cambridge University Press, 2012.

[27] J.N. MATHER, *Differentiability of the minimal average action as a function of the rotation number*, Bol. Soc. Bras. Mat. **21** (1990), 59–70.

[28] F. MORGAN, "Regularity of Isoperimetric Hypersurfaces in Riemannian Manifolds", Trans. AMS, Vol. 355 n. 12, 2003.

[29] F. MORGAN and A. PRATELLI, *Existence of isoperimetric regions in $\mathbb{R}^n$ with density*, preprint.

[30] J. MOSER, *Minimal solutions of variational problems on a torus*, Ann. IHP Analyse non linéaire **3** (1986), 229–272.

[31] J. MOSER, *A stability theorem for minimal foliations on a torus*, Ergo. Theorem Dynam. Syst. **8** (1988), 251–281.

[32] W. SENN, *Differentiability properties of the minimal average action*, Calc. Var. and PDE **3** (1995), 343–384.

[33] W. SENN, *Strikte Konvexität für Variationsprobleme auf dem n-dimensionalen Torus*, Manuscripta MaTheorem **71** (1991), 45–65.

Minimizing movements and level set approaches to nonlocal variational geometric flows

Antonin Chambolle, Massimiliano Morini and Marcello Ponsiglione

Abstract. This contribution describes recent results on a variational approach for the geometric gradient flow of perimeter-like functionals, which include a class of non-local perimeters. In particular, the consistency of the variational approach with viscosity solutions of an appropriate level set equation is established.

1 Introduction

In this contribution we present a variational approach to geometric flows associated with nonlocal perimeters, as developed in [10]. The purpose of this note, rather than giving rigorous results (and proofs of them) is to expose in an informal way the main outcomes of [10], discussing some examples, connections with other works and possible applications to variational segmentation models.

We first introduce a rather general class of (possibly nonlocal) perimeters. By *generalized perimeter* we mean a set function that satisfies some natural structural assumptions; namely, the perimeter of smooth sets with compact boundary is finite, the perimeter is lower semicontinuous with respect to the L^1-convergence, and satisfies a convexity condition, called *submodularity*. Examples of generalized perimeters include the classical perimeter, fractional perimeters (see [3, 17]) as well as the pre-Minkowsky content: the last is a peculiar perimeter, recently proposed in [2] as a variant of the standard perimeter in variational models for image segmentation and denoising problems.

Given a generalized perimeter, one is naturally led to consider its first variation, and call it (whenever exists) curvature. Classically, this is done taking into account smooth inner variations of the set E. For our purposes, it is convenient to consider all possible measurable variations of E. Having a notion of generalized perimeter and of the corresponding curvature in hands, the question we address in [10] is: do minimizing movements of the perimeter converge to the corresponding geometric curvature flow?

To explain this point let us go back to the Euclidean case. The geometric evolution in this case is nothing but the mean curvature flow: it consists in finding a family of evolving sets $t \mapsto E_t$, such that each point of the boundary of E moves with normal velocity proportional to the mean curvature of the set at that point. The variational minimizing movements approach consists in a time discretization procedure (a kind of Euler implicit scheme), and a step by step minimization of a total energy which is the sum of the perimeter plus a dissipation that approximates an L^2-distance between sets. In [1,15] it is proved that this scheme approximates the mean curvature flow in a suitable sense.

The geometric flow can also be studied through the so called level set method, where E_t is identified with the (spatial) superlevel set of some time-dependent function u. In this framework, the flow is described by a parabolic equation in u, interpreted in the viscosity sense (see [11,12]). In [6] it is proved that the two approaches described above are consistent: more precisely, if one let evolve all the superlevel sets of an initial function u_0 by minimizing movements, then the resulting time-dependent function coincides with the unique viscosity solution to the level set equation.

In [9] we have proved that the same is true for the pre-Minkowski perimeter considered in [2]; in [10] we prove that this *consistency principle* in fact holds for a rather large class of generalized perimeters. This is obtained by introducing a suitable notion of viscosity solutions for nonlocal curvatures and by exploiting the consistency of such notion with the variational definition of curvature as first variation of the generalized perimeter.

In conclusion, the approach proposed in [10] and described here provides a unifying point of view for a large class of variational geometric evolutions.

2 Generalized perimeters and curvatures

We will consider the following notion of generalized (possibly nonlocal) perimeters.

2.1 Notion of generalized perimeter

Let $\mathfrak{C}^2$ be the class of closed subsets of $\mathbb{R}^N$ with compact C^2 boundary, and let $\mathfrak{M}$ be the class of all measurable subsets of $\mathbb{R}^N$. We will say that a functional $J : \mathfrak{M} \to [0, +\infty]$ is a generalized perimeter if it satisfies the following properties:

i) $J(E) < +\infty$ for every $E \in \mathfrak{C}^2$;
ii) $J(\emptyset) = J(\mathbb{R}^N) = 0$;
iii) $J(E) = J(E')$ if $|E \Delta E'| = 0$;
iv) J is L^1_{loc}-l.s.c.: if $|(E_n \Delta E) \cap B_R| \to 0$ for every $R > 0$, then $J(E) \leq \liminf_n J(E_n)$;
v) J is submodular: for any $E,\ F \in \mathfrak{M}$,

$$J(E \cup F) + J(E \cap F) \leq J(E) + J(F)\,; \tag{2.1}$$

vi) J is translational invariant:

$$J(x + E) = J(E) \quad \text{for all } E \in \mathfrak{M},\ x \in \mathbb{R}^N\,. \tag{2.2}$$

We can extend the functional J to $L^1_{\mathrm{loc}}(\mathbb{R}^N)$ enforcing the following *generalized co-area formula*:

$$J(u) := \int_{-\infty}^{+\infty} J(\{u > s\})\, ds. \tag{2.3}$$

It can be shown that under the assumptions above, J is a convex l.s.c. functional in $L^1_{\mathrm{loc}}(\mathbb{R}^N)$ (see [7]). We remark that the use of coarea formula as a tool for the analysis of general phase transition problems was already investigated by Visintin in [18, 19].

Remark 2.1. Let $J : \mathfrak{C}^2 \to [0, \infty]$ be a functional satisfying the assumptions i), ii), iii), iv), vi). Then, J can be extended to $\mathfrak{M}$ by relaxation in L^1_{loc}:

$$J(E) := \inf\{\liminf J(E_n),\ \{E_n\} \subset \mathfrak{C}^2,\ |(E_n \Delta E) \cap B_R| \to 0 \text{ for all } R > 0\}.$$

In order to fit our definition of generalized perimeter, J has to satisfy property v). This in ensured whenever its extension to L^1_{loc} (2.3) is convex. Such a relaxation procedure is now a day classical: starting from the classical perimeter defined on regular sets, it provides the Caccioppoli-De Giorgi definition of perimeter for measurable sets.

2.2 Definition of the curvature

Definition 2.2. Let $E \in \mathfrak{C}^2$ and $x \in \partial E$. We set

$$\kappa^+(x, E) := \\ := \inf\left\{\liminf \frac{J(E \cup W_n) - J(E)}{|W_n \setminus E|} : \overline{W_n} \overset{\mathcal{H}}{\to} \{x\},\ |W_n \setminus E| > 0\right\} \tag{2.4}$$

and

$$\kappa^-(x,E) :=$$
$$:= \sup\left\{\limsup \frac{J(E) - J(E \setminus W_n)}{|W_n \cap E|} : \overline{W_n} \overset{\mathcal{H}}{\to} \{x\}, |W_n \cap E| > 0\right\}. \tag{2.5}$$

Here $\overset{\mathcal{H}}{\to}$ denotes the convergence in the Hausdorff sense. We say that $\kappa(x, E) \in [-\infty, \infty]$ is the curvature of E at x (corresponding to the perimeter J) if $\kappa^+(x, E) = \kappa^-(x, E) =: \kappa(x, E)$.

2.3 Assumptions on the curvature κ

We assume the following properties on the perimeter J and the corresponding curvature:

A) Existence of the curvature: Every set $E \in \mathfrak{C}^2$ admits a finite curvature $\kappa(x, E) \in \mathbb{R}$ at each point $x \in \partial E$.
B) Continuity: If $E_n \to E$ in $\mathfrak{C}^2$ and $x_n \in \partial E_n \to x \in \partial E$, then $\kappa(x_n, E_n) \to \kappa(x, E)$.
C) Curvature of the balls: By assumption B), for any $\rho > 0$ we can define the quantities

$$\overline{c}(\rho) := \max_{x\in\partial B_\rho} \max\{\kappa(x, B_\rho), -\kappa(x, \mathbb{R}^N \setminus B_\rho)\}, \tag{2.6}$$

$$\underline{c}(\rho) := \min_{x\in\partial B_\rho} \min\{\kappa(x, B_\rho), -\kappa(x, \mathbb{R}^N \setminus B_\rho)\}, \tag{2.7}$$

which are continuous functions of $\rho > 0$. We will assume that there exists $K > 0$ such that

$$\underline{c}(\rho) > -K > -\infty. \tag{2.8}$$

Assumption C) will guarantee that the curvature flow starting from a bounded set remains bounded at all time.

2.4 Semi-continuous extensions

We now introduce two suitable lower and upper semicontinuous extensions of κ, which will be instrumental in developing the level set formulation of the geometric flow. To this purpose, let us recall the notions of subjet and superjet of a set.

Definition 2.3. Let $E \subset \mathbb{R}^N$, $x_0 \in \partial E$, $p \in \mathbb{R}^N$, and $X \in \mathbb{M}^{N\times N}_{\text{sym}}$. We say that (p, X) is in the superjet $J^{2,+}_E(x_0)$ of E at x_0 if for every $\delta > 0$ there exists a neighborhood U of x_0 such that, for every $x \in E \cap U$

$$\langle x - x_0, p\rangle + \frac{1}{2}\langle (X + \delta I)(x - x_0), (x - x_0)\rangle \geq 0. \tag{2.9}$$

Moreover, we say that (p, X) is in the subjet $J^{2,-}_{E}(x_0)$ of E at x_0 if $(-p, -X)$ is in the superjet $J^{2,+}_{E^c}(x_0)$ of E^c at x_0.

Before defining suitable extensions of κ, we state a monotonicity property of the curvature κ with respect to the inclusion of sets, which is a consequence of the submodularity property of the perimeter J.

Proposition 2.4. *Let $E, F \in \mathfrak{C}^2$ with $E \subset F$, and assume that $x \in \partial F \cap \partial E$: then $\kappa(x, F) \leq \kappa(x, E)$.*

Proposition 2.4 justifies the following definition, inspired by a similar notion in [5]. For every $F \subset \mathbb{R}^N$ with compact boundary and $(p, X) \in J^{2,+}_{F}(x)$, we define

$$\kappa_*(x,p,X,F):=\sup\left\{\kappa(x,E) : E\in\mathfrak{C}^2,\, E\supseteq F,\, (p,X)\in J^{2,-}_{E}(x)\right\}. \quad (2.10)$$

Analogously, for any $(p, X) \in J^{2,-}_{F}(x)$ we set

$$\kappa^*(x,p,X,F)=\inf\left\{\kappa(x,E) : E\in\mathfrak{C}^2,\, \mathring{E}\subseteq F,\, (p,X)\in J^{2,+}_{E}(x)\right\}. \quad (2.11)$$

Remark 2.5. One can show that κ_* and κ^* are the l.s.c. and the u.s.c envelope of κ with respect to a suitable notion of convergence of sets, involving convergence in the sense of Hausdorff, plus a notion of uniform convergence of superjets. For more details we refer the interested reader to [10].

3 Examples of generalized curvatures

Let us present some examples of generalized perimeters and corresponding curvature, the first one being the Euclidean perimeter.

3.1 The Euclidean perimeter

Let J be the Euclidean perimeter. More precisely, let J be its lower semicontinuous extension to measurable sets introduced by Caccioppoli and De Giorgi. Then, J satisfies all the assumptions i)-vi).

Moreover, for every $E\in\mathfrak{C}^2$, $x\in\partial R$ we have that $\kappa(x,E)=\kappa_{\mathrm{mean}}(x,\partial E)$, where $\kappa_{\mathrm{mean}}(x,\partial E)$ denotes the sum of the principal curvatures of ∂E at x. Indeed, it is classical that

$$\frac{d}{d\varepsilon}\mathcal{H}^{N-1}\big(\partial\Phi_\varepsilon(E)\big)_{|\varepsilon=0} = \int_{\partial E}\kappa_{\mathrm{mean}}(x,\partial E)\varphi(x)d\mathcal{H}^{N-1}, \quad (3.1)$$

where Φ_ε is a diffeomorphism such that $\Phi_\varepsilon(x) = x + \varepsilon\varphi(x)\nu_E(x)$ for $x \in \partial E$. Now, let $\varphi^h : \partial E \to \mathbb{R}$ be a sequence of smooth functions with support concentrating on x, let $\varepsilon_n \to 0$ and let $\Phi^h_{\varepsilon_n}$ be defined in the obvious way. Looking at the identity (2.4) with $W^h_n = \Phi^h_{\varepsilon_n}(E) \setminus E$ we can easily deduce $\kappa(x, E) \le \kappa_{\mathrm{mean}}(x, E)$ (assuming that $\kappa(x, E)$ exists). The opposite inequality means that smooth perturbations of the set E are indeed the most convenient for the perimeter; this is clearly the case for the Euclidean perimeter.

3.2 The fractional perimeter

For every $\alpha \in (0, 1)$ consider the square of the $\frac{\alpha}{2}$−fractional seminorm of the characteristic function of any measurable set E:

$$J(E) := C_\alpha \int_{\mathbb{R}^N \times \mathbb{R}^d} \frac{|\chi_E(x) - \chi_E(y)|}{|x - y|^{N+\alpha}}\, dxdy = [\chi_E]^2_{H^{\frac{\alpha}{2}}},$$

where $C_\alpha \in \mathbb{R}$ and $J = +\infty$ whenever $\chi_E \notin H^{\frac{\alpha}{2}}(\mathbb{R}^N)$.

Since the work in [3], this nonlocal perimeter has attracted much attention; we refer the interested reader to [17]. A first notion of curvature corresponding to J has been introduced in [8, 13]: let $\rho(x) := 1/|x|^{N+\alpha}$, $\rho_\delta(x) = (1 - \chi_{B(0,\delta)}(x))\rho(x)$. Then, for every $E \in \mathfrak{C}^2$ set

$$\kappa_\delta(E, x) = 2 \int_{\mathbb{R}^N} (\chi_E(y) - \chi_{\mathbb{R}^N \setminus E}(y))\rho_\delta(x - y)\, dy, \qquad \kappa := \lim_{\delta \to 0} \kappa_\delta.$$

It can be proved that J satisfies all the properties i)-vi), that the curvature $\kappa(E, x)$ is well defined for smooth sets (indeed for all set E with $C^{1,1}$ boundary) and that κ is the curvature of J also in the weak sense of Definition 2.2.

3.3 The pre-Minkowski content

Let $\rho > 0$ be fixed, and consider the measure of the ρ-neighborhood of the (essential) boundary of E, *i.e.*,

$$J(E) := |\rho\text{-}\partial E| = |(\cup_{x \in \partial E} B_\rho(x))|. \tag{3.2}$$

We refer to J as the *pre-Minkowski content* of ∂E, since as $\rho \to 0$, $|\rho\text{-}\partial E|/2\rho$ approximates the Minkowski content, which coincides with the standard perimeter on smooth sets.

This functional has been proposed in [2] as a replacement of the standard perimeter penalization in variational models for image segmentation and denoising of sets with rough boundaries (indeed, with infinite perimeter as well). The idea is that such a perimeter penalizes in an additive way dilute noise, while (in contrast with the standard perimeter) it

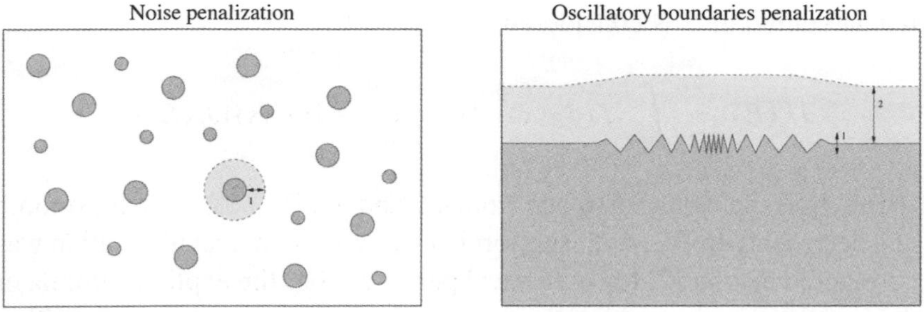

Figure 3.1.

penalizes very little rough boundaries with oscillations at scales smaller then ρ.

In [2] many numerical results based on minimization of the total energy given by the sum of a fidelity term and the nonlocal perimeter penalization (3.2) are displayed.

In view of the genaralized coarea formula (2.3), one can also consider the nonlocal total variation corresponding to J. Its flow represents a variant of the total variation flow for image regularization. The following example has been implemented by A. Chambolle using cut-graph techniques developed in [8].

Notice that the flow preserves small details of the image (for instance on the roof), while producing the desired denoising effect.

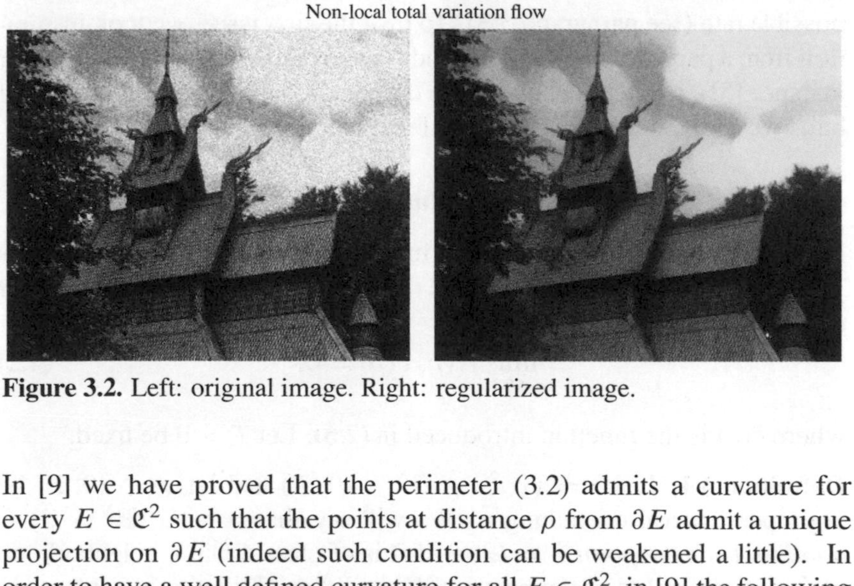

Figure 3.2. Left: original image. Right: regularized image.

In [9] we have proved that the perimeter (3.2) admits a curvature for every $E \in \mathfrak{C}^2$ such that the points at distance ρ from ∂E admit a unique projection on ∂E (indeed such condition can be weakened a little). In order to have a well defined curvature for all $E \in \mathfrak{C}^2$, in [9] the following regularization of J is considered:

$$J^f(E) = \int_{\mathbb{R}^d} f(d_E(x))\,dx = \int_0^\rho (-2sf'(s))J_s(E)\,ds,$$

where d_E is the signed distance from ∂E and $f : \mathbb{R} \to \mathbb{R}_+$ is odd, smooth and decreasing in $\mathbb{R}_+$, with support in $[-\rho, \rho]$. Such a regularization was considered also in [2] for numerical purposes. For the explicit formula of the curvature corresponding to J^f we refer the interested reader to [9].

4 The curvature flow: notion of viscosity solutions

4.1 The curvature flow

The level set approach to the geometric flow consists in solving the following parabolic Cauchy problem

$$\begin{cases} u_t(x,t) \\ \quad +|Du(x,t)|\kappa(x, Du(x,t), D^2u(x,t), \{y : u(y,t) \geq u(x,t)\}) = 0, \\ u(0,\cdot) = u_0, \end{cases} \tag{4.1}$$

in the viscosity sense. Here and in the following, D and D^2 stand for the spatial gradient and the spatial Hessian matrix, respectively.

The main technical issues are related to the fact that the Hamiltonians (or the curvatures) are nonlocal and can blow-up on small balls with any possible rate (see paragraph 3.3). To treat the first issue, we took inspiration from a paper by Slepčev [16] and some work by Cardaliaguet (see for instance [5]), while to deal with the degeneracy of the curvature of small balls we adopted the special class of test functions introduced in [14].

4.2 Definition of the viscosity solutions

Following [14] (see also [2]), we introduce the family $\mathcal{F}$ of functions $f \in C^2([0,\infty))$, such that $f(0) = f'(0) = f''(0) = 0$, $f''(r) > 0$ for all $r > 0$, and

$$\lim_{\rho \to 0^+} f'(\rho)\,\overline{c}(\rho) = 0, \tag{4.2}$$

where $\overline{c}(\rho)$ is the function introduced in (2.6). Let $T > 0$ be fixed.

Definition 4.1. Let $\hat{z} = (\hat{x}, \hat{t}) \in \mathbb{R}^N \times (0,T)$ and let $J \subset (0,T)$ be any open interval containing $\hat{t}$. We will say that $\varphi \in C^0(\mathbb{R}^N \times \overline{J})$ is *admissible at the point* $\hat{z} = (\hat{x}, \hat{t})$ if it is of class C^2 in a neighborhood of $\hat{z}$, if there exists a compact set $K \subset \mathbb{R}^N$ such that φ is constant in

$(\mathbb{R}^N \setminus K) \times J$ and, in case $D\varphi(\hat{z}) = 0$, the following holds: there exists $f \in \mathcal{F}$ and $\omega \in C^0([0, \infty))$ with $\lim_{r\to 0} \omega(r)/r = 0$, such that

$$|\varphi(x,t) - \varphi(\hat{z}) - \varphi_t(\hat{z})(t - \hat{t})| \leq f(|x - \hat{x}|) + \omega(|t - \hat{t}|)$$

for all (x, t) in $\mathbb{R}^N \times J$.

Definition 4.2. An upper semicontinuous function $u : \mathbb{R}^N \times [0, T) \to \mathbb{R}$, constant outside a compact set in $\mathbb{R}^N$, is a viscosity subsolution of (4.1) if for all $z := (x, t) \in \mathbb{R}^N \times (0, T)$ and all φ admissible at z, such that $u - \varphi$ has a maximum at z (in the domain of definition of φ) we have

$$\begin{cases} \varphi_t(z) + |D\varphi(z)|\, \kappa_* \left(x, D\varphi(z), D^2\varphi(z), \{y : \varphi(y,t) \geq \varphi(z)\}\right) \leq 0 & \\ & \text{if } D\varphi(z) \neq 0, \\ \varphi_t(z) \leq 0 & \text{otherwise.} \end{cases} \tag{4.3}$$

The definition of supersolution is analogous (with κ_* replaced by κ^*). Finally, a function u is a viscosity solution of (4.1) if its upper semicontinuous envelope is a subsolution and its lower semicontinuous envelope is a supersolution of (4.1).

As it is standard in the theory of viscosity solutions, the maximum in the definition of subsolutions can be assumed to be strict. On the other hand, it is crucial that the maximum is global (see [10]).

4.3 Existence of viscosity solutions

Let $u_0 : \mathbb{R}^N \mapsto \mathbb{R}$ be a continuous function, constant out of a compact set K. The existence of a viscosity solution to (4.1) follows by standard arguments once the existence of at least one subsolution and a stability property for subsolutions are established. To this purpose, we first need a confinement lemma.

Lemma 4.3. *Let $R, T > 0$ be fixed. There exists a constant $R' > R$ such that if $u \in USC(\overline{\mathcal{R}_T})$ is a subsolution of* (4.1) *with $u(x, 0) = C_0$ for $|x| \geq R$, then*

$$u(x,t) = C_0 \qquad \text{for } |x| > R' \text{ and } t \in [0, T]\,.$$

The crucial stability property of sub(super)solutions is stated in the following proposition.

Proposition 4.4. *Let $(u_n)_{n\geq 1}$ be a sequence of viscosity subsolutions such that $u_n = c_n$ in $(\mathbb{R}^N \setminus K) \times [0, T)$, for some constant $c_n \in \mathbb{R}$ and some compact $K \subset \mathbb{R}^N$. Let, for any $z = (x, t)$,*

$$u^*(z) = \lim_{r\downarrow 0} \sup \left\{ u_n(\zeta) : |z - \zeta| \leq r\,, n \geq \frac{1}{r} \right\}. \tag{4.4}$$

If $u^*(z) < +\infty$ for all z, then u^* is a subsolution.

Of course, similar results hold for supersolutions. We finally state the main existence theorem.

Theorem 4.5. *Let $u_0 : \mathbb{R}^N \to \mathbb{R}$ be a continuous function with $u_0 = C_0$ for $|x| \geq R$. Let moreover R' be the constant given by Lemma* 4.3. *Then, there exists a bounded uniformly continuous viscosity solution u of* (4.1) *in $[0, T]$ with $u = C_0$ for $|x| \geq R'$.*

5 Variational curvature flows

Here we present the minimizing movements scheme to solve and approximate the nonlocal κ-curvature flow, adapting to our context the approach of [1,6,15].

5.1 The time-discrete scheme

We start by introducing the incremental minimum problem. Given a bounded set E, let $d_E(x) = \operatorname{dist}(x, E) - \operatorname{dist}(x, \mathbb{R}^N \setminus E)$ be the signed distance function to ∂E. Fix a time step $h > 0$ and consider the problem

$$\min \left\{ J(F) + \frac{1}{h} \int_F d_E(x)\, dx \, : \, F \in \mathfrak{M} \right\}. \tag{5.1}$$

Recall that J satisfies the submodularity assumption v), which is a convexity assumption. It is then not surprising that the following proposition holds true.

Proposition 5.1. *The problem* (5.1) *admits a minimal and a maximal solution.*

5.2 Level by level minimizing movements

Proposition 5.1 defines a transformation $T_h E$ that associates to any bounded set E the minimal solution to problem (5.1) (we could consider the maximal solution as well). Now let us define this transformation on a bounded, uniformly continuous function u, constant outside a compact set. The main observation is that, as a consequence of the monotonicity property stated in Proposition 2.4, and since for any couple of levels $s > s' \in \mathbb{R}$ we have $\{u > s\} \subseteq \{u > s'\}$, then, $T_h\{u > s\} \subseteq T_h\{u > s'\}$. It follows that the sets $T_h\{u > s\}$ are themselves the level sets $\{v > s\}$ of a (uniformly continuous) function v. We are in a position to state the main achievement of [10].

Theorem 5.2. *Let u_0 be a bounded continuous function constant outside a compact set. For $h > 0$, $t \geq 0$, let $u_h(t, x) = T_h^{[t/h]}(u_0)(x)$. Then, up to subsequences, u_h converges locally uniformly (as $h \downarrow 0$) to a viscosity solution of* (4.1).

6 Shrinking zebras

This section contains numerical experiments (see [9] for more details) for the minimizing movements corresponding to the pre-Minkowski content. The picture we have chosen is a zebra, and animal covered of stripes very suited to show the nonlocal character of the curvature!

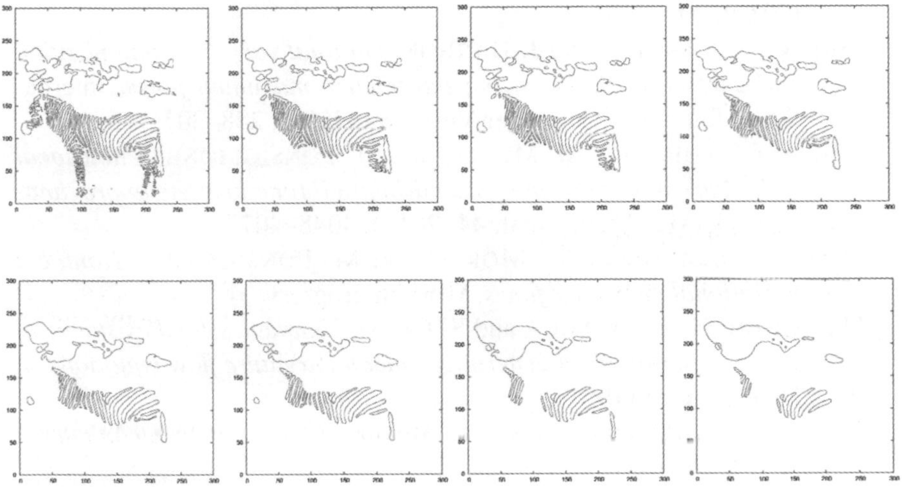

Figure 6.1. Shrinking the zebra.

Simulations show that strips shrink slower than for the standard perimeter. Somehow the end points of the strips see each other and their curvature is considerably lower than the euclidean one.

References

[1] F. ALMGREN, J. TAYLOR and L. WANG, *Curvature-driven flows: a variational approach*, SIAM J. Control Optim **31** (1993), 387–438.

[2] M. BARCHIESI, S. H. KANG, M. L. TRIET, M. MORINI and M. PONSIGLIONE, *A variational model for infinite perimeter segmentations based on Lipschitz level set functions: denoising while keeping finely oscillatory boundaries*, Multiscale Model. Simul. **8** (2010).

[3] L. A. CAFFARELLI, J.-M. ROQUEJOFFRE and O. SAVIN, *Nonlocal minimal surfaces*, Comm. Pure Appl. Math. **63** (2010), 1111–1144.

[4] L. A. CAFFARELLI and P. E. SOUGANIDIS, *Convergence of nonlocal threshold dynamics approximations to front propagation*, ARMA **195** (2010).

[5] P. CARDALIAGUET, *Front propagation problems with nonlocal terms*, II, J. Math. Anal. Appl. **260** (2001), 572–601.

[6] A. CHAMBOLLE, *An algorithm for mean curvature motion*, Interfaces Free Bound. **6** (2004), 195–218.

[7] A. CHAMBOLLE, A. GIACOMINI and L. LUSSARDI, *Continuous limits of discrete perimeters*, M2AN Math. Model. Numer. Anal. **44** (2010), 207–230.

[8] A. CHAMBOLLE and J. DARBON, *On total variation minimization and surface evolution using parametric maximum flows*, International Journal of Computer Vision **84** (2009), 288–307.

[9] A. CHAMBOLLE, M. MORINI and M. PONSIGLIONE, *A non-local mean curvature flow and its semi-implicit time-discrete approximation*, SIAM J. Math. Anal. **44** (2012), 4048–4077.

[10] A. CHAMBOLLE, M. MORINI and M. PONSIGLIONE, *Nonlocal variational curvature flows*, Work in progress.

[11] CHEN, Y.-G., Y. GIGA and S. GOTO, *Uniqueness and existence of viscosity solutions of generalized mean curvature flow equations*, J. Di. Geom. **33** (1991).

[12] L. C. EVANS and J. SPRUCK, *Motion of level sets by mean curvature*, J. Di Geom. **33** (1991).

[13] C. IMBERT, *Level set approach for fractional mean curvature flows*, Interfaces Free Bound **11** (2009).

[14] H. ISHII and P. SOUGANIDIS, *Generalized motion of noncompact hypersurfaces with velocity having arbitrary growth on the curvature tensor*, Tohoku Math. J. (2) **47** (1995), 227–250.

[15] S. LUCKHAUS and T. STURZENHECKER, *Implicit time discretization for the mean curvature flow equation*, Calc. Var. Partial Differential Equations **3** (1995), 253–271.

[16] D. SLEPČEV, *Approximation schemes for propagation of fronts with nonlocal velocities and Neumann boundary conditions*, Nonlinear Anal. **52** (2003), 79–115.

[17] E. VALDINOCI, *A fractional framework for perimeters and phase transitions*, preprint 2012.

[18] A. VISINTIN, *Nonconvex functionals related to multiphase systems*, SIAM J. Math. Anal. **21** (1990), 1281–1304.

[19] A. VISINTIN, *Generalized coarea formula and fractal sets*, Japan J. Indust. Appl. Math. **8** (1991), 175–201.

Homogenization with oscillatory Neumann boundary data in general domain

Sunhi Choi and Inwon C. Kim

Abstract. In this article we summarize recent progress on understanding averaging properties of fully nonlinear PDEs in bounded domains, when the boundary data is oscillatory. Our result on the Neumann problem is the nonlinear version of the classical result in [4] for divergence-form operators with co-normal boundary data. We also discuss the Dirichlet boundary problem.

1 Introduction

Let us consider a bounded domain Ω in $\mathbb{R}^n$ containing the closed unit ball $K = \{x : |x| \leq 1\}$ (see Figure 1.1). Let $g : \mathbb{R}^n \to \mathbb{R}$ be a Hölder continuous function which is periodic with respect to the orthonomal basis $\{(e_1, ..., e_n)\}$ of $\mathbb{R}^n$. More precisely, g satisfies

$$g(x + e_i) = g(x) \text{ for } i = 1, ..., n \text{ and } g \in C^\beta(\mathbb{R}^n) \text{ for some } 0 < \beta < 1.$$

With Ω, K and g as given above, we are interested in the limiting behavior of the following problem:

$$(N)_\varepsilon \qquad \begin{cases} F(D^2u^\varepsilon, \frac{x}{\varepsilon}) = 0 & \text{in } \Omega - K, \\ u^\varepsilon = 1 & \text{on } K, \\ \frac{\partial}{\partial \nu} u^\varepsilon = g(\frac{x}{\varepsilon}) & \text{on } \partial\Omega. \end{cases}$$

Here $\nu = \nu_x$ is the outward normal vector at $x \in \partial\Omega$ and F is a uniformly elliptic, fully nonlinear operator (see Section 2 for the precise definitions and conditions on F).

UCLA. I. K's research is partially supported by NSF DMS-0970072.

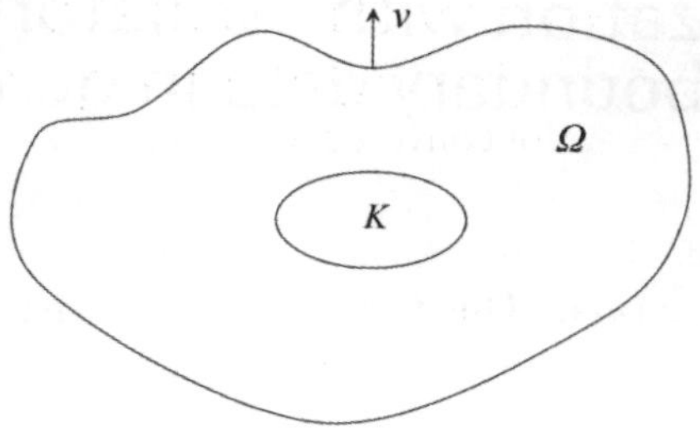

Figure 1.1.

The set K is introduced to avoid discussions of compatibility conditions on the Neumann boundary data. The operators under discussion include, for example, the divergence or non-divergence form operator

$$-\Sigma_{i,j}a_{ij}(\frac{x}{\varepsilon})\partial_{x_ix_j}u = 0, \tag{1.1}$$

where $a_{ij}(x)$ are Hölder continuous and $\lambda I_{n\times n} \leq (a_{ij})(x) \leq \Lambda I_{n\times n}$ for positive constants λ and Λ.

Let us define

$$\mathcal{S}^{n-1} = \{\nu \in \mathbb{R}^n : |\nu| = 1\} \text{ and } \mathcal{M}^n := \{M : n\times n \text{ symmetric matrix }\}.$$

Due to [8], there exists a uniformly elliptic operator $\bar{F} : \mathcal{M}^n \to \mathbb{R}$ such that any solution of $F(D^2u^\varepsilon, \frac{x}{\varepsilon}) = 0$ in Ω with *fixed* boundary data on $\partial\Omega$ converges uniformly to the solution of $\bar{F}(D^2u) = 0$ in Ω. Therefore the question under investigation is whether the oscillatory boundary data changes the averaging behavior of u^ε. As we will discuss below, several difficulties arise in answering this question due to the nonlinear (or non-divergence form) nature of the problem as well as the geometry of the domain.

1.1 Main results and the discussion of main ideas

First we introduce the following "cell problem": for given $\nu \in \mathcal{S}^{n-1}$ and $\lambda \in \mathbb{R}$

$$(P)_{\varepsilon,\nu,\lambda} \qquad \begin{cases} F(D^2u^\varepsilon, \frac{x}{\varepsilon}) = 0 & \text{in } \Pi_\nu := \{\lambda - 1 \leq x\cdot\nu \leq \lambda\}; \\ u^\varepsilon = 1 & \text{on } \Gamma_D := \{x\cdot\nu = \lambda - 1\}; \\ \partial_\nu u^\varepsilon = g(\frac{x}{\varepsilon}) & \text{on } \Gamma_N := \{x\cdot\nu = \lambda\}. \end{cases}$$

Note that when $\nu \in \mathbb{R}\mathbb{Z}^n$, *i.e.* when ν is a *rational* direction, u^ε is periodic and thus the subsequential convergence of u^ε can be relatively easily obtained. However u^ε may not converge to a unique limit, since for nonzero λ the distribution of g on Γ_N changes a lot depending on the choice of ε. Due to this difficulty most previous studies on $(P)_{\varepsilon,\nu,\lambda}$ only cover the case $\nu \in \mathbb{R}\mathbb{Z}^n$ and $\lambda = 0$. This fact serves as the main obstacle to study $(N)_\varepsilon$ for domains with general geometry, and thus also motivates us to study the remaining cases of *irrational* directions, $\nu \in \mathbb{R}^n - \mathbb{R}\mathbb{Z}^n$.

Our first main observation is that when ν is an irrational direction then due to Weyl's equi-distribution theorem g uniformly covers its entire distribution in $\mathbb{R}^n$ on Γ_N as $\varepsilon \to 0$, with the convergence rate (*discrepancy*) depending on ν. Using this fact we deduce our first main theorem:

Theorem 1.1 ([10], Theorem 3.1). *Let u^ε solve $(P)_{\varepsilon,\nu,\lambda}$ as given above. Then, for given normal direction $\nu \in S^{n-1} - \mathbb{R}\mathbb{Z}^n$ and any $\lambda \in \mathbb{R}$ there exists a unique $\mu(\nu)$ such that u^ε uniformly converges to the linear profile $l_\nu(x) := \mu(\nu)(x \cdot \nu + 1 - \lambda) + 1$ in Π_ν. Moreover, the rate of convergence of u^ε to l_ν only depends on the discrepancy depending on the direction ν.*

Our plan now is to use the result obtained above to study the general problem $(N)_\varepsilon$, adopting the perturbed test function method introduced by [17]. To be able to apply such perturbation method, we would first need to make sure that the limiting problem to have some stability properties. In other words, we would like the following problem to have a unique solution:

$$(N) \qquad \begin{cases} \bar{F}(D^2 u) = 0 & \text{in } \Omega - K, \\ u = 1 & \text{on } K, \\ \dfrac{\partial}{\partial \nu} u = \mu(\nu_x) & \text{on } \Gamma_1 \\ \left|\dfrac{\partial}{\partial \nu} u\right| \leq \max |g| & \text{on } \Gamma_2 := \partial\Omega - \Gamma_1 \end{cases}$$

where Γ_1 denotes the set of points on $\partial\Omega$ where the corresponding outer normal is irrational, $\mu(\nu_x)$ is the homogenized slope obtained from the above theorem, and ν_x is the (outer) normal at $x \in \Gamma_1$.

It is clear that the uniqueness property of (N) would be false if Γ_2 is of positive $n-1$ dimensional measure, *i.e.* if $\partial\Omega$ has a flat facet pointing toward a rational direction. This fact is well-known in the linear case (see Section 4 of [4]). Indeed in the linear case with co-normal boundary data one can show by integration by parts that $\mu(\nu_x)$ is the average of g in the unit periodic box, and thus it follows that the solution of (N) is unique if and only if $\partial\Omega$ does not have any flat boundary parts pointing toward a rational direction.

In the nonlinear case showing the uniqueness of (N) is not as straightforward, due to the likely dependence of $\mu(\nu_x)$ on the normal direction ν_x. Even in the situation where Γ_2 is only countable, to ensure the uniqueness of solutions for (N) it turns out to be necessary to obtain both continuity of $\mu(\nu_x)$ and certain stability properties of the solution of $(P)_{\varepsilon,\nu,\lambda}$, as ν varies (see Theorem 3.1). Showing this "continuity" property turns out to be rather involved, since the ergodic feature of g on Γ_N degenerates as the normal (irrational) direction ν approaches a rational direction. We indeed prove a property which is perhaps stronger than what is needed for the homogenization purposes, that is the existence of a continuous extension of $\mu(\nu)$ to all unit vectors ν, under the extra assumption on the operator $\bar{F}(M)$.

Theorem 1.2 ([10],Theorem 5.1). *Suppose that $\bar{F}(M)$ only depends on the eigenvalues of $M \in \mathcal{M}^n$. Then there exists a continuous function $\bar{\mu}(\nu) : \mathcal{S}^{n-1} \to \mathbb{R}$, given as the continuous extension of $\mu(\nu)$ over $\nu \in \mathcal{S}^{n-1} - \mathbb{R}\mathbb{Z}^n$, such that the following holds:*

Suppose Ω is a bounded domain in $\mathbb{R}^n$ such that $\partial\Omega$ is C^2 and does not contain any flat part. Let u^ε solve $(N)_\varepsilon$. Then u^ε converges locally uniformly to u, which is the unique solution of the following problem:

$$(\bar{N}) \qquad \begin{cases} \bar{F}(D^2u) = 0 & \text{in } \Omega - K; \\ u = 1 & \text{on } K; \\ \partial_\nu u = \bar{\mu}(\nu) & \text{on } \partial\Omega. \end{cases}$$

The assumption on $\bar{F}$ in Theorem 1.1 (b) seems to be a necessary condition to achieve the continuity of the homogenized slope $\bar{\mu}(\nu)$ at rational directions, a counterexample can be at least heuristically constructed (see Remark 3.2). See Section 2 for more discussions on this condition. On the other hand we may be able to achieve the uniqueness result without this assumption, as indicated from the Dirichlet case ([14]).

• *Discussions on previous results*

As mentioned before, our problem is classical for the case of uniformly elliptic, divergence-form equations

$$-\nabla \cdot \left(A\left(\frac{x}{\varepsilon}\right)\nabla u^\varepsilon\right) = 0 \text{ in } \Omega, \tag{1.2}$$

with the *co-normal* boundary data

$$\nu \cdot \left(A\left(\frac{x}{\varepsilon}\right)\nabla u^{\varepsilon}\right)(x) = g\left(\frac{x}{\varepsilon}\right) \quad \text{on } \partial\Omega. \tag{1.3}$$

For (1.2)-(1.3), a corresponding result to Theorem 1.2 was proved in the classical paper of Bensoussan, Lions and Papanicolau [4] with explicit integral formula for the limiting operator as well as the limiting boundary data.

For nonlinear or non-divergence type operators, or even for linear operators with oscillatory Neumann boundary data that is not co-normal, most available results concern half-space type domains whose boundary goes through the origin and is normal to a rational direction. In [20], Tanaka considered some model problems in half-space whose boundary is parallel to the axes of the periodicity by purely probabilistic methods. In [1] Arisawa studied special cases of problems in oscillatory domains near half spaces going through the origin, using viscosity solutions as well as stochastic control theory. Generalizing the results of [1], Barles, Da Lio and Souganidis [3] studied the problem for operators with oscillating coefficients, in half-space type domains whose boundary is parallel to the axes of periodicity, with a series of assumptions which guarantee the existence of approximate corrector. In [11] the continuity property of the averaged Neumann boundary data in half-spaces, with respect to the normal direction, was studied in the case of homogeneous operator F. We refer to Section 4 for available results on Dirichlet problem.

There are, of course, many remaining questions. One natural question is on generalization either to the stochastic (stationary ergodic) case or to other types of boundary data. Another question is on the characterization of $\mu(\nu)$: some of these questions are currently under investigation.

In Section 2 we introduce the exact conditions on the operator F and discuss the proof of Theorem 1.1. In Section 3 we give out a sketch of the proof for the existence of the continuous extension of the homogenized slope $\mu(\nu)$ to all directions ν, $\bar{\mu}(\nu)$, which is the most involved step in the proof of Theorem 1.2. In Section 4 we discuss the application of perturbed test function method to our setting and the involved difficulties.

2 The problem in the strip domain

2.1 Preliminaries

We begin with the properties of the operator F. As before, let $\mathcal{M}^n$ denote the normed space of symmetric $n \times n$ matrices. In this paper we assume that the function $F(M, y) : \mathcal{M}^n \times \mathbb{R}^n \to \mathbb{R}$ satisfies the following conditions:

(F1) [Homogeneity] $F(tM, x) = tF(M, x)$ for any $M \in \mathcal{M}^n$, $x \in \mathbb{R}^n$ and $t > 0$. In particular it follows that $F(0, x) = 0$.

(F2) [Lipschitz continuity] F is locally Lipschitz continuous in $\mathcal{M}^n \times \mathbb{R}^n$, and there exists a constant $C > 0$ such that for all $x, y \in \mathbb{R}^n$ and $M, N \in \mathcal{M}^n$

$$|F(M, x) - F(N, y)| \leq C(|x - y|(1 + \|M\| + \|N\|) + \|M - N\|).$$

(F3) [Uniform Ellipticity] There exists constants $0 < \lambda < \Lambda$ such that

$$\begin{aligned} \lambda\|N\| &\leq F(M, x) - F(M + N, x) \\ &\leq \Lambda\|N\| \text{ for any } M, N \in \mathcal{M}^n \text{ and } N \geq 0. \end{aligned} \tag{2.1}$$

(F4) [Periodicity] For any $M \in \mathcal{M}^n$ and $x \in \mathbb{R}^n$,

$$F(M, x) = F(M, x + z) \text{ if } z \in \mathbb{Z}^n.$$

A typical example of F satisfying (F1)-(F4) is the non-divergence type operator

$$F(D^2u, y) = -\Sigma_{i,j} a_{ij}(y)\partial_{x_i x_j} u, \tag{2.2}$$

where $a_{ij} : \mathbb{R}^n \to \mathbb{R}$ is periodic, Lipschitz continuous and $A = (a_{ij})$ satisfies $\lambda(Id)_{n\times n} \leq A \leq \Lambda(Id)_{n\times n}$. Another example is the Bellman-Issacs operators arising from stochastic optimal control and differential games

$$F(D^2u, y) = \inf_{\alpha\in A} \sup_{\beta\in B} \{\mathcal{L}^{\alpha\beta} u\},$$

where $\mathcal{L}^{\alpha\beta}$ is a two-parameter family of uniformly elliptic operators of the form (2.2).

2.2 The problem in the strip domain

The main ingredient in the proof of Theorem 1.1 is the following quantitative version of Weyl's equi-distribution theorem. Let $(P)_{\varepsilon,\nu,\lambda}$ be as given in the introduction.

Lemma 2.1 (Lemma 2.7, [11]). *Suppose that ν is an irrational direction. Then there exists a mode of continuity $w_\nu : [0, 1) \to \mathbb{R}^+$ and a dimensional constant $M > 0$ such that the following is true:*

(a) *For any x and x_0 on Γ_N, there exists $y \in \mathbb{R}^n$ such that*

$$|x - y| \leq M\varepsilon^{1/10}; \quad y - x_0 \in \mathbb{Z}^n$$

and

$$\text{dist}(y, \Gamma_N) < \varepsilon\omega_\nu(\varepsilon). \tag{2.3}$$

(b) *If* ν *is an irrational direction, then for any* $x \in \mathbb{R}^n$ *and* $\delta > 0$*, there exists* $y \in \Gamma_N$ *such that*

$$|x - y| \leq \delta \bmod \mathbb{Z}^n.$$

In analogy to the random settings, the first property suggests *stationarity* of g, and the second property suggests that the distribution of g on Γ_N approximates all the values of g in $\mathbb{R}^n$, *i.e.* an ergodic property.

On a technical note, we mention that the power $-9/10$ appearing in above lemma is arbitrarily chosen, and can be replaced with any $0 < \alpha < 1$. The mode of continuity ω_ν, of course, depends on the choice of α.

It should be pointed out that, if x_1 is small in mode $\varepsilon\mathbb{Z}^n$ then $g(\frac{x}{\varepsilon})$ and $g(\frac{x+x_1}{\varepsilon})$ are close, and then the regularity estimates obtained in [19] imply that $u^\varepsilon(x)$ and its shifted version $u^\varepsilon(x+x_1)$ are close as well. Thus Lemma 2.1(a) enables us to compare $u^\varepsilon(x)$ and its shifted versions along the tangential directions to ν to deduce the following "flatness" lemma:

Lemma 2.2. *Let us fix* $\nu \in S^n - \mathbb{R}\mathbb{Z}$*, and let* u^ε *solve* $(P)_{\varepsilon,\nu,\lambda}$ *for some* $\lambda \in \mathbb{R}$*. Then for any* $x_0 \in \Pi_\nu$ *with* $\mathrm{dist}(x_0, \Gamma_N) > \varepsilon^{1/20}$*, and for any* $0 < \alpha < 1$*, there exists a constant* $C = C(\alpha, n)$ *such that for any* $x \in \{(x - x_0) \cdot \nu = 0\}$

$$|u_\varepsilon(x) - u_\varepsilon(x_0)| \leq \mathcal{E}(\varepsilon), \tag{2.4}$$

where $\mathcal{E}(\varepsilon) := C\varepsilon^{\alpha/20} + C\omega_\nu(\varepsilon)^\beta$ *with* $\omega_\nu : [0, 1) \to \mathbb{R}^+$ *as given in Lemma* 2.1.

From this lemma it follows that along subsequences u^ε converges to a linear profile. It remains to show the uniqueness of the subsequential limits, when ν is irrational. Proving this requires comparing u^ε and u^η for $0 < \varepsilon < \eta$. For this we use the ergodicity of g, *i.e.* Lemma 2.1 (b), to compare u^ε with the adequately shifted (and normalized) version of u^η. Further details are referred to [11] and [10].

3 The continuity of the homogenized slope

In this section we sketch out the proof of the continuity properties of $\mu(\nu)$ as well as the mode of convergence for u^ε as the normal direction ν of the domain varies. To this end we assume the following additional condition on the homogenized operator $\bar{F}$ as given in Theorem 1.2:

$$\bar{F}(M) : \mathcal{M}^n \to \mathbb{R} \text{ only depends on the eigenvalues of } M. \tag{3.1}$$

For $\delta > 0$, we let $K_\nu(\delta) \geq 1$ be the smallest positive number ≥ 1 such that

$$|K_\nu(\delta)\nu| \leq \delta^2 \mod \mathbb{Z}^n.$$

Observe that if ν is a rational direction, there is a positive number K depending on ν such that $K\nu = 0 \mod \mathbb{Z}^n$, and hence $K_\nu(\delta) \leq K$. If ν is an irrational direction, the existence of $K_\nu(\delta)$ follows from the Weyl's equidistribution theorem. We then have the following theorem: here we adjusted the corresponding theorem in [10] so that the parameters are simplified.

Theorem 3.1. *Let $\mu(\nu) : (\mathcal{S}^{n-1} - \mathbb{R}\mathbb{Z}^n) \to \mathbb{R}$ be as given in Theorem* 1.1, *and suppose that $\bar{F}$ satisfies* (3.1). *Then μ has a continuous extension $\bar{\mu}(\nu) : \mathcal{S}^{n-1} \to \mathbb{R}$. More precisely there exists a universal constant $0 < \gamma < 1$ depending only on the λ, Λ and β such that the following holds: for any $\nu \in S^{n-1}$ and $\delta > 0$, let*

$$\theta_0 = \theta_0(\nu, \delta) = \frac{\delta^2}{K_\nu(\delta)}.$$

Then if ν_1 and ν_2 are irrational such that

$$0 < |\nu_1 - \nu|, |\nu_2 - \nu| < \theta_0, \tag{3.2}$$

then we have

(a) $|\mu(\nu_1) - \mu(\nu_2)| < \delta^\gamma$;
(b) *For the solutions u_i^ε of $(P)_{\varepsilon,\nu_i,\lambda}$ in Π_{ν_i}* $(i = 1, 2)$, *the average slope $\mu(u_i^\varepsilon)$ of u_i^ε satisfies*

$$|\mu(u_i^\varepsilon) - \mu(\nu_i)| < \delta^\gamma \quad \textit{if } \varepsilon \leq \frac{\delta|\nu_i - \nu|}{K_\nu(\delta)}.$$

Here $\mu(u_i^\varepsilon)$ denotes $u_i^\varepsilon(\lambda\nu) - u_i^\varepsilon((\lambda - 1)\nu)$. Note that our statement uses ν as the reference direction and thus is a different type of estimate than those in Theorem 1.1).

The main idea in the proof of Theorem 3.1 is to use ν as the reference direction and approximate $g(\frac{x}{\varepsilon})$ by piecewise continuous functions, where each continuous parts are "projections" of g on $\{x \cdot \nu = 0\}$ (see further description on the approximation below).

3.1 Description of the perturbation of boundary data and a sketch of the proof

First let us describe the main ideas in the proof. We begin by introducing several notations. For notational simplicity and clarity in the proof, we assume that $\nu = e_2$: we will explain in the paragraph below how to modify the notations and the proof for $\nu \neq e_2$. Let us define

$$\Omega_0 := \Pi_\nu(0) = \{x \in \mathbb{R}^2 : -1 \leq x_2 := x \cdot e_2 \leq 0\}$$

and for $i = 1, 2$

$$\Omega_i := \Pi_{\nu_i}(0) = \{x \in \mathbb{R}^2 : -1 \leq x \cdot \nu_i \leq 0\}.$$

Let us also define the family of functions

$$\begin{aligned} g_i(x_1, x_2) &= g_i(x_1) \\ &= g(x_1, \delta(i-1)), \text{ where } i = 1, ..., m := \left[\frac{1}{\delta}\right] + 1. \end{aligned} \tag{3.3}$$

(see Figure 3.1). Then g_i is a 1-periodic function with respect to x_1.

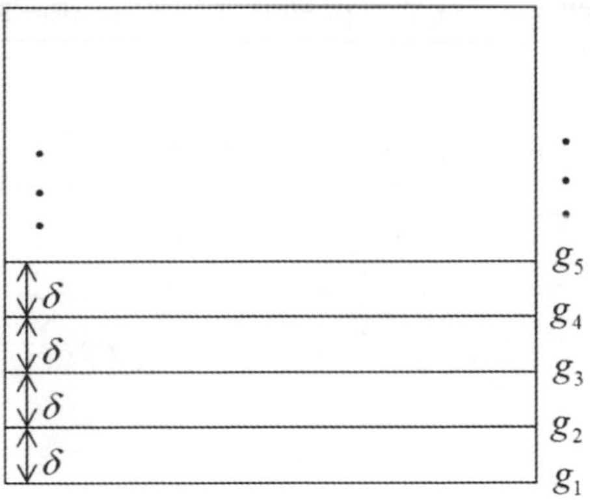

Figure 3.1.

• *A remark for* $\nu \neq e_2$

In two dimensions, if ν is a rational direction different from e_2, take the smallest $K_\nu \geq 1$ such that $K_\nu \nu = 0 \bmod \mathbb{Z}^2$. Then we define $g_i(x) = g(x' + \delta(i-1)\nu)$, where $x' = x - x\cdot\nu$, and g_i is a K_ν-periodic function. If ν is an irrational direction, take the smallest $K_\nu \geq 1$ such that $|K_\nu \nu| \leq \delta^2 \bmod \mathbb{Z}^2$. Then g_i as defined above is almost K_ν- periodic up to the order of δ^2 with respect to x'. We point out that it does not make any difference in the proof divided in the following two subsections if we replace the periodicity of g_i by the fact that g_i's are periodic up to the order δ^2.

• Proof by heuristics

Since the domains Ω_1 and Ω_2 point toward different directions ν_1 and ν_2, we cannot directly compare their boundary data, even if $\partial\Omega_1$ and $\partial\Omega_2$ cover most part of the unit cell in $\mathbb{R}^n/\mathbb{Z}^n$. To overcome this difficulty we perform a multi-scale homogenization as follows.

First we consider the functions g_i $(i = 1, .., m)$, whose profiles cover most values of g up to the order of δ^β, where β is the Hölder exponent of g. Note that most values of g are taken on $\partial\Omega_1$ and on $\partial\Omega_2$ since ν_1 and ν_2 are irrational directions. On the other hand, since ν_1 and ν_2 are very close to $\nu = e_2$ which is a rational direction, the averaging behavior of a solution u_1^ε in Ω_1 (or u_2^ε in Ω_2) would appear only after ε gets very small, as ν_1 (or ν_2) approaches $\nu = e_2$.

Let $N = [\delta/|\nu_1 - e_2|]$. If $|\nu_1 - e_2|$ is chosen much smaller than δ, then we can say that the Neumann data $g_1(\cdot/\varepsilon)$ is (almost) repeated N times on $\Gamma_1 = \{x \cdot \nu_1 = 0\}$ with period ε, up to the error $O(\delta^\beta)$. (See Figure 3.2.) Similarly, on the next piece of the boundary, $g_2(\cdot/\varepsilon)$ is (almost) repeated N times and then $g_3(\cdot/\varepsilon)$ is repeated N times: this pattern will repeat with g_k $(k \in \mathbb{N} \bmod m)$.

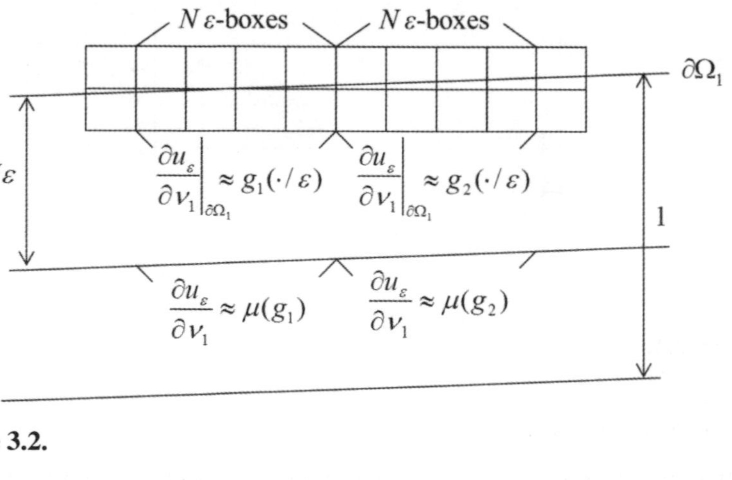

Figure 3.2.

If N is sufficiently large, *i.e.*, if $|\nu_1 - e_2|$ is sufficiently small compared to δ, the solution u_1^ε of $(P_{\varepsilon,\nu_1,0})$ in Ω_1 will exhibit averaging behavior, $N\varepsilon$-away from Γ_1. More precisely, on the hyperplane H located $N\varepsilon$-away from Γ_1, u_1^ε would be homogenized by the repeating profiles of g_i (for some fixed i) with an error of $O(\delta^\beta)$. This is the first homogenization of u_1^ε near the boundary of Ω_1: we denote, by $\mu(g_i)$, the corresponding values of the homogenized slopes of u_1^ε on H in each $N\varepsilon$-segment.

Now more than $N\varepsilon$ away from Γ_1, we obtain the second homogenization of u_1^ε, whose slope is determined by $\mu(g_i), i = 1, .., m$. In the proof, this second homogenization is divided into two parts, in the middle region which is $N\varepsilon$ to $KN\varepsilon$-away from Γ_1 and then in the rest of the domain Ω_1. The homogenization argument in the middle region is to ensure that the oscillation of the operator F in x-variable does not alter the behavior of the solution too much as ν varies. Note that, due to (3.1) after rotation and reflection we may assume that the arrangement of $\mu(g_1), ...\mu(g_m)$ is the same for ν_1 and ν_2. Therefore the second homogenization procedure applied to ν_1 and ν_2 yields that $|\mu(\nu_1) - \mu(\nu_2)|$ is small.

Remark 3.2. We finish this section by briefly discussing the necessity of the condition (3.1). The condition, as hinted above, makes sure that one obtains the same homogenization limit for $(P_{\varepsilon,\nu_i,\lambda})$ for either $g(x)$ or the reflected version $g(x-2x\cdot\nu)$. This is always true for linear operators, but for nonlinear operator it does not seem to hold, for example when F is given by maximum of two diffusion operators with different directional preferences. Further investigation is in progress [15].

4 Perturbed test function method

As given in the introduction, let Ω be a bounded domain with C^2 boundary containing a unit ball $K = B_1(0)$. Let us denote $\nu = \nu_x$ the outward normal vector of Ω at $x \in \partial\Omega$. Suppose that $\partial\Omega$ does not have any flat boundary parts in the following sense: For any $x_0 \in \partial\Omega$ and sufficiently small σ, there exists $r_0 > 0$ and $r(\sigma) > 0$ such that

$$|\nu_x - \nu_{x_0}| \geq \sigma \text{ if } x \in \partial\Omega \cap \{r(\sigma) < |x - x_0| < r_0\}. \tag{4.1}$$

The last step toward the proof of Theorem 1.2 is to use Theorem 3.1 to prove our main result. Let u^ε solve (N_ε). Following the usual argument in viscosity theory, we define the half-relaxed limits

$$\limsup{}^* u^\varepsilon(x) := \liminf_{\varepsilon\to 0}(\sup\{u^\varepsilon(y) : y \in \bar{\Omega} \text{ and } |x-y| \leq \varepsilon\})$$

and

$$\liminf{}_* u^\varepsilon(x) := \limsup_{\varepsilon\to 0}(\inf\{u^\varepsilon(y) : y \in \bar{\Omega} \text{ and } |x-y| \leq \varepsilon\}).$$

It is straightforward from the definition to check that $\limsup^* u^\varepsilon$ is upper semicontinuous and $\liminf_* u^\varepsilon$ is lower semicontinuous.

Theorem 1.2 is then a consequence of the following proposition, due to the uniquness of the solution of $(\bar{N})$.

Proposition 4.1. *The following holds true:*

(a) $\bar{u} := \limsup^* u^\varepsilon$ *is the viscosity subsolution of* $(\bar{N})$;
(b) $\underline{u} := \liminf_* u^\varepsilon$ *is the viscosity supersolution of* $(\bar{N})$.

Let us discuss the proof of (a), since the proof of (b) is parallel. The proof of above proposition is a bit different from the standard arguments for the following reason. It is previously known that the half-relaxed limits preserve the sub- and supersolution properties with respect to the operator $\bar{F}$ in the domain Ω. Thus to show (a), one proceeds by contradiction and assume that the viscosity subsolution property is violated for $\bar{u}$ at a point $x \in \partial\Omega$. Now the idea is to argue that then the viscosity subsolution property is violated for u^ε at $x_\varepsilon \in \partial\Omega$ with $x_\varepsilon \to x$ as $\varepsilon \to 0$. From here one draws a contradiction as $\varepsilon \to 0$, approximating $\partial\Omega$ with the flat surface Γ_N with normal ν_{x^ε}. Our challenge is that to apply Theorem 3.1 to obtain a contradiction we need to exclude the possibility that ν_{x_0} is rational and the irrational normal direction near x_ε is too close to ν_{x_0} compared to ε, thus in this case we would not end up with $\bar{\mu}(\nu)$ as the homogenized slope as $\varepsilon \to 0$.

This difficulty can be overcome by adding u^ε to a small perturbation which is close to zero everywhere and grows rapidly near x_0, such that one can prevent u^ε to break the subsolution property too close to x_0. This is enough to prevent the bad scenario described above, due to our assumption , *i.e.*, since $\partial\Omega$ does not have any flat boundary parts. More precisely, we consider $\tilde{u}^\varepsilon = u^\varepsilon - \varphi^\varepsilon$ instead of u^ε, where φ^ε is a smooth function with the following properties:

(1) $\varphi^\varepsilon \to 0$ as $\varepsilon \to 0$;
(2) $\mathcal{P}^-(D^2\varphi^\varepsilon) < 0$ in $B_1(x_0)$;
(3) $\partial_\nu\varphi^\varepsilon \ll 1$ on $\partial\Omega \cap B_1(x_0)$; $\quad \partial_\nu\varphi^\varepsilon \leq \min(0, \min g)$ on $\partial\Omega \cap B_{\varepsilon^{\alpha_0}}(x_0)$ with $0 \leq \alpha_0 \ll 1$.

Here $\mathcal{P}^-$ denotes the Pucci operator. Then by (2)-(3), if ϕ is a classical supersolution of $(\bar{N})$, then $\tilde{u}^\varepsilon - \phi$ can have its maximum at y in $B_r(x_0)$ only if y is more than ε^{α_0}-away from x_0. We refer to [10] for the construction of φ_ε.

The rest of the proof of Theorem 1.2 follows the standard argument.

References

[1] M. ARISAWA, *Long-time averaged reflection force and homogenization of oscillating Neumann boundary conditions,* Ann. Inst. H. Poincaré Anal. Lineaire **20** (2003), 293–332.

[2] G. BARLES and F. DA LIO, *Local $C^{0,\alpha}$ estimates for viscosity solutions of Neumann-type boundary value problems,* J. Differential Equations **225** (2006), 202–241.

[3] G. BARLES, F. DA LIO, P. L. LIONS and P. E. SOUGANIDIS, *Ergodic problems and periodic homogenization for fully nonlinear equations in half-type domains with Neumann boundary conditions,* Indiana University Mathematics Journal **57** (2008), 2355–2376.

[4] A. BENSOUSSAN, J. L. LIONS and G. PAPANICOLAOU, "Asymptotic Analysis for Periodic Structures", Vol. 5, Studies in Mathematics and its Applications, North- Holland Publishing Co., Amsterdam, 1978.

[5] G. BARLES and E. MIRONESCU, *On homogenization problems for fully nonlinear equations with oscillating Dirichlet boundary conditions*, arXiv: 1205.4496

[6] L. A. CAFFARELLI and X. CABRE, "Fully Nonlinear Elliptic Equations", AMS Colloquium Publications, Vol. 43, Providence, RI.

[7] L. CAFFARELLI, M. G. CRANDALL, M. KOCAN and A. SWIECH, *Viscosity solutions of Fully Nonlinear Equations with Measurable Ingredients,* CPAM **49** (1997), 365–397.

[8] L. A. CAFFARELLI and P. E. SOUGANIDIS, *Rates of convergence for the homogenization of fully nonlinear uniformly elliptic pde in random media,* Inventiones Mathematicae **180** (2010), 301–360.

[9] L. A. CAFFARELLI, P. E., SOUGANIDIS and L. WANG, *Homogenization of fully nonlinear, uniformly elliptic and parabolic partial differential equations in stationary ergodic media*, Comm. Pure. Applied Math. **58** (2005), 319–361.

[10] S. CHOI and I. KIM, *Homogenization for nonlinear PDEs in general domains with oscillatory Neumann data,* arxiv:1302.5386

[11] S. CHOI, I. KIM and LEE, *Homogenization of Neumann boundary data with fully nonlinear PDEs,* Analysis & PDE, to appear.

[12] M. CRANDALL, H. ISHIII and P. L. LIONS, *Users' Guide to viscosity solutions of second order partial differential equations*, Bull. Amer. Math. Soc. **27** (1992), 1–67.

[13] J. P. DANIEL, *A game interpretation of the Neumann problem for fully nonlinear parabolic and elliptic equations,* arXiv: 1204.1459.

[14] W. FELDMAN, *Homogenization of the oscillating Dirichlet boundary condition in general domains*, preprint.

[15] W. FELDMAN and C. SMART, personal communication.
[16] D. GERARD-VARET and N. MASMOUDI, *Homogenization and boundary layer,* Acta. Math., to appear.
[17] L. C. EVANS, *The perturbed test function method for viscosity solutions of nonlinear PDE,* Proc. Roy. Soc. of Edinburgh: Section A. **111** (1989), 359–375.
[18] H. ISHII and P. L. LIONS, *Viscosity solutions of fully nonlinear second-order elliptic partial differential equations,* J. Differential Equations **83** (1990), 26–78.
[19] E. MILAKIS and L. SILVESTRE, *Regularity for Fully Nonlinear Elliptic Equations with Neumann boundary data,* Comm. PDE. **31** (2006), 1227-1252.
[20] H. TANAKA, *Homogenization of diffusion processes with boundary conditions,* In: "Stochastic Analysis and Applications", Adv. Probab. Related Topics, Vol. 7, Dekker, New York, 1984, 411-437.
[21] A. SWIECH, *$W^{1,p}$ interior estimates for solutions of fully nonlinear, uniformly elliptic equations,* Adv. Differential Equations **2** (1997), 1005–1027.

The analysis of shock formation in 3-dimensional fluids

Demetrios Christodoulou

Abstract. In this lecture I shall discuss the ideas of my monograph "The Formation of Shocks in 3-Dimensonal Fluids". The monograph studies the relativistic Euler equations in 3 space dimensions for a perfect fluid with an arbitrary equation of state.

Part 1

The mechanics of a perfect fluid are described in the framework of special relativity by a future-directed timelike vectorfield u of unit magnitude relative to the Minkowski metric g, the *fluid 4-velocity*, and two positive functions n and s, the *number of particles per unit volume* and the *entropy per particle*. The mechanical properties of a perfect fluid are determined once we give an *equation of state*, which expresses the mass-energy density ρ as a function of n and s:

$$\rho = \rho(n, s)\,. \tag{1}$$

According to the laws of thermodynamics, the *pressure* p and the *temperature* θ are then given by:

$$p = n\frac{\partial \rho}{\partial n} - \rho, \qquad \theta = \frac{1}{n}\frac{\partial \rho}{\partial s}\,. \tag{2}$$

The *particle current* is the vectorfield I given by:

$$I^\mu = nu^\mu\,. \tag{3}$$

The *energy-momentum-stress tensor* is the symmetric 2-contravariant tensorfield T given by:

$$T^{\mu\nu} = (\rho + p)u^\mu u^\nu + p(g^{-1})^{\mu\nu} \tag{4}$$

and the *equations of motion* are the *differential conservation laws*:

$$\nabla_\mu I^\mu = 0, \qquad \nabla_\nu T^{\mu\nu} = 0 \tag{5}$$

There is a substantial gain in geometric insight in working with the relativistic equations because of the spacetime geometry viewpoint of special relativity. As an example consider the equation:

$$i_u\omega = -\theta ds\,. \tag{6}$$

Here ω is the vorticity 2-form:

$$\omega = d\beta \tag{7}$$

β being the 1-form defined by:

$$\beta_\mu = -\sqrt{\sigma}u_\mu, \qquad u_\mu = g_{\mu\nu}u^\nu \tag{8}$$

with $\sqrt{\sigma}$ the relativistic *enthalpy per particle*:

$$\sqrt{\sigma} = \frac{\rho + p}{n}. \tag{9}$$

In (6) i_u denotes contraction on the left by the vectorfield u. Equation (6) is equivalent to the differential energy-momentum conservation laws and is arguably the simplest explicit form of these equations.

At each point p in the spacetime manifold M, H_p, the *local simultaneous space* of the fluid at p, is the g-orthogonal complement of the linear span of u_p, the fluid velocity at p, in T_pM. The obstruction to integrability of the distribution of local simultaneous spaces is the *vorticity vector* ϖ given by:

$$\varpi^\mu = \frac{1}{2}(\epsilon^{-1})^{\mu\alpha\beta\gamma}u_\alpha\omega_{\beta\gamma} \tag{10}$$

where ϵ^{-1} is the reciprocal volume form of (M, g), or volume form in T_p^*M at each $p \in M$.

The 1-form β plays a fundamental role in my monograph. In the irrotational case it is given by $\beta = d\phi$, where ϕ is the *wave function*. In this case, the equations of motion (5) reduce to a *nonlinear wave equation*:

$$\nabla_\mu(G\partial^\mu\phi) = 0, \qquad \partial^\mu\phi = (g^{-1})^{\mu\nu}\partial_\nu\phi \tag{11}$$

where

$$G = \frac{n}{\sqrt{\sigma}} = G(\sigma), \qquad \sigma = -(g^{-1})^{\mu\nu}\partial_\mu\phi\partial_\nu\phi\,. \tag{12}$$

Equation (11) derives from the Lagrangian

$$L = p = L(\sigma) \tag{13}$$

the pressure as a function of the squared enthalpy.

Returning to the general case, the sound speed η is defined by:

$$\left(\frac{dp}{d\rho}\right)_s = \eta^2 \tag{14}$$

it being assumed that the left hand side is positive. The causality condition:

$$0 < \eta < 1 \tag{15}$$

is imposed, the right inequality meaning that the sound speed is less than the universal constant represented by the speed of light in vacuum.

The *acoustical metric* h is another Lorentzian metric on M such that at each $p \in M$ the simultaneous space H_p is also h- orthogonal to u_p, h agrees with g on H_p, and h assigns magnitude η to u. In terms of a formula:

$$h_{\mu\nu} = g_{\mu\nu} + (1 - \eta^2)u_\mu u_\nu, \qquad u_\mu = g_{\mu\nu}u^\nu . \tag{16}$$

The null cones of h are called *sound cones.* By the right inequality above, they are contained within the null cones of g, namely the light cones. What is important from the physical point of view is the *conformal geometry* induced by h on the underlying manifold. It determines the *acoustical causal structure.* That is, given any event $p \in M$ it determines $\mathcal{J}^+(p)$ the acoustical causal future of p, the set of events which are acoustically influenced by p, and $\mathcal{J}^-(p)$ the causal past of p, the set of events which acoustically influence p.

Choosing a *time function* t in Minkowski spacetime, equal to the coordinate x^0 of some rectangular coordinate system, we denote by Σ_t an arbitrary level set of the function t. The Σ_t are parallel spacelike hyperplanes relative to the Minkowski metric g.

Initial data for the equations of motion (5) is given on a domain in the hyperplane Σ_0, which may be the whole of Σ_0. It consists in the specification of the triplet (n, s, u) on this domain. In the irrotational case, where we have the nonlinear wave equation (11), initial data consists in the specification of the pair $(\phi, \partial_t\phi)$ on such a domain. To any given initial data set there corresponds a unique *maximal classical solution* of the equations of motion (5), or of the nonlinear wave equation (11) in the irrotational case. The notion of maximal classical solution or maximal development of an initial data set is the following.

Given an initial data set, the local existence theorem asserts the existence of a *development* of this set, namely of a domain $\mathcal{D}$ in Minkowski spacetime, whose past boundary is the domain of the initial data, and of a solution defined in $\mathcal{D}$ and taking the given data at the past boundary, such that the following condition holds. If we consider any point $p \in \mathcal{D}$ and any curve issuing at p with the property that its tangent vector at any point q belongs to the interior or the boundary of the past component of the sound cone at q, then the curve terminates at a point of the domain of the initial data. [Drawing 1].

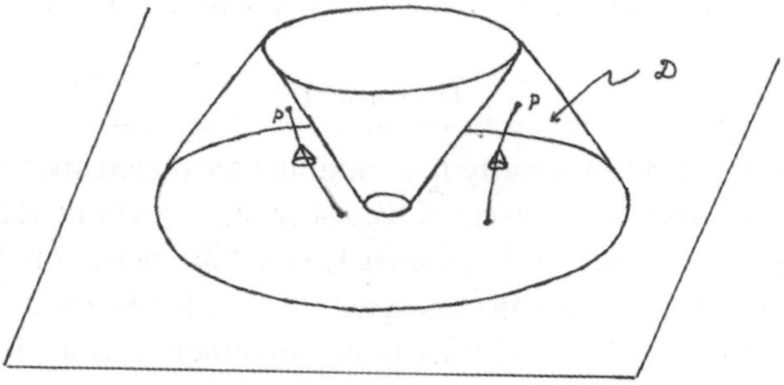

Drawing 1

The local uniqueness theorem asserts that if $(\mathcal{D}_1, (n_1, s_1, u_1))$ and $(\mathcal{D}_2, (n_2, s_2, u_2))$ are two developments of the same initial data [$(\mathcal{D}_1, \phi_1)$ and $(\mathcal{D}_2, \phi_2)$ in the irrotational case], then (n_1, s_1, u_1) coincides with (n_2, s_2, u_2) in $\mathcal{D}_1 \bigcap \mathcal{D}_2$ [ϕ_1 coincides with ϕ_2 in $\mathcal{D}_1 \bigcap \mathcal{D}_2$ in the irrotational case]. It follows that the union of all developments of a given initial data set is itself a development, the unique *maximal development* of the initial data set.

In the monograph I consider regular initial data on Σ_0 which outside a sphere coincide with the data corresponding to a constant state. That is, outside that sphere n and s are constant and u coincides with the future-directed unit normal to Σ_0. Under a suitable restriction on the size of the departure of the initial data from those of the constant state, I prove certain theorems which give a complete description of the maximal classical development. In particular, the theorems give a detailed description of the geometry of the boundary of the domain of the maximal classical development and a detailed analysis of the behavior of the solution at this boundary. A complete picture of shock formation in 3-dimensional fluids is thereby obtained.

I shall confine myself in this talk to the case that the initial data are irrotational hence so is the maximal classical development.

Let H be the function defined by:

$$1 - \eta^2 = \sigma H \tag{17}$$

where η is the sound speed. I denote by ℓ the value of $(dH/d\sigma)_s$ in the surrounding constant state. This constant determines the character of the shocks for small initial departures from the constant state. In particular when $\ell = 0$, no shocks form and the domain of the maximal classical solution is complete.

Consider the function $(dH/d\sigma)_s$ as a function of the thermodynamic variables p and s. Suppose that we have an equation of state such that at some value s_0 of s the function $(dH/d\sigma)_s$ vanishes everywhere along the adiabat $s = s_0$. In this case the irrotational fluid equations corresponding to the value s_0 of the entropy are equivalent to the minimal surface equation, the wave function ϕ defining a minimal graph in a Minkowski spacetime of one more spatial dimension. In fact in this case the Lagrangian (13) is:

$$L = 1 - \sqrt{1 - \sigma} \tag{18}$$

and the action associated to a domain is the area of the domain minus the area of the graph over the domain.

Let O be the center of the sphere $S_{0,0}$ in Σ_0 outside which we have the constant state. Let us confine ourselves to the maximal development of the restriction of the initial data to $\Sigma_0 \setminus O$. Let u be a smooth function without critical points in $\Sigma_0 \setminus O$ such that the the restriction of u to the exterior of $S_{0,0}$ is equal to minus the Euclidean distance from $S_{0,0}$. We extend u to the spacetime manifold by the condition that its level sets are outgoing null hypersurfaces relative to the acoustical metric h. Then u satisfies the h-eikonal equation:

$$(h^{-1})^{\mu\nu} \partial_\mu u \partial_\nu u = 0\,. \tag{19}$$

We call u an *acoustical function* and we denote by C_u an arbitrary level set of u.

Each C_u being a null hypersurface is generated by null geodesics of h. Let L be the tangent vectorfield to these geodesic generators parametrized not affinely but by t. We then define the wave fronts $S_{t,u}$ to be the surfaces of intersection $C_u \bigcap \Sigma_t$. Finally we define the vectorfield T to be tangential to the Σ_t and so that the flow generated by T on each Σ_t is the normal, relative to the induced on Σ_t acoustical metric $\overline{h}$, flow of the foliation of Σ_t by the surfaces $S_{t,u}$. This flow takes each wave front onto another wave front. [Drawing 2].

The structure introduced on the spacetime manifold by an acoustical function u, or, what is the same, the geometry of a foliation of spacetime

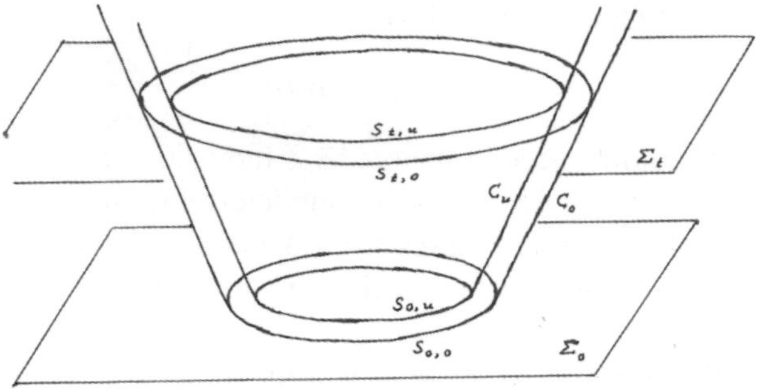

Drawing 2

by outgoing null hypersurfaces C_u, the level sets of u, plays a fundamental role in the problem. The most important geometric property of this foliation from the point of view of the study of shock formation is the density of the packing of its leaves C_u. One measure of this density is the *inverse spatial density* of the wave fronts, that is, the inverse density of the foliation of each spatial hyperplane Σ_t by the surfaces $S_{t,u}$. This is the function κ, given in arbitrary coordinates on Σ_t by:

$$\kappa^{-2} = (\overline{h}^{-1})^{ij} \partial_i u \partial_j u \tag{20}$$

where $\overline{h}_{ij}$ is the induced acoustical metric on Σ_t. An equivalent definition of κ is that it is the magnitude of the vectorfield T with respect to h. [Drawing 3].

Another measure is the *inverse temporal density* of the wave fronts, the function μ, given in arbitrary coordinates in spacetime by:

$$\frac{1}{\mu} = -(h^{-1})^{\mu\nu} \partial_\mu t \partial_\nu u \,. \tag{21}$$

The two measures are related by:

$$\mu = \alpha\kappa \tag{22}$$

where α is the inverse density, with respect to the acoustical metric, of the foliation of spacetime by the hyperplanes Σ_t. [This inverse density is of course 1 when referred to the Minkowski metric.] The function α is given in arbitrary coordinates in spacetime by:

$$\alpha^{-2} = -(h^{-1})^{\mu\nu} \partial_\mu t \partial_\nu t \,. \tag{23}$$

It is expressed directly in terms of the 1-form $\beta = d\phi$. It turns out moreover, that it is bounded above and below by positive constants. Consequently μ and κ are equivalent measures of the density of the packing of the leaves of the foliation of spacetime by the C_u.

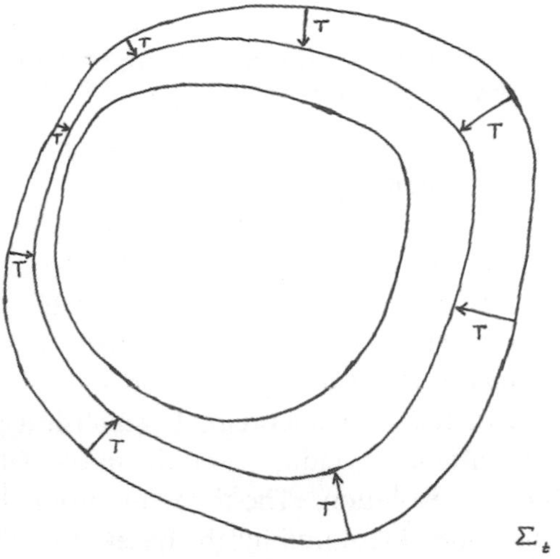

Drawing 3

Shock formation is characterized by the blow up of this density or equivalently by the vanishing of κ or μ.

The maximal development being a domain in Minkowski spacetime, which by a choice of rectangular coordinates is identified with $\mathcal{R}^4$, inherits the subset topology and the standard differential structure induced by the rectangular coordinates x^α. Choosing an acoustical function u we introduce *acoustical coordinates* (t, u, ϑ), $\vartheta \in \mathcal{S}^2$, the coordinate lines corresponding to a given value of u and to constant values of ϑ being the generators of the null hypersurface C_u. The rectangular coordi nates x^α are smooth functions of the acoustical coordinates (t, u, ϑ) and the Jacobian of the transformation is, up to a multiplicative factor which is bounded above and below by positive constants, the inverse temporal density function μ. The acoustical coordinates induce another differential structure on the same underlying topological manifold. However since $\mu > 0$ in the interior of the maximal development, the two differential structures coincide in the interior of the maximal development.

The main theorem of the monograph asserts that relative to the differential structure induced by the acoustical coordinates the maximal classical solution extends smoothly to the boundary of its domain. This boundary contains however a singular part B where the function μ vanishes. The rectangular coordinates themselves extend smoothly to the boundary but the Jacobian vanishes on the singular part of the boundary. The mapping from acoustical to rectangular coordinates has a continuous but not differentiable inverse at B. As a result, the two differential structures no longer coincide when the singular boundary B is included.

With respect to the standard differential structure the solution is continuous but not differentiable at B, the derivative $\hat{T}^{\mu}\hat{T}^{\nu}\partial_{\mu}\partial_{\nu}\phi$ blowing up as we approach B. Here $\hat{T} = \kappa^{-1}T$, is the vectorfield of unit magnitude with respect to h corresponding to T.

With respect to the standard differential structure, the acoustical metric h is everywhere in the closure of the domain of the maximal solution non-degenerate and continuous, but it is not differentiable at B, while with respect to the differential structure induced by the acoustical coordinates h is everywhere smooth, but it is degenerate at B.

After the proof of the main theorem, I establish a general theorem which gives sharp sufficient conditions on the initial data for the formation of a shock in the evolution. The theorem also gives a sharp upper bound on the time interval required for the onset of shock formation.

The last part of the work is concerned with the structure of the boundary of the domain of the maximal classical solution and the behavior of the solution at this boundary. The boundary of the maximal development consists of a regular part $\underline{C}$ and a singular part B. Each component of $\underline{C}$ is a regular incoming acoustically null hypersurface with a singular past boundary which coincides with the past boundary of an associated component of B. The union of these singular past boundaries we denote by $\partial^{-}B$. [Drawing 4].

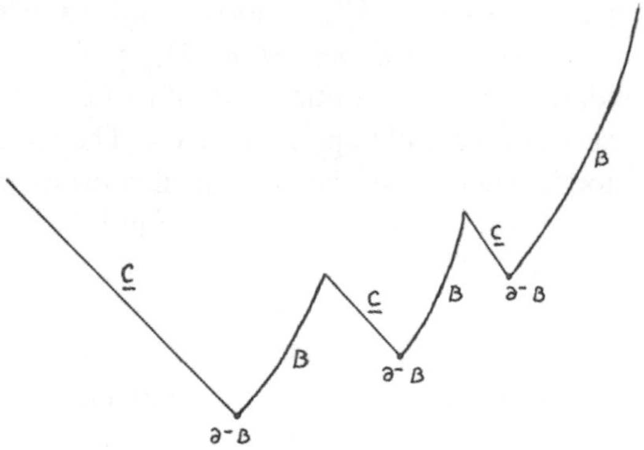

Drawing 4

Each component of B is a hypersurface which is smooth relative to both differential structures and has the intrinsic geometry of a regular null hypersurface in a regular spacetime and, like the latter, is ruled by invariant curves of vanishing arc length. [Drawing 5].

On the other hand, the extrinsic geometry of each component of B is

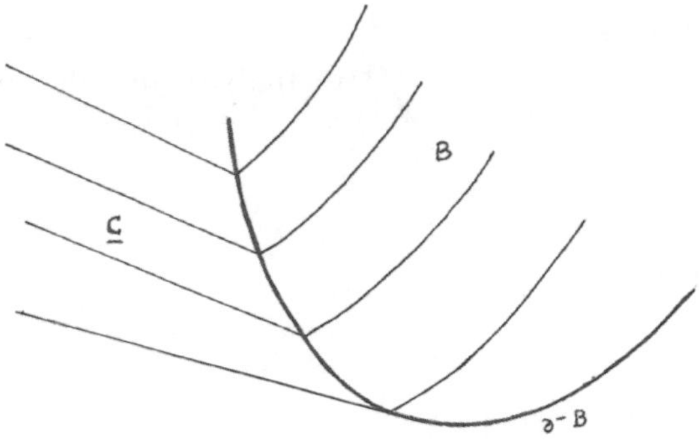

Drawing 5

that of an acoustically spacelike hypersurface which becomes acoustically null at its past boundary, an associated component of $\partial^- B$. This means that at each point $q \in B$ the past null geodesic conoid of q does not intersect B. Each component of $\partial^- B$ is an acoustically spacelike surface which is smooth relative to both differential structures.

The main result of the last part of the work is the trichotomy theorem. This theorem shows that for each point q of the singular boundary, the intersection of the past null geodesic conoid of q with any Σ_t in the past of q splits into three parts, the parts corresponding to the outgoing and to the incoming sets of null geodesics ending at q being embedded discs with a common boundary, an embedded circle, which corresponds to the set of the remaining null geodesics ending at q. *All outgoing null geodesics ending at q have the same tangent vector at q.* This vector is then an invariant null vector associated to the singular point q. [Drawing 6]. This striking result is in fact the reason why the considerable freedom in the choice of the acoustical function does not matter in the end. For, considering the transformation from one acoustical function to another, I show that the foliations corresponding to different families of outgoing null hypersurfaces have equivalent geometric properties and degenerate in precisely the same way on the same singular boundary.

Now, the components of $\partial^- B$ are physically the surfaces where shocks begin to form. The maximal classical solution is the physical solution of the problem up to $\underline{C} \bigcup \partial^- B$, but not up to B. In the epilogue of the monograph the problem of the physical continuation of the solution is set up as the *shock development problem.* This is a free-boundary problem associated to each component of $\partial^- B$. In this problem it is required to construct a hypersurface of discontinuity K, the shock hypersurface, lying in the past of the associated component of B but having the same

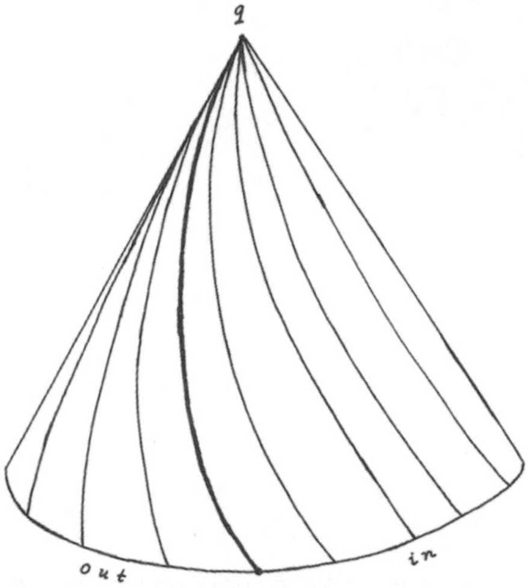

Drawing 6

past boundary as the latter, namely the given component of $\partial^- B$, the tangent hyperplanes to K and B coinciding along $\partial^- B$.

Moreover, it is required to construct a solution of the differential conservation laws (5) in the domain in Minkowski spacetime bounded in the past by $\underline{C} \bigcup K$, agreeing with the maximal classical solution on $\underline{C} \bigcup \partial^- B$, while having jumps across K relative to the data induced on K by the maximal classical solution, jumps satisfying the jump conditions which follow from the integral form of the conservation laws. Finally K is required to be acoustically spacelike with respect to the acoustical metric induced by the maximal classical solution, and timelike with respect to the acoustical metric induced by the new solution, which holds in the future of K. The maximal classical solution thus provides the boundary conditions on $\underline{C} \bigcup \partial^- B$ as well as a barrier at B. [Drawing 7].

Note that a component of $\partial^- B$ is a surface which is acoustically spacelike but not necessarily spacelike relative to the Minkowski metric g. The intersection $\Sigma_t \bigcap K$ represents the instantaneous shock surface in the Lorentz frame defined by the time function t and the intersection $\Sigma_t \bigcap \partial^- B$ represents the boundary curve of the instantaneous shock surface.

Let me close with a formula for the jump in vorticity across K which shows that even though the flow before the shock may be irrotational the flow acquires vorticity immediately after. By virtue of the last condition of the shock development problem K is a timelike hypersurface relative to the Minkowski metric g. Let N be its unit normal relative to g, point-

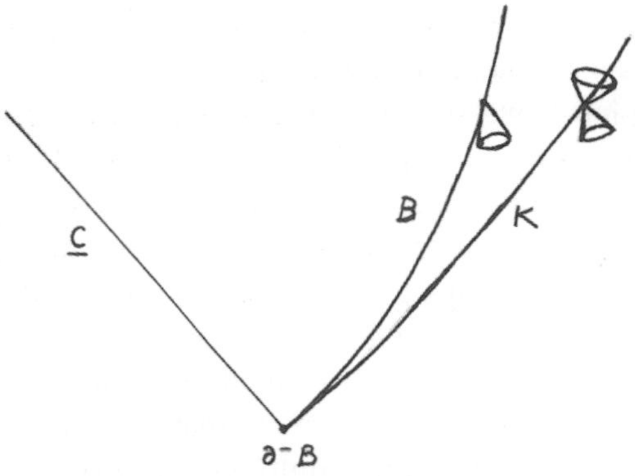

Drawing 7

ing to the future of K. Let u_0 be the fluid velocity at K induced by the past solution, u_1 be the fluid velocity at K induced by the future solution. Then by the jump conditions at each point $p \in K$ the three vectors N, u_0, u_1 lie in the same timelike plane. Let Π_p be the g- orthogonal complement of this plane.

Consider the restriction to Π_p of the differential of $[s]$, the jump in entropy across K. Let us denote this restriction by $\not{d}[s]$ and the corresponding (through g) vector, element of Π_p, by $\not{d}[s]^\sharp$. Then the vorticity vector at p induced by the future solution is given by:

$$\varpi_1 = \frac{\theta_1}{u_{\perp 1}} \, {}^*\not{d}[s]^\sharp \,. \tag{24}$$

Here θ_1 and $u_{\perp 1}$ are, respectively, the temperature and the normal component of the fluid velocity at K induced by the future solution [$u_{\perp 0}$ and $u_{\perp 1}$ are both positive and related by $n_0 u_{\perp_0} = n_1 u_{\perp 1}$]. Also, for any $V \in \Pi_p$ we denote by ${}^*V$ the result of rotating V counterclockwise by a right angle.

Part 2

The starting point of my approach is the notion of variation through solutions. In my monograph "The Action Principle and Partial Differential Equations", which treats general systems of partial differential equations arising from an action principle, I showed that such "first order" variations are associated to a linearized Lagrangian, on the basis of which energy currents are constructed. It is through energy currents and their associated integral identities that the estimates, essential to the approach, are derived. Here we consider the first order variations which correspond to the one-parameter subgroups of the Poincaré group, the isometry group of Minkowski spacetime, extended by the one-parameter scaling or dilation group, which leave the surrounding constant state invariant. The higher order variations correspond to the one-parameter groups of diffeomorphisms generated by a set of vectorfields, the commutation fields.

The construction of an energy current requires a multiplier vectorfield which at each point belongs to the closure of the positive component of the inner characteristic core in the tangent space at that point. In the case of irrotational fluid mechanics the characteristic in the tangent space at a point consists only of the sound cone at that point and this requirement becomes the requirement that the multiplier vectorfield be non-spacelike and future directed with respect to the acoustical metric h.

I use two multiplier vectorfields. The first multiplier field is the vectorfield K_0:

$$K_0 = (\eta_0^{-1} + \alpha^{-1}\kappa)L + \underline{L}, \qquad \underline{L} = \alpha^{-1}\kappa L + 2T \,. \tag{25}$$

The second multiplier field is the vectorfield K_1 defined by:

$$K_1 = (\omega/\nu)L \,. \tag{26}$$

Here ν is the mean curvature of the wave fronts $S_{t,u}$ relative to their null normal L. However ν is defined not relative to the acoustical metric $h_{\mu\nu}$ but rather relative to a conformally related metric $\tilde{h}_{\mu\nu}$:

$$\tilde{h}_{\mu\nu} = \Omega h_{\mu\nu} \,. \tag{27}$$

It turns out that there is a choice of conformal factor Ω such that a first order variation $\dot{\phi}$ of the wave function ϕ satisfies the linear wave equation relative to the metric $\tilde{h}_{\mu\nu}$:

$$\Box_{\tilde{h}}\dot{\phi} = 0 \,. \tag{28}$$

This choice defines Ω and the definition makes Ω the ratio of a function of σ to the value of this function in the surrounding constant state, thus Ω is equal to unity in the constant state. It turns out moreover that Ω is bounded above and below by positive constants. The function ω appearing in (27) is required to have linear growth in t and to be such that $\Box_{\tilde{h}}\omega$ is suitably bounded. To each variation ψ, of any order, there are energy currents associated to ψ and to K_0 and K_1 respectively.

These currents define the energies $\mathcal{E}_0^u[\psi](t)$, $\mathcal{E}_1'^u[\psi](t)$, and the fluxes $\mathcal{F}_0^t[\psi](u)$, $\mathcal{F}_1'^t[\psi](u)$. For given t and u the energies are integrals over the exterior of the surface $S_{t,u}$ in the hyperplane Σ_t, while the fluxes are integrals over the part of the outgoing null hypersurface C_u between the hyperplanes Σ_0 and Σ_t. It is these energy and flux integrals, together with a spacetime integral $K[\psi](t, u)$ associated to K_1, to be discussed below, which are used to control the solution.

Evidently, the means by which the solution is controlled depend on the choice of the acoustical function u, the level sets of which are the outgoing null hypersurfaces C_u. The function u is determined by its restriction to the initial hyperplane Σ_0.

The divergence of the energy currents, which determines the growth of the energies and fluxes, itself depends on ${}^{(K_0)}\tilde{\pi}$, in the case of the energy current associated to K_0, and ${}^{(K_1)}\tilde{\pi}$, in the case of the energy current associated to K_1. Here for any vectorfield X in spacetime, I denote by ${}^{(X)}\tilde{\pi}$ the Lie derivative of the conformal acoustical metric $\tilde{h}$ with respect to X. I call ${}^{(X)}\tilde{\pi}$ the *deformation tensor* corresponding to X. In the case of higher order variations, the divergences of the energy currents depend also on the ${}^{(Y)}\tilde{\pi}$, for each of the commutation fields Y to be discussed below.

All these deformation tensors ultimately depend on the acoustical function u, or, what is the same, on the geometry of the foliation of spacetime by the outgoing null hypersurfaces C_u. Recall from the previous lecture that the most important geometric property of this foliation from the point of view of the study of shock formation is the density of the packing of its leaves C_u. One measure of this density is the inverse spatial density of the wave fronts, that is, the inverse density of the foliation of each spatial hyperplane Σ_t by the surfaces $S_{t,u}$. This is the function κ. Another measure is the inverse temporal density of the wave fronts, the function μ.

The two measures are related by:

$$\mu = \alpha\kappa \tag{29}$$

where α is the inverse density, with respect to the acoustical metric, of the foliation of spacetime by the hyperplanes Σ_t. The function α is expressed directly in terms of the 1-form $\beta = d\phi$. It turns out moreover, that it is bounded above and below by positive constants. Consequently μ and κ are equivalent measures of the density of the packing of the leaves of the foliation of spacetime by the C_u.

Recall from the previous lecture that shock formation is characterized by the blow up of this density or equivalently by the vanishing of κ or μ.

The other entity, besides κ or μ which describes the geometry of the foliation by the C_u is the second fundamental form of the C_u. Since the C_u are null hypersurfaces with respect to the acoustical metric h their tangent hyperplane at a point is the set of all vectors at that point which are h-orthogonal to the generator L, and L itself belongs to the tangent hyperplane, being h-orthogonal to itself. Thus the second fundamental form χ of C_u is intrinsic to C_u and in terms of the metric $\not{h}$ induced by the acoustical metric on the $S_{t,u}$ sections of C_u, it is given by:

$$\not{\mathcal{L}}_L \not{h} = 2\chi \tag{30}$$

where $\not{\mathcal{L}}_X \vartheta$ for a covariant $S_{t,u}$ tensorfield ϑ denotes the restriction of $\mathcal{L}_X \vartheta$ to $TS_{t,u}$.

The acoustical structure equations are:

The propagation equation for χ along the generators of C_u.

The Codazzi equation which expresses $\not{d}\mathrm{iv}\,\chi$, the divergence of χ intrinsic to $S_{t,u}$, in terms of $\not{d}\mathrm{tr}\chi$, the differential on $S_{t,u}$ of $\mathrm{tr}\chi$, and a component of the acoustical curvature and of k, the second fundamental form of the Σ_t relative to h.

The Gauss equation which expresses the Gauss curvature of $(S_{t,u}, \not{h})$ in terms of χ and a component of the acoustical curvature and of k.

An equation which expresses $\not{\mathcal{L}}_T \chi$ in terms of the Hessian of the restriction of μ to $S_{t,u}$ and another component of the acoustical curvature and of k.

These acoustical structure equations seem at first sight to contain terms which blow up as κ or μ tend to zero. The analysis of the acoustical curvature then shows that the terms which blow up as κ or μ tend to zero cancel.

The most important acoustical structure equation from the point of view of the formation of shocks is the propagation equation for μ along the generators of C_u:

$$L\mu = m + \mu e \tag{31}$$

where the function m given by:

$$m = \frac{1}{2}(\beta_L)^2 \left(\frac{dH}{d\sigma}\right)_s (T\sigma) \tag{32}$$

and the function e depends only on the derivatives of the β_α, the rectangular components of the 1-form $\beta = d\phi$, tangential to the C_u.

It is the function m which determines shock formation, when being negative, causing μ to decrease to zero.

I first establish a theorem, the fundamental energy estimate, which applies to a solution of the homogeneous wave equation in the acoustical spacetime, in particular to any first order variation. The proof of this theorem relies on certain bootstrap assumptions on the acoustical entities. The most crucial of these assumptions concern the behavior of the function μ.

To give an idea of the nature of these assumptions, one of the assumptions required to obtain the fundamental energy estimate up to time s is:

$$\mu^{-1}(T\mu)_+ \leq B_s(t) \quad : \text{for all } t \in [0, s] \tag{33}$$

where $B_s(t)$ is a function such that:

$$\int_0^s (1+t)^{-2}[1+\log(1+t)]^4 B_s(t)dt \leq C \tag{34}$$

with C a constant independent of s. Here T is the vectorfield defined above and we denote by f_+ and f_-, respectively the positive and negative parts of an arbitrary function f.

This assumption is then established by a certain proposition with $B_s(t)$ the following function:

$$B_s(t) = C\sqrt{\delta_0}\frac{(1+\tau)}{\sqrt{\sigma-\tau}} + C\delta_0(1+\tau) \tag{35}$$

where $\tau = \log(1+t)$, $\sigma = \log(1+s)$, and δ_0 is a small positive constant appearing in the final bootstrap assumption.

The spacetime integral $K[\psi](t, u)$ mentioned above, is essentially the integral of

$$-\frac{1}{2}(\omega/\nu)(L\mu)_-|\not{d}\psi|^2$$

in the spacetime exterior to C_u and bounded by Σ_0 and Σ_t.

Another assumption states that there is a positive constant C independent of s such that in the region below Σ_s where $\mu < \eta_0/4$ we have:

$$L\mu \leq -C^{-1}(1+t)^{-1}[1+\log(1+t)]^{-1}\,. \tag{36}$$

In view of this assumption, the integral $K[\psi](t, u)$ gives effective control of the derivatives of the variations tangential to the wave fronts in the region where shocks are to form. The same assumption, which is then established by a certain proposition, also plays an essential role in the study of the singular boundary.

The final stage of the proof of of the fundamental energy estimate is the analysis of system of integral inequalities in two variables t and u satisfied by the five quantities $\mathcal{E}_0^u[\psi](t)$, $\mathcal{E}_1'^u[\psi](t)$, $\mathcal{F}_0^t[\psi](u)$, $\mathcal{F}_1'^t[\psi](u)$, and $K[\psi](t, u)$.

After this, the commutation fields Y, which generate the higher order variations, are defined. They are five: the vectorfield T which is transversal to the C_u, the field $Q = (1+t)L$ along the generators of the C_u and the three rotation fields R_i : $i = 1, 2, 3$ which are tangential to the

$S_{t,u}$ sections. The latter are defined to be $\Pi \overset{\circ}{R}_i : i = 1, 2, 3$, where the $\overset{\circ}{R}_i \ i = 1, 2, 3$ are the generators of spatial rotations associated to the background Minkowskian structure, while Π is the h-orthogonal projection to the $S_{t,u}$.

Expressions for the deformation tensors ${}^{(T)}\tilde{\pi}$, ${}^{(Q)}\tilde{\pi}$, and ${}^{(R_i)}\tilde{\pi} : i = 1, 2, 3$ are then derived, which show that these depend on the acoustical entities μ and χ.

The higher order variations satisfy inhomogeneous wave equations in the acoustical spacetime, the source functions depending on the deformation tensors of the commutation fields. These source functions give rise to error integrals, that is to spacetime integrals of contributions to the divergence of the energy currents.

The expressions for the source functions and the associated error integrals show that the error integrals corresponding to the energies of the $n + 1$st order variations contain the nth order derivatives of the deformation tensors, which in turn contain the nth order derivatives of χ and $n + 1$st order derivatives of μ. Thus to achieve closure, we must obtain estimates for the latter in terms of the energies of up to the $n + 1$st order variations. Now, the propagation equations for χ and μ give appropriate expressions for $\not{\mathcal{L}}_L \chi$ and $L\mu$. However, if these propagation equations, which may be thought of as ordinary differential equations along the generators of the C_u, are integrated with respect to t to obtain the acoustical entities χ and μ themselves, and their spatial derivatives are then taken, a loss of one degree of differentiability would result and closure would fail.

I overcome this difficulty in the case of χ by considering the propagation equation for $\mu \mathrm{tr}\chi$. I show that, by virtue of a wave equation for σ, which follows from the wave equations satisfied by the first variations corresponding to the spacetime translations, the principal part on the right hand side of this propagation equation can be put into the form $-L\check{f}$ of a derivative of a function $-\check{f}$ with respect to L. This function is then brought to the left hand side and we obtain a propagation equation for $\mu \mathrm{tr}\chi + \check{f}$. In this equation $\hat{\chi}$, the trace-free part of χ enters, but the propagation equation in question is considered in conjunction with the Codazzi equation, which constitutes an elliptic system on each $S_{t,u}$ for $\hat{\chi}$, given $\mathrm{tr}\chi$. We thus have an ordinary differential equation along the generators of C_u coupled to an elliptic system on the $S_{t,u}$ sections.

More precisely, the propagation equation which is considered at the same level as the Codazzi equation is a propagation equation for the $S_{t,u}$ 1-form $\mu \not{d} \mathrm{tr}\chi + \not{d}\check{f}$, which is a consequence of the equation just discussed. To obtain estimates for the angular derivatives of χ of order l we similarly consider a propagation equation for the $S_{t,u}$ 1-form:

$$^{(i_1...i_l)}x_l = \mu \not{d}(R_{i_l}...R_{i_1}\mathrm{tr}\chi) + \not{d}(R_{i_l}...R_{i_1}\check{f})$$

In the case of μ the aforementioned difficulty is overcome by considering the propagation equation for $\mu\not{\Delta}\mu$, where $\not{\Delta}\mu$ is the Laplacian of the restriction of μ to the $S_{t,u}$. I show that by virtue of a wave equation for $T\sigma$, which is a consequence of the wave equation for σ, the principal part on the right hand side of this propagation equation can again be put into the form $L\check{f}'$ of a derivative of a function $\check{f}'$ with respect to L. This function is then likewise brought to the left hand side and we obtain a propagation equation for $\mu\not{\Delta}\mu - \check{f}'$. In this equation $\hat{\not{D}}^2\mu$, the trace-free part of the Hessian of the restriction of μ to the $S_{t,u}$ enters, but the propagation equation in question is considered in conjunction with the elliptic equation on each $S_{t,u}$ for μ, which the specification of $\not{\Delta}\mu$ constitutes. Again we have an ordinary differential equation along the generators of C_u coupled to an elliptic equation on the $S_{t,u}$ sections.

To obtain estimates of the spatial derivatives of μ of order $l+2$ of which m are derivatives with respect to T we similarly consider a propagation equation for the function:

$$\begin{aligned} ^{(i_1...i_{l-m})}x'_{m,l-m} = \mu R_{i_{l-m}}...R_{i_1}(T)^m\not{\Delta}\mu \\ -R_{i_{l-m}}...R_{i_1}(T)^m\check{f}' \end{aligned}$$

This allows us to obtain estimates for the top order spatial derivatives of μ of which at least two are angular derivatives. A remarkable fact is that the missing top order spatial derivatives do not enter the source functions, hence do not contribute to the error integrals.

The paradigm of an ordinary differential equation along the generators of a null hypersurface coupled to an elliptic system on the sections of the hypersurface was first encountered in my work with Sergiu Klainerman on the stability of Minkowski spacetime in general relativity.

Here, the appearance of the factor of μ, which vanishes where shocks originate, in front of

$$\not{d}R_{i_l}...R_{i_1}\mathrm{tr}\chi \text{ and } R_{i_{l-m}}...R_{i_1}(T)^m\not{\Delta}\mu$$

in the definitions of

$$^{(i_1...i_l)}x_l \text{ and } {}^{(i_1...i_{l-m})}x'_{m,l-m}$$

above, makes the analysis quite delicate. This is compounded with the difficulty of the slow decay in time which the addition of the terms $-\not{d}R_{i_l}...R_{i_1}\check{f}$ and $R_{i_{l-m}}...R_{i_1}(T)^m\check{f}'$ forces.

The analysis requires a precise description of the behavior of μ itself, given by certain propositions, and a separate treatment of the condensation regions, where shocks are to form, from the rarefaction regions, the terms referring not to the fluid density but rather to the density of the stacking of the wave fronts. To overcome the difficulties the following weight function is introduced:

$$\overline{\mu}_{m,u}(t) = \min\left\{\frac{\mu_{m,u}(t)}{\eta_0}, 1\right\}, \qquad \mu_{m,u}(t) = \min_{\Sigma_t^u} \mu \tag{37}$$

where Σ_t^u is the exterior of $S_{t,u}$ in Σ_t, and the quantities $\mathcal{E}_0^u[\psi](t)$, $\mathcal{E}_1^{\prime u}[\psi](t)$, $\mathcal{F}_0^t[\psi](u)$, $\mathcal{F}_1^{\prime t}[\psi](u)$, and $K[\psi](t,u)$ corresponding to the highest order variations are weighted with a power, $2a$, of this weight function.

The following lemma then plays a crucial role here as well as in the proof of the main theorem where everything comes together. Let:

$$\begin{aligned} M_u(t) &= \max_{\Sigma_t^u}\left\{-\mu^{-1}(L\mu)_-\right\}, \\ I_{a,u} &= \int_0^t \overline{\mu}_{m,u}^{-a}(t') M_u(t') dt' . \end{aligned} \tag{38}$$

Then under certain bootstrap assumptions in the past of Σ_s, for any constant $a \geq 2$, there is a positive constant C *independent of* s*,* u *and* a such that for all $t \in [0, s]$ we have:

$$I_{a,u}(t) \leq C a^{-1} \overline{\mu}_{m,u}^{-a}(t) \tag{39}$$

The acoustical assumptions on which the previous results depend are established, using the method of continuity, on the basis of the final bootstrap assumption, which consists only of pointwise estimates for the variations up to certain order.

The analysis of the structure of the terms containing the top order spatial derivatives of the acoustical entities shows that these terms can be expressed in terms of the 1-forms ${}^{(i_1\ldots i_l)}x_l$ and the functions ${}^{(i_1\ldots i_{l-m})}x'_{m,l-m}$. These contribute *borderline error integrals*, the treatment of which is the main source of difficulties in the problem. The borderline integrals are all proportional to the constant ℓ mentioned above, hence are absent in the case $\ell = 0$.

I should make clear here that the only variations which are considered up to this point are the variations arising from the first order variations corresponding to the group of spacetime translations. In particular the final bootstrap assumption involves only variations of this type, and each of the five quantities $\mathcal{E}_{0,[n]}^u(t)$, $\mathcal{F}_{0,[n]}^t(u)$, $\mathcal{E}_{1,[n]}^{\prime u}(t)$, $\mathcal{F}_{1,[n]}^{\prime t}(u)$, and $K_{[n]}(t,u)$, which together control the solution, is defined to be the sum of the corresponding quantity $\mathcal{E}_0^u[\psi](t)$, $\mathcal{F}_0^t[\psi](u)$, $\mathcal{E}_1'^u[\psi](t)$, $\mathcal{F}_1'^t[\psi](u)$, and $K[\psi](t,u)$, over all variations ψ of this type, up to order n.

To estimate the borderline integrals however, I introduce an additional assumption which concerns the first order variations corresponding to the scaling or dilation group and to the rotation group, and the second order variations arising from these by applying the commutation field T. This assumption is later established through energy estimates of order 4 arising from these first order variations and derived on the basis of the final bootstrap assumption, just before the recovery of the final bootstrap assumption itself. It turns out that the borderline integrals all contain the factor $T\psi_\alpha$, where $\psi_\alpha : \alpha = 0, 1, 2, 3$ are the first variations corresponding to spacetime translations and the additional assumption is used to obtain an estimate for $\sup_{\Sigma_t^u}\left(\mu^{-1}|T\psi_\alpha|\right)$ in terms of $\sup_{\Sigma_t^u}\left(\mu^{-1}|L\mu|\right)$, which involves on the right the factor $|\ell|^{-1}$.

Upon substituting this estimate in the borderline integrals, the factors involving ℓ cancel, and the integrals are estimated using the inequality (39). The above is an outline of the main steps in the estimation of the borderline integrals associated to the vectorfield K_0. The estimation of the borderline integrals associated to the vectorfield K_1, is however still more delicate. In this case I first perform an integration by parts on the outgoing null hypersurfaces C_u, obtaining hypersurface integrals over Σ_t^u and Σ_0^u and another spacetime volume integral. In this integration by parts the terms, including those of lower order, must be carefully chosen to obtain appropriate estimates, because here the long time behavior, as well as the behavior as μ tends to zero, is critical. Another integration by parts, this time on the surfaces $S_{t,u}$, is then performed to reduce these integrals to a form which can be estimated.

The estimates of the hypersurface integrals over Σ_t^u are the most delicate (the hypersurface integrals over Σ_0^u only involve the initial data) and require separate treatment of the condensation and rarefaction regions, in which the properties of the function μ, established by the previous propositions, all come into play.

In proceeding to derive the energy estimates of top order, $n = l + 2$, the power $2a$ of the weight $\overline{\mu}_{m,u}(t)$ is chosen suitably large to allow us to transfer the terms contributed by the borderline integrals to the left hand side of the inequalities resulting from the integral identities associated to the multiplier fields K_0 and K_1. The argument then proceeds along the lines of that of the fundamental energy estimate, but is more complex because here we are dealing with weighted quantities.

Once the top order energy estimates are established, I revisit the lower order energy estimates using at each order the energy estimates of the next order in estimating the error integrals contributed by the highest spatial derivatives of the acoustical entities at that order. I then establish a descent scheme, which yields, after finitely many steps, estimates for the five quantities $\mathcal{E}^u_{0,[n]}(t)$, $\mathcal{F}^t_{0,[n]}(u)$, $\mathcal{E}'^u_{1,[n]}(t)$, $\mathcal{F}'^t_{1,[n]}(u)$, and $K_{[n]}(t,u)$, for $n = l + 1 - [a]$, where $[a]$ is the integral part of a, in which weights no longer appear.

It is these unweighted estimates which are used to close the bootstrap argument by recovering the final bootstrap assumption. This is accomplished by the method of continuity through the use of the isoperimetric inequality on the wave fronts $S_{t,u}$, and leads to the main theorem.

References

[1] D. CHRISTODOULOU, "The Action Principle and Partial Differential Equations", Ann. Math. Stud., Vol. 46, Princeton University Press, 2000.

[2] D. CHRISTODOULOU, "The Formation of Shocks in 3-Dimensional Fluids", EMS Monographs in Mathematics, EMS Publishing House, 2007.

The analogous problem for the classical, non-relativistic Euler equations, is studied in the following monograph:

[3] D. CHRISTODOULOU and S. MIAO, "Compressible Flow and Euler's Equations", http://arxiv.org/abs/1212.2867.

Regularity of the extremal solutions for the Liouville system

Louis Dupaigne, Alberto Farina and Boyan Sirakov

Abstract. We study the smoothness of the extremal solutions to the Liouville system.

In this short note, we study the smoothness of the extremal solutions to the following system of equations:

$$\begin{cases} -\Delta u = \mu e^{v} & \text{in } \Omega, \\ -\Delta v = \lambda e^{u} & \text{in } \Omega, \\ \quad u = v = 0 & \text{on } \partial\Omega, \end{cases} \tag{1}$$

where $\lambda, \mu > 0$ are parameters and Ω is a smoothly bounded domain of $\mathbb{R}^N$, $N \geq 1$. As shown by M. Montenegro (see [6]), there exists a limiting curve Υ in the first quadrant of the (λ, μ)-plane serving as borderline for existence of classical solutions of (1). He also proved the existence of a weak solution u^* for every (λ^*, μ^*) on the curve Υ and left open the question of its regularity. Following standard terminology (see e.g. the books [3], [5] for an introduction to this vast subject), u^* is called an extremal solution. Our result is the following.

Theorem 1. *Let* $1 \leq N \leq 9$. *Then, extremal solutions to* (1) *are smooth.*

Remark 2. C. Cowan ([1]) recently obtained the same result under the further assumption that $(N-2)/8 < \lambda/\mu < 8/(N-2)$.

Any extremal solution u^* is obtained as the increasing pointwise limit of a sequence of regular solutions (u_n) associated to parameters $(\lambda_n, \mu_n) = (1 - 1/n)(\lambda^*, \mu^*)$. In addition, see [6], u_n is stable in the sense that the principal eigenvalue of the linearized operator associated to (1) is non-negative. In other words, there exist $\lambda_1 \geq 0$ and two positive functions

$\varphi_1, \psi_1 \in C^2(\overline{\Omega})$ such that

$$\begin{cases} -\Delta\varphi_1 - g'(v)\psi_1 = \lambda_1\varphi_1 & \text{in } \Omega, \\ -\Delta\psi_1 - f'(u)\varphi_1 = \lambda_1\psi_1 & \text{in } \Omega. \\ \varphi_1 = \psi_1 = 0 & \text{on } \partial\Omega, \end{cases} \tag{2}$$

where, in the context of (1), $g(v) = \mu e^v$ and $f(u) = \lambda e^u$. This motivates the following useful inequality.

Let f, g denote two nondecreasing C^1 functions and consider the more general system

$$\begin{cases} -\Delta u = g(v) & \text{in } \Omega, \\ -\Delta v = f(u) & \text{in } \Omega, \\ u = v = 0 & \text{on } \partial\Omega. \end{cases} \tag{3}$$

Lemma 3. *Let $N \geq 1$ and let $(u, v) \in C^2(\overline{\Omega})^2$ denote a stable solution of* (3). *Then, for all $\varphi \in C^1_c(\Omega)$, there holds*

$$\int_\Omega \sqrt{f'(u)g'(v)}\varphi^2 \, dx \leq \int_\Omega |\nabla\varphi|^2 \, dx \tag{4}$$

Remark 4. As we just learnt, the same inequality has been obtained independently by C. Cowan and N. Ghoussoub. See [2].

Proof. Since (u, v) is stable, there exist $\lambda_1 \geq 0$ and two positive functions $\varphi_1, \psi_1 \in C^2(\overline{\Omega})$ solving (2). Given $\varphi \in C^1_c(\Omega)$, multiply the first equation in (2) by φ^2/φ_1 and integrate. Then,

$$\begin{aligned} \int_\Omega g'(v)\frac{\psi_1}{\varphi_1}\varphi^2 \, dx &\leq \int_\Omega \frac{\varphi^2}{\varphi_1}(-\Delta\varphi_1) \\ &= -\int_\Omega |\nabla\varphi_1|^2\left(\frac{\varphi}{\varphi_1}\right)^2 + 2\int_\Omega \frac{\varphi}{\varphi_1}\nabla\varphi\nabla\varphi_1 \\ &= -\int_\Omega \left|\frac{\varphi}{\varphi_1}\nabla\varphi_1 - \nabla\varphi\right|^2 + \int_\Omega |\nabla\varphi|^2 \\ &\leq \int_\Omega |\nabla\varphi|^2. \end{aligned} \tag{5}$$

Working similarly with the second equation, we also have

$$\int_\Omega f'(u)\frac{\varphi_1}{\psi_1}\varphi^2 \, dx \leq \int_\Omega |\nabla\varphi|^2 \, dx \tag{6}$$

(4) then follows by combining the Cauchy-Schwarz inequality and (5)- (6). □

Thanks to the inequality (4), we obtain the following estimate.

Lemma 5. *Let* $N \geq 1$. *There exists a universal constant* $C > 0$ *such that any stable solution of* (1) *satisfies*

$$\int_\Omega e^{u+v}\,dx \leq C\,|\Omega|\left(\frac{\lambda}{\mu}+\frac{\mu}{\lambda}\right). \tag{7}$$

Proof. Multiply the second equation in (1) by $e^v - 1$ and integrate.

$$\begin{aligned}\lambda\int_\Omega e^{u+v}dx \geq \lambda\int_\Omega e^u(e^v-1)dx &= \int_\Omega \nabla v \nabla(e^v-1)\,dx\\ &= 4\int_\Omega \left|\nabla(e^{v/2}-1)\right|^2\,dx.\end{aligned} \tag{8}$$

Using (4) with test function $\varphi = e^{v/2}-1$, it follows that

$$\begin{aligned}\lambda\int_\Omega e^{u+v}\,dx &\geq 4\sqrt{\lambda\mu}\int_\Omega e^{\frac{u+v}{2}}(e^{v/2}-1)^2\,dx\\ &\geq 4\sqrt{\lambda\mu}\int_\Omega e^{\frac{u+v}{2}}e^v\,dx - 8\sqrt{\lambda\mu}\int_\Omega e^{\frac{u+v}{2}}e^{v/2}\,dx.\end{aligned} \tag{9}$$

By Young's inequality, $e^{v/2} = \frac{1}{\sqrt{2}}e^{v/2}\cdot\sqrt{2} \leq \frac{1}{4}e^v + 1$. So,

$$\int_\Omega e^{\frac{u+v}{2}}e^{v/2}\,dx \leq \frac{1}{4}\int_\Omega e^{\frac{u+v}{2}}e^v\,dx + \int_\Omega e^{\frac{u+v}{2}}\,dx.$$

Plugging this in (9), we obtain

$$\lambda\int_\Omega e^{u+v}\,dx + 8\sqrt{\lambda\mu}\int_\Omega e^{\frac{u+v}{2}}\,dx \geq 2\sqrt{\lambda\mu}\int_\Omega e^{\frac{u+v}{2}}e^v\,dx. \tag{10}$$

Similarly,

$$\mu\int_\Omega e^{u+v}\,dx + 8\sqrt{\lambda\mu}\int_\Omega e^{\frac{u+v}{2}}\,dx \geq 2\sqrt{\lambda\mu}\int_\Omega e^{\frac{u+v}{2}}e^u\,dx. \tag{11}$$

Multiply (10) and (11) to get

$$\begin{aligned}\lambda\mu\left(\int_\Omega e^{u+v}\,dx\right)^2 &+ 64\lambda\mu\left(\int_\Omega e^{\frac{u+v}{2}}\,dx\right)^2\\ &+ 8\sqrt{\lambda\mu}(\lambda+\mu)\int_\Omega e^{u+v}\,dx\int_\Omega e^{\frac{u+v}{2}}\,dx\\ &\geq 4\lambda\mu\int_\Omega e^{\frac{u+v}{2}}e^u\,dx\int_\Omega e^{\frac{u+v}{2}}e^v\,dx.\end{aligned} \tag{12}$$

Using Young's inequality, the left-hand side in the above inequality is bounded above by

$$2\lambda\mu\left(\int_\Omega e^{u+v}\,dx\right)^2 + C(\lambda+\mu)^2\left(\int_\Omega e^{\frac{u+v}{2}}\,dx\right)^2, \tag{13}$$

where C is a universal constant. In addition, by the Cauchy-Schwarz inequality,

$$\int_\Omega e^{\frac{u+v}{2}}e^u\,dx\int_\Omega e^{\frac{u+v}{2}}e^v\,dx \geq \left(\int_\Omega e^{u+v}\,dx\right)^2. \tag{14}$$

Plugging (14) in (13) and remembering that (13) is an upper bound of the left-hand side in (12), we obtain

$$C(\lambda+\mu)^2\left(\int_\Omega e^{\frac{u+v}{2}}\,dx\right)^2 \geq 2\lambda\mu\int_\Omega e^{\frac{u+v}{2}}e^u\,dx\int_\Omega e^{\frac{u+v}{2}}e^v\,dx. \tag{15}$$

By the Cauchy-Schwarz inequality and (14), we have

$$\begin{aligned}\left(\int_\Omega e^{\frac{u+v}{2}}\,dx\right)^2 &\leq |\Omega|\int_\Omega e^{u+v}\,dx \qquad (16)\\ &\leq |\Omega|\left(\int_\Omega e^{\frac{u+v}{2}}e^u\,dx\int_\Omega e^{\frac{u+v}{2}}e^v\,dx\right)^{1/2}.\end{aligned}$$

Using (16) in (15), we obtain

$$C\frac{(\lambda+\mu)^2}{\lambda\mu}|\Omega| \geq \left(\int_\Omega e^{\frac{u+v}{2}}e^u\,dx\int_\Omega e^{\frac{u+v}{2}}e^v\,dx\right)^{1/2}. \tag{17}$$

Applying once more (14), we obtain the desired estimate. □

We can now prove Theorem 1.

Step 1. Case $1 \leq N \leq 3$. It is enough to treat the case $N = 3$, the cases $N = 1, 2$ being easier. By (8) and (7), $e^{v/2} - 1$ is bounded in $H_0^1(\Omega)$ (with a uniform bound with respect to λ and μ). By the Sobolev embedding, it follows that e^v is bounded in $L^{\frac{N}{N-2}}(\Omega)$. By (8) and elliptic regularity, u is bounded in $W^{2,\frac{N}{N-2}}$. For $N = 3$, $\frac{N}{N-2} > \frac{N}{2}$. By Sobolev's embedding, we deduce that u is bounded, and so must be v. This implies the desired conclusion for the corresponding extremal solution.

Step 2. General case. We adapt a method introduced in [4]. Fix $\alpha > 1/2$ and multiply the first equation in (1) by $e^{\alpha u} - 1$. Integrating over Ω, we obtain

$$\mu \int_\Omega \left(e^{\alpha u} - 1\right) e^v \, dx = \alpha \int_\Omega e^{\alpha u} |\nabla u|^2 \, dx = \frac{4}{\alpha} \int_\Omega \left|\nabla \left(e^{\frac{\alpha u}{2}} - 1\right)\right|^2 \, dx$$

By (4),

$$\sqrt{\lambda\mu} \int_\Omega e^{\frac{u+v}{2}} \left(e^{\frac{\alpha u}{2}} - 1\right)^2 \, dx \le \int_\Omega \left|\nabla \left(e^{\frac{\alpha u}{2}} - 1\right)\right|^2 \, dx.$$

Combining these two inequalities, we deduce that

$$\sqrt{\lambda\mu} \int_\Omega e^{\frac{u+v}{2}} \left(e^{\frac{\alpha u}{2}} - 1\right)^2 \, dx \le \frac{\alpha}{4} \mu \int_\Omega \left(e^{\alpha u} - 1\right) e^v \, dx \qquad (18)$$

Hence,

$$\sqrt{\lambda\mu} \int_\Omega e^{\frac{2\alpha+1}{2} u} e^{\frac{v}{2}} \, dx \le \frac{\alpha}{4} \mu \int_\Omega e^{\alpha u} e^v \, dx + 2\sqrt{\lambda\mu} \int_\Omega e^{\frac{\alpha+1}{2} u} e^{\frac{v}{2}} \, dx \quad (19)$$

Let us estimate the terms on the right-hand side. By Hölder's inequality,

$$\int_\Omega e^{\alpha u} e^v \, dx \le \left(\int_\Omega e^{\frac{2\alpha+1}{2} u} e^{\frac{v}{2}} \, dx\right)^{\frac{2\alpha-1}{2\alpha}} \left(\int_\Omega e^{\frac{u}{2}} e^{\frac{2\alpha+1}{2} v} \, dx\right)^{\frac{1}{2\alpha}} \qquad (20)$$

Given $\varepsilon > 0$, it also follows from Young's inequality that

$$\int_\Omega e^{\frac{\alpha+1}{2} u} e^{\frac{v}{2}} \, dx \le \frac{\varepsilon}{2} \sqrt{\frac{\mu}{\lambda}} \int_\Omega e^{\alpha u} e^v \, dx + \frac{1}{2\varepsilon} \sqrt{\frac{\lambda}{\mu}} \int_\Omega e^u \, dx.$$

Using (7), we deduce that

$$\int_\Omega e^{\frac{\alpha+1}{2} u} e^{\frac{v}{2}} \, dx \le \frac{\varepsilon}{2} \sqrt{\frac{\mu}{\lambda}} \int_\Omega e^{\alpha u} e^v \, dx + \frac{1}{2\varepsilon} \sqrt{\frac{\lambda}{\mu}} C \, |\Omega| \left(\frac{\lambda}{\mu} + \frac{\mu}{\lambda}\right). \quad (21)$$

where C is the universal constant of Lemma 3.1.

So, gathering (19), (20), (21), and letting

$$X = \int_\Omega e^{\frac{2\alpha+1}{2} u} e^{\frac{v}{2}} \, dx \quad \text{and} \quad Y = \int_\Omega e^{\frac{2\alpha+1}{2} v} e^{\frac{u}{2}} \, dx,$$

we obtain

$$\sqrt{\lambda\mu}\, X \le \left(\frac{\alpha}{4} + \varepsilon\right) \mu \, X^{\frac{2\alpha-1}{2\alpha}} Y^{\frac{1}{2\alpha}} + C \frac{\lambda}{\varepsilon} |\Omega| \left(\frac{\lambda}{\mu} + \frac{\mu}{\lambda}\right).$$

By symmetry, we also have

$$\sqrt{\lambda\mu}\, Y \le \left(\frac{\alpha}{4}+\varepsilon\right)\lambda\, Y^{\frac{2\alpha-1}{2\alpha}} X^{\frac{1}{2\alpha}} + C\frac{\mu}{\varepsilon}\,|\Omega|\left(\frac{\lambda}{\mu}+\frac{\mu}{\lambda}\right).$$

Multiplying these inequalities, we deduce that

$$\left(1-\left(\frac{\alpha}{4}+\varepsilon\right)^2\right) X\, Y \le C_1\left(\frac{\lambda}{\mu}+\frac{\mu}{\lambda}\right)^2\left(1+X^{\frac{2\alpha-1}{2\alpha}}Y^{\frac{1}{2\alpha}}+Y^{\frac{2\alpha-1}{2\alpha}}X^{\frac{1}{2\alpha}}\right).$$

where $C_1 > 0$ depends on $\Omega, \alpha, \varepsilon$ only. Hence, for every $\alpha < 4$, either X or Y must be bounded (with a uniform bound with respect to λ and μ). Without loss of generality, $\lambda \ge \mu$ and by the maximum principle, $v \ge u$. It follows that e^u is bounded in $L^p(\Omega)$ for every $p = \alpha + 1 < 5$. Using standard elliptic regularity, the result follows. □

Remark. O. Goubet informed us that J. Dávila and him obtained a simpler proof of Theorem 1.

References

[1] C. COWAN, *Regularity of the extremal solutions in a Gelfand systems problem*, Advanced Nonlinear Studies **11** (2011), n. 3, 695–700.

[2] C. COWAN and N. GHOUSSOUB, *Regularity of semi-stable solutions to fourth order nonlinear eigenvalue problems on general domains*, Calc. Var., to appear.

[3] L. DUPAIGNE, "Stable Solutions of Elliptic Partial Differential Equations", Chapman & Hall/CRC Monographs and Surveys in Pure and Applied Mathematics, Vol. 143, Chapman & Hall/CRC, Boca Raton, FL, 2011, xiv+321.

[4] L. Dupaigne, M. Ghergu, O. Goubet and G. Warnault, *The Gel'fand problem for the biharmonic operator*, Arch. Ration. Mech. Anal. **208** (2013), n. 3, 725–752.

[5] P. ESPOSITO, N. GHOUSSOUB and Y. GUO, "Mathematical Analysis of Partial Differential Equations Modeling Electrostatic MEMS", Courant Lecture Notes in Mathematics, Vol. 20, Courant Institute of Mathematical Sciences, New York, 2010, xiv+318.

[6] M. MONTENEGRO, *Minimal solutions for a class of elliptic systems*, Bull. London Math. Soc. **37** (2005), 405–416.

On general existence results for one-dimensional singular diffusion equations with spatially inhomogeneous driving force

Mi-Ho Giga, Yoshikazu Giga and Atsushi Nakayasu

Abstract. A general anisotropic curvature flow equation with singular interfacial energy and spatially inhomogeneous driving force is considered for a curve given by the graph of a periodic function. We prove that the initial value problem admits a unique global-in-time viscosity solution for a general periodic continuous initial datum. The notion of a viscosity solution used here is the same as proposed by Giga, Giga and Rybka, who established a comparison principle. We construct the global-in-time solution by careful adaptation of Perron's method.

1 Introduction

In this paper we study a one-dimensional nonlinear degenerate parabolic equation whose diffusion effect is very strong at particular slopes of unknown functions. We are in particular interested in an equation, where the driving force term is spatially inhomogeneous. A typical example is a quasilinear equation

$$u_t = a(u_x)[(W'(u_x))_x + \sigma(t, x)], \tag{1.1}$$

where W is a given convex function on $\mathbf{R}$ but may not be of class $C^1(\mathbf{R})$ so that its derivative W' may have jump discontinuities and σ is a given Lipschitz function depending on the space variable x as well as the time variable t; here a is a given nonnegative continuous function, and u_t and u_x denote the time and the space derivative of an unknown function $u = u(t, x)$.

In order to explain the motivation of this work, let us consider an evolution law of a curve $\Gamma_t \subset \mathbf{R}^2$ moved by an anisotropic curvature flow

$$V = M_0(\mathbf{n})(\kappa_{\gamma_0} + \sigma) \quad \text{on } \Gamma_t, \tag{1.2}$$

The work of the second author has been partly supported by Japan Society for the Promotion of Science (JSPS) through grants for scientific research Kiban (S) 21224001, Kiban (A) 23244015 and 25610025. The work of the third author was supported by Leading Graduate School Doctoral Program from The Ministry of Education, Culture, Sports, Science and Technology, and by a Grant-in-Aid for JSPS Fellows No. 25-7077.

where V is the normal velocity of the evolving curve in the direction of the normal vector $\mathbf{n}$ and let the mobility M_0 and the surface energy density γ_0 be positive functions on the unit circle; the term κ_{γ_0} called a nonlocal curvature is the first variation of surface energy. We note that if γ_0 is the constant 1, then κ_{γ_0} is nothing but usual curvature κ; the quantity κ_{γ_0} formally equals $((\gamma_0)_{\theta\theta} + \gamma_0)\kappa$ if one writes γ_0 as a function of the argument θ of $\mathbf{n} = (\cos\theta, \sin\theta)$. The equation (1.2) appears in crystal growth as an equation to describe the interface of two phases; see, *e.g.*, [2].

If the curve Γ_t is given as a graph of a function $u = u(t,x)$, the equation (1.2) then becomes of the form (1.1) with

$$a(p) = M(p,-1), \qquad M(p,q) = \sqrt{p^2+q^2}\,M_0\left(\frac{(p,q)}{\sqrt{p^2+q^2}}\right),$$

$$W(p) = \gamma(p,-1), \qquad \gamma(p,q) = \sqrt{p^2+q^2}\,\gamma_0\left(\frac{(p,q)}{\sqrt{p^2+q^2}}\right).$$

Assume that the Frank diagram $F := \{(p,q) \in \mathbf{R}^2 \mid \gamma(p,q) \le 1\}$ is convex so that W is a convex function. If F has a smooth (C^2) boundary ∂F, the theory of (1.2) is well developed [6, 9, 16]. Indeed, since W is $C^2(\mathbf{R})$, we are able to apply the classical theory of viscosity solutions [7] to the equation (1.1). We are concerned with the case that ∂F is of class C^2 except finitely many points. A typical example of F is a polygon so that W is a piecewise linear function. For examples if γ is a crystalline energy of the form

$$\gamma(p,q) = |p| + |q|,$$

then $W''(p)$ is twice the Dirac delta function δ and so the equation (1.1) formally becomes

$$u_t = a(u_x)[2\delta(u_x)u_{xx} + \sigma],$$

which is not a classical partial differential equation.

Admissible curves such as polygons moving by a crystalline energy with no driving force have been studied by Taylor [19, 20] and by Angenent and Gurtin [1]. For the evolution law of graphs (1.1) a notion of solutions is introduced by adapting the subdifferential theory [10] ($\sigma = 0$) and [11]. Elliott, Gardiner and Schätzle [8] study relationship between the solutions in the sense of [10] and admissible curves. When σ is independent of x, the theory of viscosity solutions to (1.1) and (1.2) is established in a series of papers [12–14].

The goal of this paper is to establish a global-in-time existence theorem of a viscosity solution for a class of equations including (1.1) with a given continuous periodic initial condition. Our result is a generalization of [12, Section 8, 9] to the equation with spatially inhomogeneous driving force. Notion of viscosity solutions to (1.1) with σ depending on x is introduced in [15], where a comparison theorem is established. The authors of [15] also show some existence results by showing that a special semi-explicit variational solution studied in [17] is a viscosity solution but their initial data is very restrictive. We also point out that in a recent paper by Chambolle and Novaga [4] the authors establish short-time existence for (1.2) by time-discrete implicit scheme, which is introduced in [3,5]. Our argument based on the theory of viscosity solutions is completely different from theirs and can be applied to a fully nonlinear equation.

Following [15], let us consider an energy functional which formally equals

$$\Phi[f] = \int_{\mathbf{T}} (W(f_x) - \sigma f)dx$$

for a smooth function f; we assumed a periodic boundary condition so that $\mathbf{T} = \mathbf{R}/\omega\mathbf{Z}$ with $\omega > 0$. Let $\partial^0\Phi[f]$ be the canonical restriction of the subdifferential $\partial\Phi[f]$ in the Hilbert space $H := L^2(\mathbf{T})$, *i.e.*

$$\partial^0\Phi[f] = \arg\min\{\|\lambda\|_H \mid \lambda \in \partial\Phi[f]\}.$$

As mentioned in [11], the above minimizing problem is equivalent to an obstacle problem: The condition $\lambda \in -\partial\Phi[f]$ holds if and only if λ is of the form $\lambda = \xi'$ such that $\xi \in \partial W(f_x) + Z$ a.e. on $\mathbf{T}$, where Z is a primitive function of σ, *i.e.* $Z_x = \sigma$. Therefore, we minimize

$$\left\{\int_{\mathbf{T}} |\xi'|^2 dx \mid \xi \in \partial W(f_x) + Z \text{ a.e. on } \mathbf{T}\right\}. \tag{1.3}$$

There might be a chance that there is no such ξ satisfying $\xi \in \partial W(f_x) + Z$ a.e. on $\mathbf{T}$. We need to require special structure to guarantee the existence of such ξ. A sufficient condition is that f is flat (facet) on a nontrivial interval (called a faceted region) containing each fixed point x whenever ∂W has a jump at $f_x(x)$. Such a function f is called a faceted function and we see that (1.3) admits a unique minimizer $\bar{\xi}$ for a faceted function f since the problem is convex. It is natural to guess that $\bar{\xi}'$ gives a candidate for the value of the nonlocal curvature

$$\Lambda_W^\sigma(f)(x) = (W'(f_x))_x + \sigma(x).$$

Based on this observation we establish a notion of viscosity solutions to (1.1).

We prove the existence theorem by Perron's method, which is standard in the theory of viscosity solutions for regular equations; we refer the reader to [7, 18]. In our problem, however, it is necessary to modify a smooth faceted test function keeping its property. In the previous work [12] it suffices to modify the test function outside the faceted region. However, this method heavily relies on the fact that the nonlocal curvature $\Lambda^\sigma_W(f)$ is constant on a faceted region when σ is independent of x.

The main idea to solve this problem is to find a small effective region which determines the quantity of the nonlocal curvature. We construct a modification as in the previous work [12] using the effective region instead of the faceted region. Then the argument works well for our setting with the spatially inhomogeneous driving force term σ.

This paper is organized as follows. In Section 2 we recall the definition of faceted functions and the nonlocal curvature Λ^σ_W and define generalized solutions for the equations. In Section 3 we describe how to construct an effective region and modifications for test functions. In Section 4 we prove Perron type existence theorems and Section 5 is devoted to proving the existence theorem for periodic initial data.

2 Definitions of generalized solutions

In this section we recall some notions of functions and the nonlocal curvature Λ^σ_W introduced in [15, Section 2] and define generalized solutions for fully nonlinear equations of the form

$$u_t + F(t, u_x, \Lambda^\sigma_W(u)) = 0 \quad \text{in } Q := (0, T) \times \Omega, \tag{2.1}$$

where $T > 0$ and Ω is an open set in $\mathbf{R}$. We assume the following conditions throughout this paper.

(W) Assume W is a convex function on $\mathbf{R}$ with values in $\mathbf{R}$ of class C^2 outside a closed discrete set P and that its second derivative W'' is bounded in any compact set except all points in P.

(S) The continuous function $\sigma = \sigma(t, x)$ on $[0, T] \times \Omega$ is Lipschitz continuous in x uniformly with respect to t, *i.e.* there exists a constant L such that

$$|\sigma(t, x) - \sigma(t, y)| \le L|x - y| \quad \text{for all } t \in [0, T), x, y \in \Omega.$$

(F1) F is continuous on $[0, T] \times \mathbf{R} \times \mathbf{R}$ with values in $\mathbf{R}$.

(F2) $F(t, p, X) \le F(t, p, Y)$ for all $t \in [0, T]$, $p \in \mathbf{R}$, $X \ge Y$.

The discrete set P in (W) is either a finite set or a countable set having no accumulation point in $\mathbf{R}$. If P is nonempty, P is of form $\{p_j\}_{j=1}^{m}$, $\{p_j\}_{j=1}^{\infty}$, $\{p_j\}_{j=-1}^{-\infty}$ or $\{p_j\}_{j=-\infty}^{\infty}$, where $\{p_j\}$ is a strictly increasing sequence $p_j < p_{j+1}$ with $\lim_{j\to\infty} p_j = \infty$ and $\lim_{j\to-\infty} p_j = -\infty$, and m is a positive integer. We often let $\sigma(t)$ denote the function $\sigma(t)(x) = \sigma(t,x)$ for $t \in [0,T)$. We say that a family of a functions σ_t on Ω is equi-Lipschitz continuous if there exists a constant L such that $|\sigma_t(x) - \sigma_t(y)| \leq L|x-y|$ for all t and $x, y \in \Omega$. Our assumption (S) is equivalent to saying that $\sigma(t)$ on Ω is equi-Lipschitz continuous.

2.1 Faceted functions

We first define a notion of a faceted function.

Definition 2.1 (Faceted function). A function $f \in C^1(\Omega)$ is *faceted* at a point $\hat{x} \in \Omega$ with *slope* $p \in \mathbf{R}$ (or *p-faceted* at $\hat{x}$) if there exists a closed nontrivial finite interval $I = [c_l, c_r] \subset \Omega$ containing $\hat{x}$ (*i.e.* $c_l, c_r \in \Omega$ satisfy $c_l < c_r$ and $c_l \leq \hat{x} \leq c_r$) such that

$$\begin{aligned} f'(x) &= p \quad \text{for all } x \in I, \\ f'(x) &\neq p \quad \text{for all } x \in J \setminus I \end{aligned}$$

with some neighborhood $J = (b_l, b_r) \subset \Omega$ of I. The closed interval I is called a *faceted region* of f containing $\hat{x}$. We say that a function f is P-faceted at $\hat{x}$ if f is p-faceted at $\hat{x}$ for some $p \in P$ and let

$$C_P^2(\Omega) := \left\{ f \in C^2(\Omega) \mid f \text{ is } P\text{-faceted at } \hat{x} \text{ whenever } f'(\hat{x}) \in P \right\}.$$

We also define the *left transition number* $\chi_l = \chi_l(f,\hat{x})$ and the *right transition number* $\chi_r = \chi_r(f,\hat{x})$ for a p-faceted function f at $\hat{x}$ by

$$\chi_l = \begin{cases} +1 & \text{if } f' < p \text{ on } (b_l, c_l), \\ -1 & \text{if } f' > p \text{ on } (b_l, c_l), \end{cases}$$

$$\chi_r = \begin{cases} +1 & \text{if } f' > p \text{ on } (c_r, b_r), \\ -1 & \text{if } f' < p \text{ on } (c_r, b_r). \end{cases}$$

Let $R(f,\hat{x}) = [c_l, c_r]$ denote a maximal closed interval containing $\hat{x}$ on which f' is constant, *i.e.*

$$\begin{aligned} c_l &:= \inf\{x \in \Omega \mid f'(y) = f'(\hat{x}) \text{ for all } y \in [x, \hat{x}]\}, \\ c_r &:= \sup\{x \in \Omega \mid f'(y) = f'(\hat{x}) \text{ for all } y \in [\hat{x}, x]\}. \end{aligned}$$

The interval $R(f,\hat{x})$ is nothing but the faceted region if f is a P-faceted function at $\hat{x}$.

Remark 2.2. We note that a p-faceted function f at $\hat{x}$ agrees with an affine function

$$\ell_p(x) := p(x - \hat{x}) + f(\hat{x})$$

on $I = R(f, \hat{x})$ and that

$$\chi_l = \begin{cases} +1 & \text{if } f > \ell_p \text{ on } (b_l, c_l), \\ -1 & \text{if } f < \ell_p \text{ on } (b_l, c_l), \end{cases}$$

$$\chi_r = \begin{cases} +1 & \text{if } f > \ell_p \text{ on } (c_r, b_r), \\ -1 & \text{if } f < \ell_p \text{ on } (c_r, b_r). \end{cases}$$

2.2 Nonlocal curvature with a nonuniform driving force

We next recall the definition of the nonlocal curvature for a smooth faceted function. Assume (W) and that

$$\sigma \text{ is a Lipschitz function on } \Omega. \tag{2.2}$$

For $f \in C^2_P(\Omega)$ and $\hat{x} \in \Omega$ define the *nonlocal curvature* $\Lambda^\sigma_W(f)(\hat{x})$ as below.

On one hand, if $f'(\hat{x}) \notin P$, we set

$$\Lambda^\sigma_W(f)(\hat{x}) = W''(f'(\hat{x}))f''(\hat{x}) + \sigma(\hat{x})$$

as expected. On the other hand, if $p := f'(\hat{x}) \in P$, *i.e.* f is p-faceted at $\hat{x}$, the definition is more involved since it is based on the obstacle problem (1.3).

Let Z be a primitive function of σ and let

$$\Delta = |\partial W(p)| = \lim_{q \downarrow p} W'(q) - \lim_{q \uparrow p} W'(q).$$

We also take the faceted region $I = R(f, \hat{x}) = [c_l, c_r]$ and the transition numbers $\chi_l = \chi_l(f, \hat{x})$, $\chi_r = \chi_r(f, \hat{x})$. We note that

$$Z \in C^{1,1}(I),\ \Delta > 0,\ I \text{ is a nontrivial closed interval and } \chi_l,\ \chi_r \in [-1, 1]. \tag{2.3}$$

For later convenience we have defined K for χ_l, χ_r whose values are in $[-1, 1]$ not necessarily in $\{\pm 1\}$. Let $K = K^{Z,\Delta,I}_{\chi_l \chi_r}$ be the set of all $\xi \in H^1(I)$ satisfying an obstacle condition

$$Z(x) - \Delta/2 \le \xi(x) \le Z(x) + \Delta/2 \quad \text{for all } x \in I$$

and a boundary condition

$$\xi(c_l) = Z(c_l) - \chi_l \Delta/2, \quad \xi(c_r) = Z(c_r) + \chi_r \Delta/2.$$

We now consider the functional $J = J^{Z,\Delta,I}_{\chi_l \chi_r}$ on $L^2(I)$ defined by

$$J[\xi] = \begin{cases} \int_I |\xi'(x)|^2 dx & \text{if } \xi \in K, \\ \infty & \text{otherwise.} \end{cases}$$

It is easy to see that K is a closed convex set with respect to H^1 norm and thus J admits a unique minimizer denoted by $\bar{\xi} = \xi^{Z,\Delta,I}_{\chi_l \chi_r}$.

An equivalent condition to being a minimizer of the obstacle problem is known. Assume (2.3). For $\xi \in K$ define the *upper coincidence set* D_+ and the *lower coincidence set* D_- by

$$D_\pm = D_\pm(\xi) = \{x \in I \mid \xi(x) = Z(x) \pm \Delta/2\}.$$

We say that ξ satisfies *concave-convex condition* if ξ is concave outside the upper coincidence set D_+ and convex outside the lower coincidence set D_-.

Proposition 2.3 (Characterization of minimizer). *A function $\xi \in K$ is the minimizer of J if and only if ξ satisfies the concave-convex condition.*

This proposition is proved in the same way as in [15, Proposition 2.2], which shows the equivalence with the assumption $\chi_l, \chi_r = \pm 1$, and so we omit it. Noting that Proposition 2.3 in particular implies that the minimizer of the obstacle problem $\bar{\xi}$ belongs to $C^{1,1}(I)$, so we define

$$\Lambda^{Z'}_{\chi_l \chi_r}(x; I, \Delta) = \bar{\xi}'(x) \quad \text{for } x \in I.$$

The reason we write Z' instead of Z is that the derivative $\bar{\xi}'$ depends on Z only through its derivative. Proposition 2.3 also shows that restriction of $\bar{\xi}$ is also a minimizer of an obstacle problem on the restricted domain:

Corollary 2.4. *Let $M = [c_l, c_r] \subset I$ be a nontrivial closed interval. Then,*

$$\xi^{Z,\Delta,I}_{\chi_l \chi_r} = \xi^{Z,\Delta,M}_{\chi_l' \chi_r'} \quad \textit{on } M.$$

with

$$\chi_l' = 2(\bar{\xi}(c_l) - Z(c_l))/\Delta, \quad \chi_r' = 2(\bar{\xi}(c_r) - Z(c_r))/\Delta.$$

Definition 2.5 (Nonlocal curvature). Assume (W) and (2.2). Let $f \in C^2_P(\Omega)$ and $\hat{x} \in \Omega$.

(i) If $f'(\hat{x}) \notin P$, then define

$$\Lambda^\sigma_W(f)(\hat{x}) = W''(f'(\hat{x}))f''(\hat{x}) + \sigma(\hat{x}).$$

(ii) If f is P-faceted at $\hat{x}$, then define

$$\Lambda^\sigma_W(f)(\hat{x}) = \Lambda^\sigma_{\chi_l \chi_r}(\hat{x}; I, \Delta)$$

with $\Delta = |\partial W(p)|$, $I = R(f, \hat{x})$, $\chi_l = \chi_l(f, \hat{x})$, $\chi_r = \chi_r(f, \hat{x})$.

We prepare several propositions on the nonlocal curvature.

Proposition 2.6 (Comparison). *Assume* (W) *and* (2.2). *Let* $f, g \in C^2_P(\Omega)$ *and* $\hat{x} \in \Omega$. *If* $\max_\Omega(f - g) = (f - g)(\hat{x})$, *then*

$$\Lambda^\sigma_W(f)(\hat{x}) \le \Lambda^\sigma_W(g)(\hat{x}).$$

Proposition 2.7 (Continuity with respect to σ and x). *Assume* (W) *and let* $f \in C^2_P(\Omega)$ *and* $\hat{x} \in \Omega$. *Let* $y, y_k \in R(f, \hat{x})$ *and equi-Lipschitz continuous functions* σ, σ_k *on* Ω *satisfy* $y_k \to y$ *and* $\sigma_k \to \sigma$ *uniformly. Then*

$$\Lambda^{\sigma_k}_W(f)(y_k) \to \Lambda^\sigma_W(f)(y).$$

Proposition 2.8 (Continuity with respect to I). *Assume* (2.2), $\chi_l, \chi_r = \pm 1$, $\Delta > 0$. *Let nontrivial intervals* $I = [c_l, c_r]$, $I^k = [c^k_l, c^k_r]$ *of* Ω *satisfy* $I^k \to I$, i.e. $c^k_l \to c_l$ *and* $c^k_r \to c_r$, *and let* $y \in I$, $y_k \in I_k$ *satisfy* $y_k \to y$. *Then*

$$\Lambda^\sigma_{\chi_l \chi_r}(y_k; I_k, \Delta) \to \Lambda^\sigma_{\chi_l \chi_r}(y; I, \Delta).$$

Proposition 2.6–2.8 are immediate consequence of [15, Theorem 2.8, 2.9, 2.12].

2.3 Admissible functions and definition of a generalized solution

We recall a natural class of test function.

Definition 2.9 (Admissible function). Let I and J be open intervals in $\mathbf{R}$. An *admissible* function on $\mathcal{Q} := J \times I$ is a function φ of the form

$$\varphi(t, x) = f(x) + g(t) \quad \text{on } \mathcal{Q} \tag{2.4}$$

with some functions $f \in C^2_P(I)$ and $g \in C^1(J)$. Let $A_P(\mathcal{Q})$ be the set of all admissible functions on $\mathcal{Q}$.

We are now able to define a generalized solution in the viscosity sense for the singular parabolic equation (2.1). For a real-valued function u recall the *upper semicontinuous envelope* and the *lower semicontinuous envelope*

$$u^*(t, x) := \lim_{\varepsilon \downarrow 0} \sup\{u(s, y) \mid (s, y) \in \mathcal{Q}, |s - t| + |y - x| < \varepsilon\},$$

$$u_*(t, x) := \lim_{\varepsilon \downarrow 0} \inf\{u(s, y) \mid (s, y) \in \mathcal{Q}, |s - t| + |y - x| < \varepsilon\}$$

for $(t, x) \in \overline{\mathcal{Q}}$.

Definition 2.10 (Viscosity solution). A real-valued function u on $\mathcal{Q}$ is a *viscosity subsolution* of (2.1) in $\mathcal{Q}$ if $u^* < \infty$ in $[0, T) \times \overline{\Omega}$ and

$$\varphi_t(\hat{t}, \hat{x}) + F(\hat{t}, \varphi_x(\hat{t}, \hat{x}), \Lambda_W^{\sigma(\hat{t})}(\varphi(\hat{t}, \cdot))(\hat{x})) \leq 0 \tag{2.5}$$

whenever $(\hat{t}, \hat{x}) \in \mathcal{Q}$ and $\varphi \in A_P(\mathcal{Q})$ satisfy

$$\max_{\mathcal{Q}}(u^* - \varphi) = (u^* - \varphi)(\hat{t}, \hat{x}). \tag{2.6}$$

A real-valued function u on $\mathcal{Q}$ is a *viscosity supersolution* of (2.1) in $\mathcal{Q}$ if $u_* > -\infty$ in $[0, T) \times \overline{\Omega}$ and

$$\varphi_t(\hat{t}, \hat{x}) + F(\hat{t}, \varphi_x(\hat{t}, \hat{x}), \Lambda_W^{\sigma(\hat{t})}(\varphi(\hat{t}, \cdot))(\hat{x})) \geq 0 \tag{2.7}$$

whenever $(\hat{t}, \hat{x}) \in \mathcal{Q}$ and $\varphi \in A_P(\mathcal{Q})$ satisfy

$$\min_{\mathcal{Q}}(u_* - \varphi) = (u_* - \varphi)(\hat{t}, \hat{x}). \tag{2.8}$$

If u is both a subsolution and a supersolution, u is called a *viscosity solution*.

Hereafter we suppress the word "viscosity". A function φ satisfying (2.6) or (2.8) is called a *test function* of u at $(\hat{t}, \hat{x})$.

The following propositions are easily derived.

Proposition 2.11 (Smooth solution and viscosity solution). *We assume* (W), (S), (F2). *If* $\varphi \in A_P(\mathcal{Q})$ *of the form* (2.4) *with* $f \in C_P^2(\Omega)$ *and* $g \in C^1(0, T)$ *satisfies* (2.5) *(resp.* (2.7)*) for each* $(\hat{t}, \hat{x}) \in \mathcal{Q}$*, then* φ *is a subsolution (resp. supersolution) of* (2.1) *in* $\mathcal{Q}$.

Proof. We only show that φ is a subsolution. Fix $\psi \in A_P(\mathcal{Q})$ of the form

$$\psi(t, x) = \tilde{f}(x) + \tilde{g}(t) \quad \text{on } \mathcal{Q}$$

with $\tilde{f} \in C_P^2(\Omega)$ and $\tilde{g} \in C^1(0, T)$, and suppose that

$$\varphi(t, x) - \psi(t, x) = f(x) - \tilde{f}(x) + g(t) - \tilde{g}(t)$$

attains a maximum at a point $(\hat{x}, \hat{t}) \in \mathcal{Q}$. We then see that $f'(\hat{x}) = \tilde{f}'(\hat{x})$ and $g'(\hat{t}) = \tilde{g}'(\hat{t})$. Moreover, Proposition 2.6 yields

$$\Lambda_W^{\sigma(\hat{t})}(f)(\hat{x}) \leq \Lambda_W^{\sigma(\hat{t})}(\tilde{f})(\hat{x}).$$

Therefore, we have

$$\tilde{g}'(\hat{t}) + F(\hat{t}, \tilde{f}'(\hat{x}), \Lambda_W^{\sigma(\hat{t})}(\tilde{f})(\hat{x})) \leq g'(\hat{t}) + F(\hat{t}, f'(\hat{x}), \Lambda_W^{\sigma(\hat{t})}(f)(\hat{x})) \leq 0$$

by (F2) and (2.5). □

Proposition 2.12 (Addition by affine functions). *Let u be a subsolution (resp. supersolution) of* (2.1) *in $\mathcal{Q}$ and $a, b \in \mathbf{R}$. Then $v(t,x) = u(t,x) - ax - b$ is a subsolution (resp. supersolution) of*

$$v_t + F(t, v_x + a, \Lambda^{\sigma}_{W_a}(v)) = 0 \text{ in } \mathcal{Q},$$

where $W_a(p) = W(p+a)$.

In order to show the existence of a solution by Perron's method we define a local version of the notion of solutions. We say that a function $\varphi \in C(\mathcal{Q})$ is *locally admissible* at a point $(\hat{t}, \hat{x}) \in \mathcal{Q}$ if φ is admissible on $J \times I$ with some bounded open intervals I and J such that $\hat{t} \in J \subset (0,T)$ and $\hat{x} \in I \subset \Omega$.

Definition 2.13. A real-valued function u on $\mathcal{Q}$ is a *subsolution in the local sense* of (2.1) in $\mathcal{Q}$ if $u^* < \infty$ in $[0,T) \times \overline{\Omega}$ and (2.5) holds for all locally admissible $\varphi \in C(\mathcal{Q})$ at $(\hat{t}, \hat{x}) \in \mathcal{Q}$ satisfying (2.6). A *supersolution in the local sense* is defined by replacing $u^* < \infty$ by $u_* > -\infty$, the inequality (2.5) by (2.7) and the equality (2.6) by (2.8) as before.

Lemma 2.14. *A real-valued function u on $\mathcal{Q}$ is a subsolution (resp. supersolution) of* (2.1) *in $\mathcal{Q}$ if and only if u is a subsolution (resp. supersolution) in the local sense of* (2.1) *in $\mathcal{Q}$.*

These facts can be shown by the same argument as in [12, Section 6].

3 Effective region and canonical modification

In this section we construct an upper and lower modification $f^{\#,\varepsilon}$ and $f_{\#,\varepsilon}$ for a faceted function f and a small number $\varepsilon > 0$. These modifications play an important role in order to prove a Perron type existence theorem in the next section.

Definition 3.1. Let $f \in C(\Omega) \cap C^2_P(\Omega_1)$ satisfy $f'(\hat{x}) = 0$ with an open interval $\Omega_1 = (a_l, a_r) \subset \Omega$ and $\hat{x} \in \Omega_1$. Let

$$p_1 = \sup\{p \in P \cup \{-\infty\} \mid p < 0\} \in [-\infty, 0),$$
$$p_2 = \inf\{p \in P \cup \{\infty\} \mid p > 0\} \in (0, \infty].$$

Consider the case (i) $f'(\hat{x}) = 0 \notin P$. We then define $M = [d_l, d_r]$ by

$$d_l = d_r = \hat{x}, \qquad \text{i.e. } M = \{\hat{x}\}$$

and set

$$f^{\#,\varepsilon}(x) = f^{\#}(x) = f(x) + (x - \hat{x})^4 \quad \text{for } x \in \Omega.$$

Let us note that there exists an open neighborhood $\Omega_2 = (b_l, b_r) \subset \Omega_1$ of $\hat{x}$ such that

$$\frac{p_1}{2} < f'(x) < \frac{p_2}{2} \quad \text{for all } x \in \Omega_2, \tag{3.1}$$

$$d_l + \frac{\sqrt[3]{p_1}}{2} \le b_l < d_l, \quad d_r < b_r \le d_r + \frac{\sqrt[3]{p_2}}{2}. \tag{3.2}$$

Consider the case (ii) $f'(\hat{x}) = 0 \in P$, *i.e.* f is P-faceted at $\hat{x}$. Take the faceted region $[c_l, c_r] = R(f, \hat{x})$ and the minimizer of the obstacle problem ξ. Define $M = [d_l, d_r]$ by

$$\begin{aligned} d_l &= \max\{x \le \hat{x} \mid x \in D_-(\xi) \cup \{c_l\}\}, \\ d_r &= \min\{x \ge \hat{x} \mid x \in D_+(\xi) \cup \{c_r\}\}. \end{aligned}$$

Take an open interval $\Omega_2 = (b_l, b_r) \subset \Omega_1 \cap J$ such that (3.1) and (3.2) hold, where J is the neighborhood of $R(f, \hat{x})$ appearing in Definition 2.1. Define $f^{\#,\varepsilon}$ for each $\varepsilon > 0$ as below: First set

$$f^{\#,\varepsilon}(x) = f(x) = f(\hat{x}) \quad \text{for } x \in M = [d_l, d_r].$$

If $d_l \in D_-(\xi)$, set

$$f^{\#,\varepsilon}(x) = f(x) + (x - d_l)^4 \quad \text{for } x \in \Omega, x \le d_l.$$

If $d_l \notin D_-(\xi)$, that is $d_l = c_l$ and $d_l \in D_+(\xi)$, set

$$f^{\#,\varepsilon}(x) = \begin{cases} f(d_l) = f(\hat{x}) & \text{for } x \in [d_l - \varepsilon, d_l], \\ f(x + \varepsilon) & \text{for } x \in [b_l, d_l - \varepsilon], \\ f(x) + f(b_l + \varepsilon) - f(b_l) & \text{for } x \in \Omega, x \le b_l. \end{cases}$$

If $d_r \in D_+(\xi)$, set

$$f^{\#,\varepsilon}(x) = f(x) + (x - d_r)^4 \quad \text{for } x \in \Omega, x \ge d_r.$$

If $d_r \notin D_+(\xi)$, that is $d_r = c_r$ and $d_r \in D_-(\xi)$, set

$$f^{\#,\varepsilon}(x) = \begin{cases} f(d_r) = f(\hat{x}) & \text{for } x \in [d_r, d_r + \varepsilon], \\ f(x - \varepsilon) & \text{for } x \in [d_r + \varepsilon, b_r], \\ f(x) + f(b_r - \varepsilon) - f(b_r) & \text{for } x \in \Omega, x \ge b_r. \end{cases}$$

We call the function $f^{\#,\varepsilon}$ an *upper canonical modification* of f at $\hat{x}$ with an *effective region* M and a *canonical neighborhood* Ω_2. By a similar way we are able to construct a *lower canonical modification* $f_{\#,\varepsilon}$ with an *effective region* M and a *canonical neighborhood* Ω_2: Let $-f_{\#,\varepsilon}$ be an upper canonical modification of $-f$ at $\hat{x}$.

The figures below illustrate how to construct the effective region M and the upper canonical modification $f^{\#} = f^{\#,\varepsilon}$ when f is P-faceted at $\hat{x}$ and $\chi_l = \chi_r = -1$. While Figure 3.1 indicates the cases $d_l \in D_-(\xi)$ and $d_r \in D_+(\xi)$, Figure 3.2 shows the cases $d_l \in D_-(\xi)$ and $d_r \notin D_+(\xi)$.

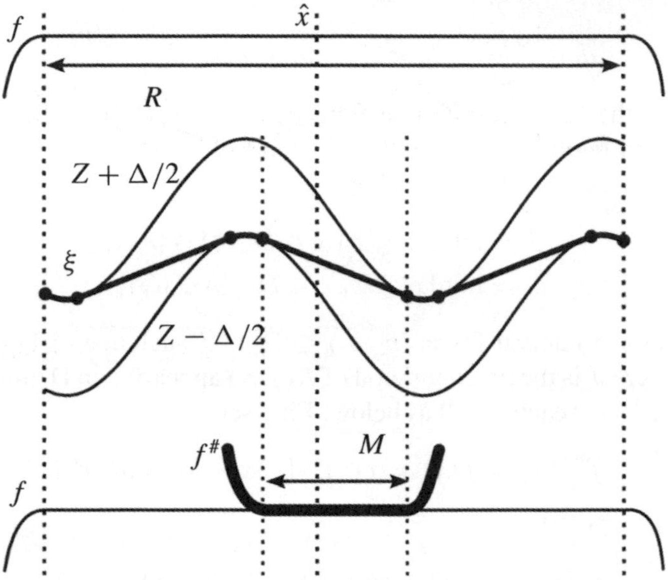

Figure 3.1. Construction of M and $f^{\#} = f^{\#,\varepsilon}$ (case $d_l \in D_-(\xi)$ and $d_r \in D_+(\xi)$)

The upper and lower canonical modification fulfills

Proposition 3.2. *Assume (W). Let $\Omega_1 = (a_l, a_r) \subset \Omega$ be an open interval. For $f \in C(\Omega) \cap C^2_P(\Omega_1)$ and $\hat{x} \in \Omega_1$ satisfying $f'(\hat{x}) = 0$, let f^ε be an upper canonical modification $f^{\#,\varepsilon}$ (resp. lower canonical modification $f^{\#,\varepsilon}$) with effective region $M = [d_l, d_r]$ and a canonical neighborhood $\Omega_2 = (b_l, b_r)$ and let $s = 1$ (resp. $s = -1$). Let $y, y_\varepsilon \in M$, $y_k \in \Omega$ and equi-Lipschitz functions σ, σ_k satisfy $y_\varepsilon \to y$, $y_k \to y$ and $\sigma_k \to \sigma$ uniformly. Then the conditions*

$$f^\varepsilon \in C(\Omega) \cap C^2_P(\Omega_2), \tag{3.3}$$

$$sf^\varepsilon > sf \quad on\ \Omega \setminus M, \tag{3.4}$$

$$\inf_{\Omega \setminus \Omega_2} s(f^\varepsilon - f) > 0, \tag{3.5}$$

$$f^\varepsilon(y) = f(y) = f(\hat{x}), \tag{3.6}$$

$$\lim_k (f^\varepsilon)'(y_k) = (f^\varepsilon)'(y), \tag{3.7}$$

$$(f^\varepsilon)'(y) = f'(y) = f'(\hat{x}) = 0, \tag{3.8}$$

$$\limsup_k s\Lambda_W^{\sigma_k}(f^\varepsilon)(y_k) \le s\Lambda_W^{\sigma}(f^\varepsilon)(y) \tag{3.9}$$

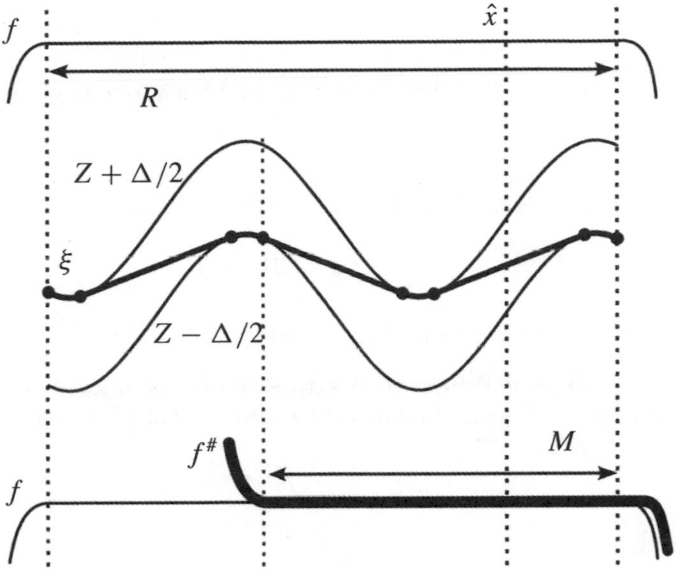

Figure 3.2. Construction of M and $f^{\#} = f^{\#,\varepsilon}$ (case $d_l \in D_-(\xi)$ and $d_r \notin D_+(\xi)$)

hold for all $\varepsilon > 0$ *small enough, and*

$$\Lambda_W^{\sigma}(f^{\varepsilon})(y_{\varepsilon}) \to \Lambda_W^{\sigma}(f)(y) \quad \text{as } \varepsilon \to 0, \tag{3.10}$$

$$s\Lambda_W^{\sigma}(f)(y) \le s\Lambda_W^{\sigma}(f)(\hat{x}) \tag{3.11}$$

hold.

Proof. We only consider the case $f^{\varepsilon} = f^{\#,\varepsilon}$ and $s = 1$. Since it is easy to verify the conditions (3.3)–(3.11) in the case (i) $f'(\hat{x}) = 0 \notin P$, we only consider the case (ii) f is P-faceted at $\hat{x}$. The conditions (3.3)–(3.8) are shown by the definition of the canonical modification.

Show (3.9). Take a subsequence k_j such that

$$\Lambda_W^{\sigma_{k_j}}(f^{\#,\varepsilon})(y_{k_j}) \to \limsup_k \Lambda_W^{\sigma_k}(f^{\#,\varepsilon})(y_k).$$

Since Proposition 2.7 implies

$$\Lambda_W^{\sigma_{k_j}}(f^{\#,\varepsilon})(y_{k_j}) \to \Lambda_W^{\sigma}(f^{\#,\varepsilon})(y)$$

provided that $y_{k_j} \in R^{\varepsilon} = [c_l^{\varepsilon}, c_r^{\varepsilon}] := R(f^{\#,\varepsilon}, \hat{x})$ for each j, we may assume that $y_{k_j} \notin R^{\varepsilon}$. Also it is enough to consider the case $y_{k_j} < c_l^{\varepsilon}$.

Hence,

$$\Lambda_W^{\sigma_{k_j}}(f^{\#,\varepsilon})(y_{k_j}) = W''((f^{\#,\varepsilon})'(y_{k_j}))(f^{\#,\varepsilon})''(y_{k_j}) + \sigma_{k_j}(y_{k_j}) \\ \to \sigma(y).$$

Since $y_{k_j} \to y \in M \subset R^\varepsilon$, we observe that

$$y = c_l^\varepsilon = d_l \in D_-(\xi^\varepsilon),$$

where ξ^ε is the minimizer of the obstacle problem $\xi_{\chi_l'\chi_r'}^{Z,\Delta,R^\varepsilon}$ with a primitive Z of σ, $\Delta = |\partial W(0)|$, $R^\varepsilon =$, $\chi_l' = \chi_l(f^{\#,\varepsilon}, \hat{x})$, $\chi_r' = \chi_r(f^{\#,\varepsilon}, \hat{x})$. Noting that $\xi^\varepsilon - Z + \Delta/2$ attains zero minimum at y, we have

$$\sigma(y) \le \Lambda_W^\sigma(f^{\#,\varepsilon})(y),$$

and hence

$$\limsup_k \Lambda_W^{\sigma_k}(f^\varepsilon)(y_k) = \lim_j \Lambda_W^{\sigma_{k_j}}(f^\varepsilon)(y_{k_j}) \le \Lambda_W^\sigma(f^{\#,\varepsilon})(y).$$

Show (3.10). Write $R = R(f, \hat{x})$, $\chi_l = \chi_l(f, \hat{x})$, $\chi_r = \chi_r(f, \hat{x})$ so that $\xi = \xi_{\chi_l\chi_r}^{Z,\Delta,R}$. Also note that χ_l' and χ_r' are independent of ε. It follows from Corollary 2.4 that

$$\Lambda_W^\sigma(f)(y) = \Lambda_{\chi_l\chi_r}^\sigma(y; R, \Delta) = \Lambda_{\chi_l'\chi_r'}^\sigma(y; M, \Delta).$$

Since $R^\varepsilon \to M$ as $\varepsilon \to 0$, we see by Proposition 2.8 that

$$\Lambda_W^\sigma(f^{\#,\varepsilon})(y_\varepsilon) = \Lambda_{\chi_l'\chi_r'}^\sigma(y_\varepsilon; R^\varepsilon, \Delta) \to \Lambda_{\chi_l'\chi_r'}^\sigma(y; M, \Delta) = \Lambda_W^\sigma(f)(y).$$

Since $y \notin D_-(\xi)$ for all $y \in [\hat{x}, d_r)$, we see that ξ is concave on $[\hat{x}, d_r]$ by the concave-convex condition of ξ. By a similar argument we see that ξ is convex on $[d_l, \hat{x}]$. Therefore we obtain (3.11) for all $y \in M$. □

4 Perron type existence theorem

In this section we show Perron type existence theorem. Let Ω be an open set in $\mathbf{R}$ and $\mathcal{Q} = (0, T) \times \Omega$.

Theorem 4.1 (Perron type existence). *Assume* (W), (S), (F1), (F2). *Let u^- and u^+ respectively be a subsolution and a supersolution of* (2.1) *satisfying*

$$u^- \le u^+ \text{ in } \mathcal{Q}, \quad (u^-)_* > -\infty, (u^+)^* < \infty \text{ on } [0, T) \times \Omega. \tag{4.1}$$

(1) *Then, there exists a solution u of* (2.1) *satisfying*

$$u^- \le u \le u^+ \quad \text{in } Q. \tag{4.2}$$

(2) *Moreover, if*

$$\sigma(t, x + \omega) = \sigma(t, x) \tag{4.3}$$

$$u^-(t, x + \omega) = u^-(t, x), \quad u^+(t, x + \omega) = u^+(t, x) \tag{4.4}$$

for all $(t, x) \in Q$ *with* $\omega > 0$ *and* $\Omega = \mathbf{R}$, *then there exists a solution* u *of* (2.1) *satisfying* (4.2) *and*

$$u(t, x + \omega) = u(t, x) \quad \text{for all } (t, x) \in Q. \tag{4.5}$$

We divide the main part of the proof into two lemmas.

Lemma 4.2. *Assume* (W), (S), (F1), (F2). *Let* $\mathcal{S}$ *be a nonempty family of subsolutions (resp. supersolutions) of* (2.1). *Define*

$$u(t, x) = \sup\{v(t, x) \mid v \in \mathcal{S}\} \ (\textit{resp. } v(t, x) = \inf\{v(t, x) \mid v \in \mathcal{S}\})$$

for $(t, x) \in Q$. *Assume that* $u^* < \infty$ *(resp.* $v_* > -\infty$*) in* $[0, T) \times \overline{\Omega}$. *Then* u *is a subsolution (resp. supersolution) of* (2.1).

Lemma 4.3. *Assume* (W), (S), (F1), (F2). *Let* $\mathcal{S}$ *be the set of all subsolutions* u *of* (2.1) *satisfying* $v \le u^+$ *in* Q *with a supersolution* u^+ *of* (2.1). *If* $u \in \mathcal{S}$ *is not a supersolution of* (2.1) *and satisfies* $u_* > -\infty$ *in* $[0, \infty) \times \Omega$, *then there exist a function* $v \in \mathcal{S}$ *and a point* $(s, y) \in Q$ *such that* $u(s, y) < v(s, y)$.

We first show the Perron type existence theorems under the assumption that Lemma 4.2 and 4.3 hold.

Proof of Theorem 4.1. we shall show the part (1). Let $\mathcal{S}$ be the set of all subsolutions v of (2.1) satisfying $v \le u^+$ in Q. Note that $\mathcal{S}$ is not empty since $u^- \in \mathcal{S}$. Define

$$u(t, x) = \sup\{v(t, x) \mid v \in \mathcal{S}\} \quad \text{for } (t, x) \in Q.$$

We then have $u^- \le u \le u^+$ in Q, which implies $u_* \ge (u^-)_* > -\infty$ and $u^* \le (u^+)^* < \infty$ on $[0, T) \times \Omega$. We hence see that u is a subsolution by Lemma 4.2. We next claim that u is a supersolution. If u were not a supersolution, Lemma 4.3 would imply that there exist $v \in \mathcal{S}$ and $(s, y) \in Q$ such that $u(s, y) < v(s, y)$, which contradicts the maximality of u. Therefore, we conclude that u is a solution.

It remains to show (4.5). Note that for $v \in \mathcal{S}$ the periodicity conditions (4.3) and (4.4) imply that $\tilde{v}(t, x) = v(t, x \pm \omega) \in \mathcal{S}$. Hence, we see that

$$u(t, x+\omega) = \sup\{v(t, x+\omega) \mid v \in \mathcal{S}\} = \sup\{v(t, x) \mid v \in \mathcal{S}\} = u(t, x).$$

The proof is now complete. □

We next show the lemmas. We note that being a subsolution is equivalent to being a subsolution in the local sense by Lemma 2.14.

Proof of Lemma 4.2. We only show that u is a subsolution (in the local sense). Fix a point $(\hat{t}, \hat{x}) \in \mathcal{Q}$ and a locally admissible test function $\varphi \in C(\mathcal{Q})$ at $(\hat{t}, \hat{x})$ such that (2.6) holds. Our goal is to show (2.5). Since φ is locally admissible, there exist $f \in C_P^2(\Omega_1)$ and $g \in C^1(I)$ with open intervals Ω_1 and J such that

$$\begin{aligned} &\varphi(t, x) = f(x) + g(t) \quad \text{on } \mathcal{Q}_1 := J \times \Omega_1, \\ &R(f, \hat{x}) \subset \Omega_1 \subset \Omega, \quad \hat{t} \in J \subset (0, T). \end{aligned} \tag{4.6}$$

We may assume that

$$(u^* - \varphi)(\hat{t}, \hat{x}) = 0, \quad \varphi_x(\hat{t}, \hat{x}) = 0$$

by Proposition 2.12 with $a = \varphi_x(\hat{t}, \hat{x}) = f'(\hat{x})$ and $b = u^*(\hat{t}, \hat{x}) - f'(\hat{x})\hat{x}$. Therefore, the desired inequality (2.5) becomes

$$g'(\hat{t}) + F(\hat{t}, 0, \Lambda_W^{\sigma(\hat{t})}(f)(\hat{x})) \leq 0, \tag{4.7}$$

which we should show.

We now let $\psi \in C(\mathcal{Q})$ be an $A_P(\mathcal{Q}_2)$ function such that

$$\psi = \varphi \text{ on } K, \quad \psi > \varphi \text{ on } \mathcal{Q} \setminus K, \quad \inf_{\mathcal{Q}\setminus\mathcal{Q}_2} (\psi - \varphi) > 0 \tag{4.8}$$

with a closed set K and an open set $\mathcal{Q}_2$ satisfying

$$(\hat{x}, \hat{t}) \in K \subset \mathcal{Q}_2 \subset \mathcal{Q}.$$

The function ψ is to be determined later. By the definition of the upper semicontinuous envelope there exists a sequence $\{(t_k, x_k)\}_{k\in\mathbf{N}} \subset \mathcal{Q}_2$ such that

$$(t_k, x_k, u(t_k, x_k)) \to (\hat{t}, \hat{x}, u^*(\hat{t}, \hat{x})) \quad \text{as } k \to \infty.$$

By the definition of u there exists $\{v_k\}_{k\in\mathbf{N}} \subset \mathcal{S}$ such that

$$v_k(t_k, x_k) > u(t_k, x_k) - 1/k$$

and so

$$v_k(t_k, x_k) \to u^*(\hat{t}, \hat{x}) \quad \text{as } k \to \infty.$$

Taking a maximizer (s_k, y_k) of $v_k^* - \psi$ on $\overline{Q_2}$, we observe that

$$((v_k)^* - \psi)(t_k, x_k) \leq ((v_k)^* - \psi)(s_k, y_k) \leq (u^* - \psi)(s_k, y_k)$$

for each k. Sending $k \to \infty$ yields

$$(u^* - \psi)(\hat{t}, \hat{x}) \leq (u^* - \psi)(\bar{s}, \bar{y}),$$

where

$$(\bar{s}, \bar{y}) = \lim_{k \to \infty} (s_k, y_k) \in \overline{Q_2}$$

by taking a subsequence if necessary. We see that $(\bar{s}, \bar{y}) \in K$ and $(s_k, y_k) \in Q_2$ for sufficiently large k. We also note that

$$\max_{Q}((v_k)^* - \psi) = ((v_k)^* - \psi)(s_k, y_k),$$

i.e. ψ is a test function of v_k at (s_k, y_k) by the last inequality of (4.8).

Let $f^{\#,\varepsilon}$ be an upper canonical modification of f at $\hat{x}$ with effective region M and canonical neighborhood $\Omega_2 \subset \Omega_1$ for $\varepsilon > 0$. We then see that

$$\psi(x, t) = f^{\#,\varepsilon}(x) + g(t) + (t - \hat{t})^2$$

is an admissible function on a set $Q_2 = J \times \Omega_2 \subset Q_1$ and that (4.8) holds with $K = \{\hat{t}\} \times M$ by Proposition 3.2. By the above argument we have $v_k^\varepsilon \in \mathcal{S}$ and $(s_k^\varepsilon, y_k^\varepsilon) \in Q_2$ such that

$$(s_k^\varepsilon, y_k^\varepsilon) \to (\hat{t}, y^\varepsilon) \in \{\hat{t}\} \times M \quad \text{as } k \to \infty$$

and ψ is a test function of v_k^ε at $(s_k^\varepsilon, y_k^\varepsilon)$. Since v_k^ε is a subsolution, we have

$$g'(s_k^\varepsilon) + 2(s_k^\varepsilon - \hat{t}) + F(s_k^\varepsilon, (f^{\#,\varepsilon})'(y_k^\varepsilon), \Lambda_W^{\sigma(s_k^\varepsilon)}(f^{\#,\varepsilon})(y_k^\varepsilon)) \leq 0. \quad (4.9)$$

Proposition 3.2 implies that

$$\lim_{k \to \infty} (f^{\#,\varepsilon})'(y_k^\varepsilon) = f'(\hat{x}),$$

$$\lim_{\varepsilon \to 0} \limsup_{k \to \infty} \Lambda_W^{\sigma(s_k^\varepsilon)}(f^{\#,\varepsilon})(y_k^\varepsilon) \leq \Lambda_W^{\sigma(\hat{t})}(f)(\hat{x}).$$

Therefore, it follows from (4.9) that (4.7) holds by (F1) and (F2).

We conclude that u is a subsolution. □

Proof of Lemma 4.3. Since u is not a supersolution, there exist $(\hat{x}, \hat{t}) \in \mathcal{Q}$ and a locally admissible test function $\varphi \in C(\mathcal{Q})$ at $(\hat{t}, \hat{x})$ such that (2.6) and

$$\varphi_t(\hat{t}, \hat{x}) + F(\hat{t}, \varphi_x(\hat{t}, \hat{x}), \Lambda_W^{\sigma(\hat{t})}(\varphi(\hat{t}, \cdot))(\hat{x})) < 0 \tag{4.10}$$

hold. Since φ is locally admissible, there exist $f \in C_P^2(\Omega_1)$ and $g \in C^1(J)$ with open intervals Ω_1 and J such that (4.6) holds with $\mathcal{Q}_1 := J \times \Omega_1$. We may assume that

$$(u_* - \varphi)(\hat{t}, \hat{x}) = 0, \quad \varphi_x(\hat{t}, \hat{x}) = 0$$

by Proposition 2.12 with $a = \varphi_x(\hat{t}, \hat{x}) = f'(\hat{x})$ and $b = u_*(\hat{x}, \hat{t}) - f'(\hat{x})\hat{x}$. Therefore, the inequality (4.10) becomes

$$g'(\hat{t}) + F(\hat{t}, 0, \Lambda_W^{\sigma(\hat{t})}(f)(\hat{x})) < 0. \tag{4.11}$$

Take a lower canonical modification $f_{\#,\varepsilon}$ of f at $\hat{x}$ for $\varepsilon > 0$ with effective region M and canonical neighborhood $\Omega_2 \subset \Omega_1$. Set

$$\psi(x, t) = f_{\#,\varepsilon}(x) + g(t) - (t - \hat{t})^2.$$

We now claim that

$$\psi_t(t, x) + F(t, \psi_x(t, x), \Lambda_W^{\sigma(t)}(\varphi(t, \cdot))(x)) < 0, \tag{4.12}$$

i.e.

$$g'(t) - 2(t - \hat{t}) + F(t, (f_{\#,\varepsilon})'(x), \Lambda_W^{\sigma(t)}(f_{\#,\varepsilon})(x)) < 0 \tag{4.13}$$

for all (t, x) in some neighborhood of $K := \{\hat{t}\} \times M$ choosing ε small enough. Indeed, since Proposition 3.2 implies that

$$\lambda^\varepsilon(t, x) := \Lambda_W^{\sigma(t)}(f_{\#,\varepsilon})(x)$$

is lower semicontinuous at each point of the compact set K, we see that for every $m > 0$ there exists an open set $\mathcal{Q}_3 \supset K$ on which the inequality

$$\lambda^\varepsilon(t, x) > \min_K \lambda^\varepsilon - m$$

holds. Choose $y_\varepsilon \in M$ such that $(\hat{t}, y_\varepsilon)$ is a minimum point of λ^ε on $K = \{\hat{t}\} \times M$. Proposition 3.2 implies that

$$\Lambda_W^{\sigma(t)}(f_{\#,\varepsilon})(x) > \Lambda_W^{\sigma(\hat{t})}(f)(\hat{x}) - m$$

for all $(t, x) \in \mathcal{Q}_3$ with small ε and m. Since Proposition 3.2 also implies that

$$|(f_{\#,\varepsilon})'(x)| < m,$$

it follows from (4.9) that (4.13) and so (4.12) holds on $\mathcal{Q}_3$ by (F1) and (F2).

We next claim that $\psi < (u^+)_*$ in $\mathcal{Q}_3$. First note that $\psi \le \varphi \le u \le u^+$ and so $\psi \le (u^+)_*$. If $\psi(t, x) = (u^+)_*(t, x)$ at some point $(t, x) \in \mathcal{Q}_3$, then ψ would be a test function of the supersolution u^+ at (t, x). Hence,

$$\psi_t(t, x) + F(t, \psi_x(t, x), \Lambda_W^{\sigma(t)}(\psi(t, \cdot))(x)) \ge 0,$$

which contradicts to (4.12).

Take a bounded open set $\mathcal{Q}_4$ such that $K \subset \mathcal{Q}_4$ and $\overline{\mathcal{Q}_4} \subset \mathcal{Q}_3$. Letting $\sigma_1 = \inf_{\mathcal{Q}_4}((u^+)_* - \psi) > 0$, we have

$$\psi + \sigma_1 \le (u^+)_* \quad \text{in } \mathcal{Q}_4.$$

Since $f_{\#,\varepsilon} < f$ on $\Omega_2 \setminus M$ by Proposition 3.2, we also have

$$\psi + \sigma_2 \le u_* \quad \text{in } \mathcal{Q}_3 \setminus \mathcal{Q}_4$$

with $\sigma_2 = \inf_{\mathcal{Q}_3 \setminus \mathcal{Q}_4}(u_* - \psi) > 0$. Define a function v by

$$v(t, x) = \begin{cases} \max\{\psi(t, x) + \sigma, u(t, x)\} & \text{for } (t, x) \in \mathcal{Q}_3, \\ u(t, x) & \text{for } (t, x) \notin \mathcal{Q}_3. \end{cases}$$

with $\sigma = \min\{\sigma_1, \sigma_2\}$. We show that this function v is a desirable function in the statement of this lemma.

Note that $v \ge u$. In addition, since $(u_* - \psi)(\hat{t}, \hat{x}) = 0$, there exists $(s, y) \in \mathcal{Q}_4$ such that $(u - \psi)(s, y) < \sigma$, which implies

$$v(s, y) > u(s, y).$$

Since

$$\psi(t, x) + \sigma \le \begin{cases} (u^+)_*(t, x) & \text{if } (t, x) \in \mathcal{Q}_4, \\ u_*(t, x) & \text{if } (t, x) \in \mathcal{Q}_3 \setminus \mathcal{Q}_4, \end{cases}$$

and $u \le u^+$ in $\mathcal{Q}$, we have

$$v = u \quad \text{on } \mathcal{Q} \setminus \mathcal{Q}_4,$$

$$v \le u^+ \quad \text{in } \mathcal{Q}.$$

Noting that ψ is a subsolution of (2.1) in $\mathcal{Q}_3$, we see that v is a subsolution of (2.1) in $\mathcal{Q}_3$ by Lemma 2.12 and 4.2. Therefore, if we admit the next lemma, we have $w \in \mathcal{S}$ and so the proof is finished. □

Lemma 4.4. *Assume* (W), (S), (F1), (F2). *Let u be a subsolution of* (2.1) *in $\mathcal{Q}$. Let w be a function defined on $\mathcal{Q}$ such that $w \geq u$ in $\mathcal{Q}$, $w = u$ in $\mathcal{Q} \setminus \overline{N_2}$, and w is a subsolution of* (2.1) *in N_1 with open rectangle sets $N_1 = J_1 \times I_1$, $N_2 = J_2 \times I_2$ satisfying $\overline{N_2} \subset N_1$, $\overline{N_1} \subset \mathcal{Q}$. Then w is a subsolution of* (2.1) *in $\mathcal{Q}$.*

In the classical setting, say $P = \emptyset$, this is easy to prove; if a function is a solution in two domains, then it is a solution in their union. However, this assertion on locality of solutions is not true for our equation (2.1) in general.

Proof. Fix a point $(\hat{t}, \hat{x}) \in \mathcal{Q}$ and a locally admissible test function $\varphi \in C(\mathcal{Q})$ of w at $(\hat{t}, \hat{x})$, *i.e.* $\max_{\mathcal{Q}}(w^* - \varphi) = (w^* - \varphi)(\hat{t}, \hat{x})$. Since φ is locally admissible, there exist $f \in C_P^2(\Omega_1)$ and $g \in C^1(J)$ with open intervals Ω_1 and J such that (4.6) holds. We may assume that

$$(w^* - \varphi)(\hat{t}, \hat{x}) = 0, \quad \varphi_x(\hat{t}, \hat{x}) = 0$$

by Proposition 2.12 with $a = \varphi_x(\hat{t}, \hat{x}) = f'(\hat{x})$ and $b = w^*(\hat{x}, \hat{t}) - f'(\hat{x})\hat{x}$. We should show (4.7).

It is enough to consider the case

$$(\hat{t}, \hat{x}) \in N_2, \quad \varphi(\hat{t}, \hat{x}) > u(\hat{t}, \hat{x});$$

otherwise, φ is a test of the subsolution u and so we have (4.7). We may also assume that f is P-faceted at $\hat{x}$ and $R(f, \hat{x})$ is not contained by I_1; otherwise, (4.7) holds since w is a subsolution in N_1.

Let $f^\# = f^{\#,\varepsilon}$ be a upper canonical modification of f at $\hat{x}$ with effective region M and canonical neighborhood $\Omega_2 \subset \Omega_1$. Set

$$\psi(x, t) = f^\#(x) + g(t) + (t - \hat{t})^2.$$

We then observe that

$$\psi > \varphi \geq w^* \geq u^* \quad \text{in } \mathcal{Q} \setminus \{\hat{t}\} \times M.$$

Let us assume for the moment that $\psi(\hat{t}, x_0) = u^*(\hat{t}, x_0)$ at some $x_0 \in M$. Then, since ψ is a test function of the subsolution u at $(\hat{t}, x_0)$, we have

$$g'(\hat{t}) + F(\hat{t}, (f^\#)'(x_0), \Lambda_W^{\sigma(\hat{t})}(f^\#)(x_0)) \leq 0.$$

Proposition 3.2 yields (4.7) by (F2). Therefore, we have

$$\psi > u^* \quad \text{in } \mathcal{Q} \tag{4.14}$$

We now take a faceted function whose faceted region is contained in I_1; set

$$\tilde{f}^{\#}(x) = \begin{cases} f^{\#}(x) + k|x - c_l|^3(|x - c_l| - 1) & \text{for } x \in \Omega, x \le c_l \\ f^{\#}(x) & \text{for } x \in [c_l, c_r] \\ f^{\#}(x) + k|x - c_r|^3(|x - c_r| - 1) & \text{for } x \in \Omega, x \ge c_r, \end{cases}$$

where $I_2 \subset [c_l, c_r] \subset I_1$ and $k > 0$. Note that

$$\tilde{\psi}(t, x) := \tilde{f}^{\#}(x) + g(t) + (t - \hat{t})^2$$

is locally admissible in N_1. Taking k small enough, we have

$$\tilde{\psi} > u^* \quad \text{for } (t, x) \in \mathcal{Q}$$

by (4.14). Noting that

$$w = u \text{ in } \mathcal{Q} \setminus \overline{N_2}, \quad \tilde{\psi} = \psi \text{ in } I_1 \times [c_l, c_r] \supset \overline{N_2},$$

we see that $\max_{\mathcal{Q}}(w^* - \tilde{\psi}) = (w^* - \tilde{\psi})(\hat{t}, \hat{x})$. Since $\tilde{\psi}$ is a test function,

$$g'(\hat{t}) + F(\hat{t}, (\tilde{f}^{\#})'(\hat{x}), \Lambda_W^{\sigma(\hat{t})}(\tilde{f}^{\#})(\hat{x})) \le 0.$$

Note that Proposition 2.6 yields

$$\Lambda_W^{\sigma(\hat{t})}(\tilde{f}^{\#})(\hat{x}) \le \Lambda_W^{\sigma(\hat{t})}(f^{\#})(\hat{x}).$$

Therefore, we have

$$g'(\hat{t}) + F(\hat{t}, (f^{\#})'(\hat{x}), \Lambda_W^{\sigma(\hat{t})}(f^{\#})(\hat{x})) \le 0,$$

which gives (4.7). □

5 Existence theorem for periodic initial data

In this section we prove an existence theorem for the equation (2.1) with periodic boundary condition and initial condition. In order to utilize the Perron type existence theorem (Theorem 4.1) we construct a subsolution u^- and a supersolution u^+ with given initial data; for a general strategy; see [16].

Lemma 5.1 (Existence of sub- and supersolutions). *Assume* (W), (S), (F1), (F2) *with* $\Omega = \mathbf{R}$. *Also assume that* u_0 *is a bounded and uniformly continuous function on* $\mathbf{R}$ *and* σ *is bounded. Then, there exist an upper semicontinuous function* u^+ *and an lower semicontinuous function* u^- *on*

$\overline{\mathcal{Q}}$ such that u^+ and u^- respectively are a supersolution and a subsolution of (2.1) *in $\mathcal{Q}$ and*

$$u^-(0,x) = u_0(x) = u^+(0,x), \quad u^-(t,x) \le u_0(x) \le u^+(t,x)$$

holds for all $(t,x) \in \mathcal{Q}$. Moreover, if

$$u_0(x+\omega) = u_0(x), \tag{5.1}$$

then $u^\pm$ can be taken so that it is spatially periodic with period ω, i.e. (4.4) *holds.*

We show this existence theorem as in [12, Section 9].

Lemma 5.2 ([12, Lemma 9.5]). *For each $\delta \in (0, 1/2)$ and $M > 0$ there exists $V = V_{\delta,M} \in C^2_P(\mathbf{R})$ such that*

$$V \ge 0, \quad V'' \ge 0 \text{ in } \mathbf{R}, \quad V(0) = 0, \quad V(x) \ge M \text{ for } |x| > \delta, \tag{5.2}$$

$$V'(x) = \begin{cases} q & \text{for } x \le -1, \\ q' & \text{for } x \ge 1 \end{cases} \tag{5.3}$$

with some $q, q' \notin P$.

We need to show

Lemma 5.3. *Let $V \in C^2_P(\mathbf{R})$ be such that $V'' \ge 0$ and* (5.3) *holds with some $q, q' \notin P$. Then for $B \in \mathbf{R}$ large enough*

$$V^+(t,x) = Bt + V(x) \tag{5.4}$$

is a supersolution of (2.1) *in* $(0,T) \times \mathbf{R}$.

Proof. We first claim that

$$C := \sup \left\{ |\Lambda_W^{\sigma(t)}(V)(x)| \mid (t,x) \in \mathcal{Q} \right\} < \infty. \tag{5.5}$$

Note that $V'(x) \in [q, q']$ for $x \in \mathbf{R}$ and

$$\sup_{\mathbf{R}} |V''| = \sup_{[-1,1]} |V''| < \infty.$$

Moreover, we have

$$\sup_{p \in [q,q'] \setminus P} |W''(p)| < \infty, \quad \sup_{\mathcal{Q}} |\sigma| < \infty.$$

Therefore, for each $(t, x) \in \mathcal{Q}$ with $V'(x) \notin P$ we observe that

$$\begin{aligned} |\Lambda_W^{\sigma(t)}(V)(x)| &\le |W''(V'(x))||V''(x)| + |\sigma(t,x)| \\ &\le \sup_{p\in[q,q']\setminus P} |W''(p)| \sup_{\mathbf{R}} |V''| + \sup_{\mathcal{Q}} |\sigma| < \infty. \end{aligned} \tag{5.6}$$

We shall show that

$$c_p := \sup \left\{ |\Lambda_W^{\sigma(t)}(V)(x)| \mid (t,x) \in \mathcal{Q},\, V'(x) = p \right\} < \infty$$

for each $p \in P$. Indeed, since a faceted region $R = \{x \in \mathbf{R} \mid V'(x) = p\}$ is a bounded closed interval, Proposition 2.7 implies that $(t, x) \mapsto \Lambda_W^{\sigma(t)}(V)(x)$ is continuous on $[0, T] \times R$, and so $c_p < \infty$. We note that the number of faceted regions of V is finite, *i.e.* $P \cap [q, q']$ is finite by (W). Hence we have

$$\sup \left\{ |\Lambda_W^{\sigma(t)}(V)(x)| \mid (t,x) \in \mathcal{Q},\, V'(x) \in P \right\} = \sup_{p\in P\cap[q,q']} c_p < \infty. \tag{5.7}$$

Combining (5.6) and (5.7), we obtain (5.5). Moreover, we see that

$$F(t, V'(x), \Lambda_W^{\sigma(t)}(V)(x)) \ge \inf_{[0,T]\times[q,q']\times[-C,C]} F =: -B_0 > -\infty.$$

Therefore, V^+ in (5.4) is a supersolution of (2.1) for $B \ge B_0$. □

Proof of Lemma 5.1. Let δ be a modulus of continuity of u_0; δ is a continuous nondecreasing function on $[0, \infty)$ with $\delta(0) = 0$ such that

$$|u_0(x) - u_0(y)| \le \delta(|x - y|) \quad \text{for } x, y \in \mathbf{R}.$$

By Lemma 5.2 and 5.3 take $V_\delta = V_{\delta,M} \in C_P^2(\mathbf{R})$ and $B_\delta \ge 0$ for small δ and $M = \max u_0 - \min u_0$ satisfying (5.2) and that $V_\delta^+(t, x) = B_\delta t + V_\delta(x)$ is a supersolution of (2.1). Define

$$u_{\varepsilon,\xi}^+(t, x) := V_{\delta(\varepsilon)}^+(t, x - \xi) + u_0(\xi) + \varepsilon$$

for small $\varepsilon > 0$ and $\xi \in \mathbf{R}$. Note that $u_{\varepsilon,\xi}^+$ is a supersolution of (2.1) and

$$u_{\varepsilon,\xi}^+(t, x) \ge V_{\delta(\varepsilon)}(x - \xi) + u_0(\xi) + \varepsilon.$$

On the case $|\xi - x| \le \delta(\varepsilon)$ we observe that

$$u_{\varepsilon,\xi}^+(t, x) \ge u_0(\xi) + \varepsilon \ge u_0(x);$$

on the other case

$$u^+_{\varepsilon,\xi}(t,x) \geq M + u_0(\xi) \geq u_0(x).$$

Therefore, Lemma 4.2 implies that

$$u^+(t,x) := \inf_{\varepsilon>0,\xi\in\mathbf{R}} u^+_{\varepsilon,\xi}(t,x)$$

is an upper semicontinuous supersolution of (2.1) satisfying $u^+ \geq u_0$. Moreover, since

$$u^+_{\varepsilon,x}(0,x) = u_0(x) + \varepsilon \to u_0(x) \quad \text{as } \varepsilon \to 0,$$

we have $u^+(0,x) = u_0(x)$ for all $x \in \mathbf{R}$. Under the assumption that u_0 is periodic we see that

$$\begin{aligned} u^+(t,x+\omega) &= \inf_{\varepsilon>0,\xi\in\mathbf{R}} (V^+_{\delta(\varepsilon)}(t,x+\omega-\xi) + u_0(\xi) + \varepsilon) \\ &= \inf_{\varepsilon>0,\xi\in\mathbf{R}} (V^+_{\delta(\varepsilon)}(t,x-\xi) + u_0(\xi+\omega) + \varepsilon) = u^+(t,x). \end{aligned}$$

The same proof is valid for existence of a subsolution u^-. □

Combining Theorem 4.1 and Lemma5.1 we have

Theorem 5.4 (Existence theorem for periodic initial data). *Assume* (W), (S), (F1), (F2) *and* (4.3) *with* $\Omega = \mathbf{R}$ *and* $\omega > 0$. *Let* u_0 *be a continuous function satisfying* (5.1). *Then there exists a solution* u *of* (2.1) *satisfying* (4.5) *and*

$$u(0,x) = u(x) \quad \text{for all } x \in \mathbf{R}.$$

References

[1] S. Angenent and M. E. Gurtin, *Multiphase thermomechanics with interfacial structure. II. Evolution of an isothermal interface*, Arch. Rational Mech. Anal. **108** (1989), 323–391.

[2] J. W. Barrett, H. Garcke and R. Nürnberg, *Numerical computations of faceted pattern formation in snow crystal growth*, Phys. Rev. E **86** (2012), 011604.

[3] G. Bellettini, V. Caselles, A. Chambolle and M. Novaga, *Crystalline mean curvature flow of convex sets*, Arch. Ration. Mech. Anal. **179** (2006), 109–152.

[4] A. CHAMBOLLE and M. NOVAGA, *Existence and uniqueness for planar anisotropic and crystalline curvature flow*, Proc. of 'Variational Methods for Evolving Objects', (eds. Y. Giga, Y. Tonegawa and P. Rybka), Adv. Stud. Pure Math., to appear.

[5] A. CHAMBOLLE, *An algorithm for mean curvature motion*, Interfaces Free Bound. **6** (2004), 195–218.

[6] Y. G. CHEN, Y. GIGA and S. GOTO, *Uniqueness and existence of viscosity solutions of generalized mean curvature flow equations*, J. Differential Geom. **33** (1991), 749–786.

[7] M. G. CRANDALL, H. ISHII and P.-L. LIONS, *User's guide to viscosity solutions of second order partial differential equations*, Bull. Amer. Math. Soc. **27** (1992), 1–67.

[8] C. M. ELLIOTT, A. R. GARDINER and R. SCHÄTZLE, *Crystalline curvature flow of a graph in a variational setting*, Adv. Math. Sci. Appl. **8** (1998), 425–460.

[9] L. C. EVANS and J. SPRUCK, *Motion of level sets by mean curvature. I*, J. Differential Geom. **33** (1991), 635–681.

[10] T. FUKUI and Y. GIGA, *Motion of a graph by nonsmooth weighted curvature*, World Congress of Nonlinear Analysts '92, de Gruyter, Berlin, 1996, 47–56.

[11] M.-H. GIGA and Y. GIGA, *A subdifferential interpretation of crystalline motion under nonuniform driving force*, Discrete Contin. Dynam. Systems (1998), no. Added Volume I, 276–287.

[12] M.-H. GIGA and Y. GIGA, *Evolving graphs by singular weighted curvature*, Arch. Rational Mech. Anal. **141** (1998), 117–198.

[13] M.-H. GIGA and Y. GIGA, *Stability for evolving graphs by nonlocal weighted curvature*, Comm. Partial Differential Equations **24** (1999), 109–184.

[14] M.-H. GIGA and Y. GIGA, *Generalized motion by nonlocal curvature in the plane*, Arch. Ration. Mech. Anal. **159** (2001), 295–333.

[15] M.-H. GIGA, Y. GIGA and P. RYBKA, *A comparison principle for singular diffusion equations with spatially inhomogeneous driving force for graphs*, Hokkaido University Preprint Series in Mathematics #981 (2011).

[16] Y. GIGA, "Surface Evolution Equations: a Level set Approach", Birkhäuser Verlag, Basel, 2006.

[17] Y. GIGA and P. RYBKA, *Facet bending in the driven crystalline curvature flow in the plane*, J. Geom. Anal. **18** (2008), 109–147.

[18] H. ISHII, *Perron's method for Hamilton-Jacobi equations*, Duke Math. J. **55** (1987), 369–384.

[19] J. E. TAYLOR, *Constructions and conjectures in crystalline nondifferential geometry*, In: "Differential Geometry", Pitman Monogr.

Surveys Pure Appl. Math., Vol. 52, Longman Sci. Tech., Harlow, 1991, 321–336.

[20] J. E. TAYLOR, *Motion of curves by crystalline curvature, including triple junctions and boundary points*, In: "Differential Geometry: Partial Differential Equations on Manifolds" (Los Angeles, CA, 1990), Proc. Sympos. Pure Math., Vol. 54, Amer. Math. Soc., Providence, RI, 1993, 417–438.

On representation of boundary integrals involving the mean curvature for mean-convex domains

Yoshikazu Giga and Giovanni Pisante

Abstract. Given a mean-convex domain $\Omega \subset \mathbb{R}^n$ with boundary of class $C^{2,1}$, we provide a representation formula for a boundary integral of the type

$$\int_{\partial\Omega} f(k(x))\, d\mathcal{H}^{n-1}$$

where $k \geq 0$ is the mean curvature of $\partial\Omega$ and f is non-increasing and sufficiently regular, in terms of volume integrals and defect measure on the ridge set.

1 Introduction

In this note we are interested in giving an explicit representation for a particular class of curvature depending integral functionals defined on compact manifolds without boundary. We restrict our analysis to $C^{2,1}$ regular manifolds with non-negative mean curvature that can be identified as boundaries of mean-convex domains. More precisely, for a mean-convex domain $\Omega \subset \mathbb{R}^n$, denoted by $k(x)$ the mean curvature of $\partial\Omega$, we are interested in recovering the value of boundary integrals of the type

$$\int_{\partial\Omega} f(k(x))\, d\mathcal{H}^{n-1} \tag{1.1}$$

in terms of the behavior of f inside Ω when f satisfies suitable regularity assumptions. It turns out that, if f is a differentiable non-increasing function (*cf.* Theorem 4.1 for the precise assumptions), (1.1) can be expressed

This work was initiated while the second author visited the Department of Mathematics of Hokkaido University during 2010. Its hospitality is gratefully acknowledged. This work was completed when the second author visited the School of Engineering Science of Osaka University during 2012/13 under Marie Curie project IRSES-2009-247486. The work of the first author has been partly supported by the Japan Society for the Promotion of Science (JSPS) through grants for scientific research Kiban (S) 21224001 and Kiban (A) 23244015. Kiban (S) 21224001, Kiban (A) 23244015 and Houga 25610025.

as the sum of volume integrals plus a defect measure δ^f concentrated on the ridge set of Ω as follows

$$\int_{\partial\Omega} f(k(x))\, d\mathcal{H}^{n-1} = \int_{\Omega} k(x) f(k(x))\, dx - \int_{\Omega} f'(k(x))|D^2 d(x)|^2\, dx + \delta^f(\Sigma) \tag{1.2}$$

where with d and Σ we have denoted the distance function from $\partial\Omega$ and its singular set respectively and we have set $k = -\Delta d$.

Our initial motivation for this study comes from the understanding of a useful consequence of the area formula and Fubini's theorem, which is sometimes referred to as Heintze-Karcher's inequality (see [10] and [13, Theorem 6.16]). In our setting it provides an upper bound of the measure of Ω in terms of the boundary integral of $1/k$. Indeed, it states that for any regular strictly mean-convex domain $\Omega \subset \mathbb{R}^n$ we have

$$\mathcal{L}^n(\Omega) \leq \frac{1}{n} \int_{\partial\Omega} \frac{1}{k(x)}\, d\mathcal{H}^{n-1}, \tag{1.3}$$

where $\mathcal{L}^n$ denotes the n-dimensional Lebesgue measure. The case of equality in (1.3) is particularly interesting, as is known that equality occurs if and only if Ω is a ball and this property, combined with Minkowski integral formula, gives an elegant proof of the classical Alexandrov's theorem that identifies the sphere as the unique compact connected surface with constant mean curvature (see [13, Theorem 6.17]). This point of view has been recently used in [9] to prove an anisotropic version of Alexandrov's theorem for compact embedded hypersurfaces with constant anisotropic mean curvature.

Our goal was to obtain a sharp estimate for the error term in (1.3), and this led us to the study of the boundary integral

$$\int_{\partial\Omega} f(k(x))\, d\mathcal{H}^{n-1}.$$

An application of (1.2) with $f(t) = 1/t$ allows us to write

$$\int_{\partial\Omega} \frac{1}{k(x)}\, d\mathcal{H}^{n-1} - n\mathcal{L}(\Omega) = \underbrace{\int_{\Omega} \frac{|D^2 d(x)|^2 - (n-1)k^2(x)}{k^2(x)}\, dx + \delta^f(\Sigma)}_{\geq 0}$$

filling the gap in (1.3) by explicitly expressing the error term as a sum of a volume integral and a defect measure in the spirit of [2].

The idea behind the proof of the main result can be easily explained by a formal application of the divergence theorem. Indeed, recalling that, denoted by $\nu(x)$ the unit inner normal to $\partial\Omega$ at a point x, we have $\nu(x) = \nabla d(x)$, the divergence theorem should give us

$$\int_{\partial\Omega} f(k) d\mathcal{H}^{n-1} = \int_{\partial\Omega} f(k)\nabla d \cdot \nu \, d\mathcal{H}^{n-1} \approx -\int_{\Omega} \operatorname{div}(f(k)\nabla d).$$

The meaning of the last integral in the previous formula has to be clarified, since the integrand term a priori is just defined as a distribution. This is a key step of the proof of (1.2) in Theorem 4.1, whose main ingredient is indeed the identification of $-\operatorname{div}(f(k)\nabla d)$ as a non-negative Radon measure which is absolutely continuous with respect to the $(n-1)$-dimensional Hausdorff measure. The formula then follows once we identify the densities of its Lebesgue decomposition.

The paper is structured as follows. In the next section we recall some preliminary results on the regularity of the distance function and on functionals of measures. In Section 3 we prove that $-\operatorname{div}(f(k)\nabla d)$ is indeed a non-negative Radon measure in a neighborhood of Ω. The proof of the main result is presented in Section 4. In the last section the anisotropic version of the representation formula (1.2) is briefly discussed.

2 Preliminaries

We start recalling some classical regularity results concerning the distance function from the boundary of a domain that have been proved in [11, 12]. Let Ω be a domain in $\mathbb{R}^n$. We define the singular set Σ as the complement of the open set G defined as the largest open subset of Ω such that for every $x \in G$ there is a unique closest point on $\partial\Omega$ from x. The following regularity result holds true.

Theorem 2.1 (Li-Nirenberg). *Suppose that $\partial\Omega$ is of class $C^{\ell,\alpha}$, with $\ell \geq 2$ and $0 < \alpha \leq 1$, then the distance function belongs to $C^{\ell,\alpha}(G \cup \partial\Omega)$.*

Let us comment on the properties of the singular set $\Sigma = \Omega \setminus G$ of the distance function to the boundary $\partial\Omega$ (sometimes called also *ridge* of Ω or *medial axes*). It is known that Σ is always a connected set and has finite $(n-1)$-dimensional Hausdorff measure, and moreover Σ is a $(n-1)$-rectifiable set. Another useful property is resumed here. Consider $y \in \partial\Omega$, the *cut point* of y, denoted by $m(y)$ is defined as the point when, moving along the inner normal from y, the set Σ is hit for the first time.

Theorem 2.2. *Let Ω be of class $C^{2,1}$. Then from every point $y \in \partial\Omega$, the length $s(y)$ of the segment joining y and its cut point $m(y)$ is Lipschitz continuous in y.*

As pointed out in [11, Remark 1.2] the regularity condition $C^{2,1}$ is sharp for the validity of the previous result.

Definition 2.3 (mean-convex domain). Let Ω be a bounded domain in $\mathbb{R}^n$ with boundary $\partial\Omega$ of class $C^{1,1}$. We let ν be the interior unit normal to $\partial\Omega$ and $\mathbf{H}_{\partial\Omega}$ the mean curvature vector of $\partial\Omega$. We say that Ω is mean-convex if $\mathbf{H}_{\partial\Omega}$ is pointing inside Ω at every point, *i.e.* $\mathbf{H}_{\partial\Omega} \cdot \nu \geq 0$.

In the sequel we will assume that Ω is a mean-convex domain. Since the boundary of Ω is supposed to be of class C^2 this is equivalent to require that the mean curvature $k(y)$ is non-negative for any $y \in \partial\Omega$.

Denote by $d(x)$ the distance function form $\partial\Omega$. We will write, with a slight abuse of notation, $k(x) = -\Delta d(x)$ and $\nu(x) = \nabla d(x)$. One of the steps needed for the proof of the identity (1.2) is the fact that the distribution $S = -\operatorname{div}(f(k)\nabla d)$ is a non-negative Radon measure if Ω is mean-convex. Let S be a real distribution, *i.e.* $S(\phi)$ is real for any real valued $\phi \in \mathcal{D}$, we say that S is non-negative if $S(\phi) \geq 0$ for any $\phi \in \mathcal{D}$ with $\phi \geq 0$. We will make use of the following well known result (see [15, Théorème V]).

Theorem 2.4. *A non-negative distribution S can be identified as a continuous linear form on $\mathcal{D}^k$ equipped with the topology induced by C^k $(k \geq 0)$. Moreover S is identified with a non-negative Radon measure μ_S through the equality*

$$S(\phi) = \int \phi \, d\mu_S.$$

Here we recall some notations on functionals of measures, we refer to [5] and [8] for further details. In order to understand the behavior of continuous functions at infinity (needed to deal with functions of Radon measures) we will use the following notion. Let $f : \mathbb{R}^m \to \mathbb{R} \cup \{\infty\}$ with $f(0) < +\infty$, we define its *recession function* as

$$f^\infty(p) := \lim_{t \nearrow \infty} \frac{f(tp)}{t}.$$

The recession function can be used to give a meaning to functions depending on pairs of Radon measures. Let $f : \mathbb{R}^m \to [0, \infty]$ be continuous and such that f^∞ is well defined. The recession function is positively homogeneous of degree 1, it is finite along the direction of (at most) linear growth of f and is infinite along the direction of superlinear growth. Moreover it is clear that if $\|f\|_\infty < \infty$ then $f^\infty = 0$ identically. Let μ and λ be respectively an $\mathbb{R}^m$-valued and a positive measure in $\Omega \subset \mathbb{R}^n$, we can define the measure

$$G(\mu, \lambda) := f\left(\frac{\mu}{\lambda}\right)\lambda + f^\infty\left(\frac{\mu^s}{|\mu^s|}\right)|\mu^s|,$$

i.e. for a measurable set E

$$G(\mu,\lambda)(E) := \int_E f\left(\frac{\mu}{\lambda}(x)\right) d\lambda(x) + \int_E f^\infty\left(\frac{\mu^s}{|\mu^s|}(x)\right) d|\mu^s|(x),$$

where μ^s is the singular part of μ with respect to λ and $\frac{\mu}{\lambda}$ denotes the Radon-Nikodym derivative of μ with respect to λ. When λ is the Lebesgue measure we will indicate the measure $G(\mu,\lambda)$ with $f(\mu)$. It is worth to observe that if f is a convex and lower semicontinuous function, then the recession function is always well defined and the functional G turns out to be lower semicontinuous with respect to the weak-$*$ convergence of measures (*cf.* [1, Theorem 2.34]).

Lemma 2.5. *Let f be a real-valued bounded continuous function in $(0,+\infty)$ and let μ be a positive Radon measure. Then $f(\mu)$ is bounded, i.e. $f(\mu)$ is absolutely continuous with respect to $\mathcal{L}^n$ and its density is an L^∞ function.*

Proof. Using the previous notation we have

$$f(\mu) := f\left(\frac{\mu}{\mathcal{L}^n}\right)\mathcal{L}^n + f^\infty\left(\frac{\mu^s}{|\mu^s|}\right)|\mu^s|.$$

Now we observe that f^∞ is identically zero by the boundedness assumption on f. This imply the desired result since we can write

$$f(\mu) := f\left(\frac{\mu}{\mathcal{L}^n}\right)\mathcal{L}^n$$

and $\|f\left(\frac{\mu}{\mathcal{L}^n}\right)\|_\infty \leq \|f\|_\infty$. □

3 The measure $-\operatorname{div}(f(k)\nabla d)$

In this section we prove that $-\operatorname{div}(f(k)\nabla d)$ is a non-negative Radon measure. The first step is contained in Lemma 3.1 where we prove that $k := -\Delta d$ is a positive Radon measure. We recently learned that this result has been showed, using a different argument, also in [14], where is indeed proved the equivalence of the non-negativity of $-\Delta d$ and the mean-convexity for a sufficiently regular domain.

We will use the *normal coordinates* to represent points in $\Omega \setminus \Sigma$. For $z \in \Omega\setminus\Sigma$ we can write $z = x + t\nu_x$ for a unique $x \in \partial\Omega$ and $0 < t < s(x)$, where $s(x) = \langle m(x) - x, \nu_x\rangle$ and ν_x is the unit inner normal to $\partial\Omega$ at x. We recall that by Theorem 2.2 the function $s(x)$ is Lipschitz continuous. Moreover from Theorem 2.1 we know that $-\Delta d(z)$ is well-defined in $G \cup \partial\Omega$. We can explicitly compute it in terms of the principal curvatures

$k_r(x)$ of $\partial\Omega$ in the point x, we have indeed (see for instance [7, Section 14.6, Lemma 14.17])

$$-\Delta d(z) = \sum_{r=1}^{n-1} \frac{k_r(x)}{1 - tk_r(x)} \,. \tag{3.1}$$

Lemma 3.1. *Let Ω be a mean-convex domain with boundary of class $C^{2,1}$. Then $-\Delta d$ is a non-negative Radon measure. If $k(x) \geq c > 0$ for any $x \in \partial\Omega$, then $-\Delta d$ is a positive Radon measure uniformly bounded from below, in particular $-\Delta d \geq c$ in distributional sense.*

Proof. Consider $S := -\Delta d$ which is well-defined as a distribution on Ω since $\|\nabla d\|_\infty \leq 1$. Let $\varphi \geq 0$ be a test function. We use (3.1) to infer, by explicit calculations, that if φ is supported in $\Omega \setminus \Sigma$ we have $S(\varphi) \geq c$. Let θ be a cut-off function such that $\theta = 1$ on $[0, 1]$, $\theta = 0$ on $[2, \infty)$ and θ is non-increasing C^1 function. Define for $\varepsilon > 0$

$$\psi_\varepsilon(x, t) = \theta\left(\frac{s(x) - t}{\varepsilon}\right) \quad \text{with} \quad \varepsilon < \inf_{x\in\partial\Omega} \frac{s(x)}{2}.$$

We note that, by definition,

$$\langle \nabla\psi_\varepsilon, \nabla d\rangle = \frac{\partial\psi_\varepsilon}{\partial t} \geq 0. \tag{3.2}$$

We use this cut-off to write $S(\varphi) = S(\psi_\varepsilon\varphi) + S((1 - \psi_\varepsilon)\varphi)$. As noted before, since $(1 - \psi_\varepsilon)\varphi$ is supported in $\Omega \setminus \Sigma$ we have $S((1 - \psi_\varepsilon)\varphi) \geq c$. For the other term we write, using (3.2)

$$S(\psi_\varepsilon\varphi) = \int_\Omega \varphi \,\langle\nabla\psi_\varepsilon, \nabla d\rangle + \int_\Omega \psi_\varepsilon \,\langle\nabla\varphi, \nabla d\rangle \geq \int_\Omega \psi_\varepsilon \,\langle\nabla\varphi, \nabla d\rangle.$$

Here and hereafter we often suppress dx unless confusion occurs. Since the last term tends to zero for ε going to zero we get $S(\varphi) \geq 0$. The claim follows from the Theorem 2.4. □

Remark 3.2. From the previous two lemmata we easily infer that if f is bounded and if Ω is mean-convex, then the distribution $f(k)\nabla d$ can be identified with an essentially bounded map in Ω.

Using the same idea of the previous lemma we are able now to prove the following.

Theorem 3.3. *Let Ω be a mean-convex domain with boundary of class $C^{2,1}$ and let f be a non-negative and non-increasing function of class C^1 in a neighborhood of the interval $[\min_{\partial\Omega} k(x), \infty)$, then the distribution*

$$S = -\operatorname{div}[f(k)\nabla d]$$

is a non-negative Radon measure in a sufficiently small neighborhood of $\overline{\Omega}$.

Proof. We start by observing that the regularity of $\partial\Omega$ allows us to extend in a $C^{2,1}$ way the distance function in a δ-neighborhood of $\overline{\Omega}$ defined by

$$\Omega_\delta := \{x \in \mathbb{R}^n \ : \ d(x, \Omega) < \delta\}.$$

By the regularity assumptions on f, we can choose δ sufficiently small in order to have $f(k(y)) \leq c < \infty$ for any $y \in \Omega_\delta \setminus \Omega$. We will continue to denote by $d(x)$ the extended distance function from $\partial\Omega$ defined in Ω_δ. Now we observe that if φ is a test function supported in $\Omega_\delta \setminus \Sigma$ with $\varphi \geq 0$, then $S(\varphi) \geq 0$. This follows by direct calculations. Indeed, since in $\Omega_\delta \setminus \Sigma$ the distance function is $C^{2,1}$, we can differentiate two times the equality $|\nabla d| = 1$ to get (using Einstein notation of summated indices)

$$\partial_{lji} d \, \partial_i d + \partial_{ji} \, \partial_{li} d = 0,$$

that implies, for $l = j$

$$\langle \nabla d, \nabla k \rangle = -(\nabla d \cdot \nabla) \operatorname{div} \nabla d = |D^2 d|^2 \tag{3.3}$$

and consequently

$$\operatorname{div}(f(k)\nabla d) = f'(k)\langle \nabla d, \nabla k \rangle - f(k)k \leq 0.$$

Hence the last inequality follows as in the previous proof, observing that by (3.1) the function k is uniformly bounded by a positive constant in any compact set far from Σ.

For a general φ we can use the same cut-off procedure used in the proof of the previous lemma. We can therefore write as before

$$S(\varphi) = S(\psi_\varepsilon \varphi) + S((1 - \psi_\varepsilon)\varphi).$$

We observed already that the second term is positive due to the aforementioned explicit calculation. It remains to bound the quantity

$$\begin{aligned} S(\psi_\varepsilon \varphi) &= \int_{\Omega_\delta} \varphi \, f(k) \, \langle \nabla \psi_\varepsilon, \nabla d \rangle + \int_{\Omega_\delta} \psi_\varepsilon \, \langle \nabla \varphi, f(k) \nabla d \rangle \\ &\geq \int_{\Omega_\delta} \psi_\varepsilon \, \langle \nabla \varphi, f(k) \nabla d \rangle, \end{aligned}$$

where the last inequality is a consequence of (3.2) and of the non-negativity of the function f. Finally we observe that Remark 3.2 and the definition of ψ_ε allow us to infer that

$$\lim_{\varepsilon \to 0} \int_{\Omega_\delta} \psi_\varepsilon \, \langle \nabla \varphi, f(k) \nabla d \rangle = 0$$

concluding the proof by Theorem 2.4. □

We observe now that the measure $\mu := -\operatorname{div}[f(k)\nabla d]$ is absolutely continuous with respect to $\mathcal{H}^{n-1}$. To this aim it is sufficient to bound the upper $(n-1)$-density of the measure μ (see [1, Theorem 2.56])

$$\Theta_{n-1}(\mu, x) = \limsup_{\rho\to 0} \frac{\mu(B_\rho(x))}{\rho^{n-1}}$$

for any $x \in \Omega$, where $B_\rho(x)$ denotes the closed ball of radius ρ centered at x. This can be achieved using a smoothing argument and integration by parts. Indeed we recall that by Remark 3.2 $f(k)\nabla d$ is in $L^\infty(\Omega)$. Then if we consider a family of standard mollifiers ρ_ε we have (up to subsequences) $|\rho_\varepsilon * f(k)\nabla d| \leq C$ for some $C < \infty$. Moreover, by standard properties of convolutions of measures (see [1, Theorem 2.2]) we also have that $-\operatorname{div}(f(k)\nabla d) * \rho_\varepsilon$ is a C^∞ function that locally weak-$*$ converges to $-\operatorname{div}(f(k)\nabla d)$ in the sense of measures. Let $x \in \Omega$, we use the lower semicontinuity of the total variation with respect to weak-$*$ convergence of measures to ensure that (for sufficiently small ρ) we have

$$\begin{aligned}
\mu(B_\rho(x)) &\leq \lim_{\varepsilon\to 0} \int_{B_\rho(x)} -\operatorname{div}(f(k)\nabla d) * \rho_\varepsilon \, dx \\
&= -\int_{\partial B_\rho(x)} \langle f(k)\nabla d * \rho_\varepsilon, \nu\rangle \, d\mathcal{H}^{n-1} \leq C\omega_n \rho^{n-1}
\end{aligned}$$

which proves the claim.

4 Main result

In this section we will precisely state and prove the main result of the paper. To this aim, we will need to know more precise properties of the ridge set. In particular, it will be useful to understand its *stratified* structure. The singular set of the distance function is characterized by the property that if $y \in \Sigma$ then there exist at least two points in $\partial\Omega$ in which the value $d(x) := d(x, \partial\Omega)$ of the distance function is attained. Let us consider the set $\Sigma_0 \subset \Sigma$ of the points where the distance function form $\partial\Omega$ is attained exactly in two points. Namely we set

$$\Sigma_0 := \left\{ x \in \Sigma \;\middle|\; \begin{array}{l} \text{there exists a unique pair } (y_1(x), y_2(x)) \in \partial\Omega \times \partial\Omega \\ \text{such that } y_1(x) \neq y_2(x),\ d(x) = |x - y_1| = |x - y_2| \end{array} \right\}.$$

For a fixed $x \in \Sigma_0$ let us denote by ν_1 and ν_2 the inner normal directions to $\partial\Omega$ at $y_1(x)$ and $y_2(x)$ respectively. We observe that, due to the expression (3.1), when moving along ν_i from y_i toward $x \in \Sigma_0$, we can identify two limit values for the extended curvature function $k = -\Delta d$, namely for $j \in \{1, 2\}$,

$$k^j(x) = \lim_{t \to s(y_j)} \sum_{r=1}^{n-1} \frac{k_r(y_j)}{1 - tk_r(y_j)}.$$

Our main result can be stated as follows.

Theorem 4.1. *Let Ω be a mean-convex domain with boundary of class $C^{2,1}$ and let f be a non-negative and non-increasing function of class C^1 in a neighborhood of the interval $[\min_{\partial\Omega} k(x), \infty)$. Then the following formula holds*

$$\begin{aligned}\int_{\partial\Omega} f(k(x))\, d\mathcal{H}^{n-1} = &\int_{\Omega} k(x) f(k(x))\, dx \\ &- \int_{\Omega} f'(k(x)) |D^2 d(x)|^2\, dx + \delta^f(\Sigma),\end{aligned} \tag{4.1}$$

where

$$\delta^f(\Sigma) = \int_{\Sigma_0} \frac{|\nu_1(x) - \nu_2(x)|}{\sqrt{2}} \left[f(k^1(x)) + f(k^2(x))\right] d\mathcal{H}^{n-1}(x).$$

Once we know that the distribution $S = -\operatorname{div}(f(k)\nabla d)$ is a measure on Ω_δ which is absolutely continuous with respect to the $(n-1)$-dimensional Hausdorff measure, a key step toward the proof of the main result is to identify the densities of its absolutely continuous part and of its singular part. The next lemma provides the desired identification.

Lemma 4.2. *Let f be as in Theorem 4.1, then for the measure $-\operatorname{div}(f(k)\nabla d)$ the following decomposition holds*

$$\begin{aligned}-\operatorname{div}(f(k)\nabla d) = &\left(f(k)k - f'(k)|D^2 d|^2\right) d\mathcal{L}^n \\ &+ \left(\frac{|\nu_1(x) - \nu_2(x)|}{\sqrt{2}} \left[f(k^1(x)) + f(k^2(x))\right]\right) \mathcal{H}^{n-1}|_{\Sigma_0}.\end{aligned} \tag{4.2}$$

Before proving the Lemma 4.2 we state two preliminary lemmata that will be useful throughout the proof. The first one is an orthogonality result of linear algebra.

Lemma 4.3. *Let e_i for $i \in \{0, 1, 2\}$ be three different unit vectors in $\mathbb{R}^n$. Then $e_i - e_0$ for $i = 1, 2$ are linearly independent.*

Proof. Assume that c_1 and c_2 in $\mathbb{R}$ satisfy $c_1(e_1 - e_0) + c_2(e_2 - e_0) = 0$. Thus $c_1e_1 + c_2e_2 = (c_1 + c_2)e_0$. If $c_1 + c_2$ is not zero, then e_0, e_1, e_2 lie on the same line which is impossible since e_i has length one. If $c_1 + c_2 = 0$ and e_1, e_2 are linearly independent, then we necessarily have $c_1 = c_2 = 0$. If $c_1 + c_2 = 0$ and $e_1 = -e_2$ we have $c_1e_1 + c_2e_2 = 2c_1e_1 = 0$ which implies $c_1 = 0$ and consequently $c_2 = 0$. □

The second one gives us some informations on the Jacobian of the Lipschitz map m at regular points which are inverse images of conjugate points.

Lemma 4.4. *Let $x \in \partial\Omega$ be such that m and s are differentiable at x. Assume that*

$$\lim_{t\to s(x)} \sum_{r=1}^{n-1} \frac{k_r(x)}{1 - tk_r(x)} = \infty ,$$

then the Jacobian of m at x is zero.

Proof. By assumption we know that there exists $r \in \{1, \dots n-1\}$ such that $s(x) = 1/k_r(x)$ and by definition we have

$$m(x) = x + s(x)\nu(x),$$

where $\nu(x)$ is the inner unit normal to $\partial\Omega$. Let $\tau_r(x)$ be the unit tangent vector at x such that it is the eigenvector of the Weigarten map corresponding to the eigenvalue $k_r(x)$. Differentiating along the direction τ_r we get

$$D_{\tau_r} m(x) = D_{\tau_r} s(x)\nu(x) + \big(1 - s(x)k_r(x)\big)\tau_r(x) = D_{\tau_r} s(x)\nu(s).$$

Assume that $D_{\tau_r} s$ is not zero at x. By the implicit function theorem we observe that Σ is tangent to ν at $m(x)$. One often assumes C^1 regularity to get a C^1 implicit function. However, if one is interested in differentiability of the implicit function at x, the (Fréchet) differentiability of m at x is enough. Thus the set Σ is tangent to the ray from x to $m(x)$. This contradicts the fact that $s(x)$ is Lipschitz. Thus we conclude that $D_{\tau_r} m(x) = 0$. This implies that the Jacobian of m at x is zero. □

Proof of Lemma 4.2. The proof will be divided in several steps.

STEP 1: First we justify the expression of $\delta^f(\Sigma)$ analyzing the local representation of $-\operatorname{div}(f(k)\nabla d)$ near Σ_0. Let $U \subset \Omega$ be an open set such that $U \cap \Sigma_0 \neq \emptyset$ and there exist two connected, mutually disjoint and relatively open sets $\mathcal{N}_1$ and $\mathcal{N}_2$ on $\partial\Omega$ with the property that

$$U \cap \Sigma_0 = \{x \in U \mid d(x, \mathcal{N}_1) - d(x, \mathcal{N}_2) = 0\}. \tag{4.3}$$

We use the notation $d_i(x) = d(x, \mathcal{N}_i)$ for $i = \{1, 2\}$. Suppose moreover that $k \le M < \infty$ in U. Under this condition, by the regularity of $\partial\Omega$ we can write $U = U_1 \cup U_2 \cup (\Sigma_0 \cap U)$, where $\Sigma_0 \cap U$ is a $C^{2,1}$ hypersurface and

$$U_1 := \{x \in U \mid d_1(x) < d_2(x)\} \ , \qquad U_2 := \{x \in U \mid d_1(x) > d_2(x)\}.$$

Let ϕ be a test function supported in U, we have

$$\begin{aligned}
&\langle -\operatorname{div}(f(k)\nabla d), \phi\rangle \\
&= \int_U f(k)\nabla d \cdot \nabla\phi \, dx \\
&= \int_{U_1} f(k)\nabla d \cdot \nabla\phi \, dx + \int_{U_2} f(k)\nabla d \cdot \nabla\phi \, dx \\
&= \int_{U_1} f(-\Delta d_1)\nabla d_1 \cdot \nabla\phi \, dx + \int_{U_2} f(-\Delta d_2)\nabla d_2 \cdot \nabla\phi \, dx \\
&= -\int_{U_1} f'(-\Delta d_1)|D^2 d_1|^2 \phi \, dx - \int_{U_2} f'(-\Delta d_2)|D^2 d_2|^2 \phi \, dx \\
&\quad - \int_{U_1} f(-\Delta d_1)\Delta d_1 \phi \, dx - \int_{U_2} f(-\Delta d_2)\Delta d_2 \phi \, dx \\
&\quad + \int_{\partial U_1} \phi f(-\Delta d_1)\nabla d_1 \cdot \nu_1 \, d\mathcal{H}^{n-1} + \int_{\partial U_2} \phi f(-\Delta d_2)\nabla d_2 \cdot \nu_2 \, d\mathcal{H}^{n-1} \\
&= -\int_U f'(k)|D^2 d|^2 \phi \, dx + \int_U f(k) k \, \phi \, dx \\
&\quad + \int_{\Sigma_0} \frac{|\nu_1(x) - \nu_2(x)|}{\sqrt{2}} [f(k^1(x)) + f(k^2(x))] \, \phi \, d\mathcal{H}^{n-1}(x).
\end{aligned}$$

We note explicitly that, in order to justify the previous calculations, we can extend up to the boundary, in a canonical way, all the quantities we are interested in, when considered separately on U_1 and U_2. In the last equality of the previous formula we have used (3.1), (3.3) and the fact that we can explicitly calculate the expression of the normal to the boundary ∂U_i in the portion that lives on Σ_0 (which is the only one that gives a contribution to the integral since ϕ is supported in U). More precisely in view of (4.3), we have

$$\nu_1 = \frac{\nabla d_1 - \nabla d_2}{|\nabla d_1 - \nabla d_2|} = -\nu_2 \, ,$$

from which it follows that

$$\nabla d_i \cdot \nu_i = \frac{1 - \nabla d_1 \cdot \nabla d_2}{\sqrt{2}\sqrt{1 - \nabla d_1 \cdot \nabla d_2}} = \frac{|\nabla d_1 - \nabla d_2|}{\sqrt{2}} \, .$$

Note that along a geodesic line the gradient of the distance function points always in the same direction.

STEP 2: We give now an estimate of the size of the subset of $\Sigma \setminus \Sigma_0$ where the curvature remains bounded. For a given $z \in \Sigma$ we define

$$\kappa(z) := \sup\left\{ \lim_{t\to s(y)} \sum_{r=1}^{n-1} \frac{k_r(y)}{1 - tk_r(y)} \mid y \in \partial\Omega \text{ with } d(z) = |z - y| \right\}.$$

We are interested in the points of Σ where κ is bounded. For $M > 0$ define

$$G_M := \left\{ z \in \Sigma \mid \begin{array}{l} \kappa(z) \leq M \text{ and there are } y_1(z), y_2(z), y_3(z) \in \partial\Omega \\ \text{such that } d(z) = |z - y_i(z)| \text{ for any } i \in \{1, 2, 3\} \end{array} \right\}.$$

We claim that $\mathcal{H}^{n-1}(G_M) = 0$. The proof of the claim is based on the observation that for a fixed $z \in G_M$ and for any collection of neighborhoods $\mathcal{N}_i$ of $y_i(z)$ in $\partial\Omega$ such that $y_i(z) \notin \mathcal{N}_j$ if $j \neq i$, we have that z is contained in the intersection of the two hypersurfaces given by

$$\Gamma_1 := \left\{ x \in \Omega \ : \ d(x, \mathcal{N}_1) = d(x, \mathcal{N}_3) \right\}$$

and

$$\Gamma_2 := \left\{ x \in \Omega \ : \ d(x, \mathcal{N}_2) = d(x, \mathcal{N}_3) \right\}.$$

It is easy to check, as in STEP 1 that, for $i \in \{1, 2\}$, the normal vector of Γ_i is parallel to $\nabla d(x, \mathcal{N}_i) - \nabla d(x, \mathcal{N}_3)$. Lemma 4.3 ensures us that Γ_1 and Γ_2 are transversal and thus $\mathcal{H}^{n-2}(\Gamma_1 \cap \Gamma_2) < \infty$. The claim will be proved once we know that we can reduce to consider just countably many hypersurfaces of the type Γ_i to cover the set G_M.

We start by defining for any $h \in \mathbb{N}$ the set

$$G_M^h := \left\{ z \in G_M \mid \begin{array}{l} \text{there exist } y_1, y_2, y_3 \in \partial\Omega \text{ with } d(z) = |z - y_i| \\ \text{and } |y_i - y_j| > \frac{1}{h} \text{ for any } i, j \in \{1, 2, 3\} \end{array} \right\}.$$

Choose $\rho_h > 0$ such that for any $y \in \mathcal{N}_h(y_i(z)) = B_{\rho_h}(y_i) \cap \partial\Omega$ we have, independently of $z \in G_M^h$,

$$\sum_{r=1}^{n-1} \frac{k_r(y)}{1 - s(y)k_r(y)} \leq M + 1.$$

Since the set

$$L := \left\{ y \in \partial\Omega \ : \ \sum_{r=1}^{n-1} \frac{k_r(y)}{1 - s(y)k_r(y)} \leq M + 1 \right\}$$

is compact, we may find a finite collection of points $\{y_i\}_{i\in I}$ with $y_i \notin \mathcal{N}_h(y_j)$ for $i \neq j$ and

$$L \subset \bigcup_{i\in I} \mathcal{N}_h(y_i).$$

For any pair of points y_i, y_j with $i, j \in I$ we can consider the hypersurfaces

$$\Gamma_{i,j} := \left\{x \in \Omega \ : \ d(x, \mathcal{N}_h(y_i)) = d(x, \mathcal{N}_h(y_j))\right\}$$

As before, we have that $\Gamma_{i,l}$ and $\Gamma_{j,l}$ are transversal. Then the set G^h_M is contained in the finite union of the intersections of the type $\Gamma_{j,l} \cap \Gamma_{i,l}$ for $i, j, l \in I$. The claim follows by noticing that $G_M \subseteq \bigcup_{h\in\mathbb{N}} G^h_M$.

STEP 3: We analyze now the measure of the set

$$K := \{z \in \Sigma \ : \ \kappa(z) = +\infty\}.$$

We claim that $\mathcal{H}^{n-1}(K) = 0$. Since K is a subset of the image through the Lipschitz map m of the set

$$A_\infty := \left\{x \in \partial\Omega \ : \ \lim_{t\to s(x)} \sum_{r=1}^{n\ 1} \frac{k_r(x)}{1 - tk_r(x)} = \infty\right\},$$

it is sufficient to prove that

$$\mathcal{H}^{n-1}\big(m(A_\infty)\big) = 0. \tag{4.4}$$

We first observe that at $\mathcal{H}^{n-1}$-a.e. x in A_∞ the mappings m and s are differentiable by the Rademacher's theorem. Then the equality (4.4) follows from Lemma 4.4 and from the area formula for Lipschitz maps (*cf.* [1, Theorem 2.71] or [6, Theorem 3.2.3]).

STEP 4: From STEP 3, since the measure $\mu := -\operatorname{div}(f(k)\nabla d)$ is absolutely continuous with respect to $\mathcal{H}^{n-1}$, we have that the set K is negligible with respect to μ. It is then enough to prove (4.2) on the set $\Omega_\delta \setminus K$. The claim will follow by continuity of measures along increasing sequences of sets once we have proved that (4.2) holds on the set

$$W_M := \{x \in \Omega_\delta \ : \ \kappa(x) < M\},$$

for any $M > 0$. To this aim we observe that, since $z \in \Sigma_0$ is a conjugate point only if $\kappa(z) = \infty$, by [12, Theorem 1], we can find an open covering of $\Sigma_0 \cap W_M$ given by $\{U_i\}_{i\in I}$, with U_i satisfying (4.3) as in STEP 1. We can then consider the open covering of $W_M \setminus G_M$ given by $U_0 \cup \bigcup_{i\in I} U_i$ where U_0 is such that $\partial\Omega \subset U_0$ and $U_0 \cap \Sigma = \emptyset$.

If ϕ is a test function supported on U_0, we can apply the integration by parts to obtain

$$\langle -\operatorname{div}(f(k)\nabla d), \phi\rangle = -\int_{U_0} f'(k(x))|D^2 d(x)|^2 \phi(x)\, dx + \int_{U_1} f(k(x))k(x)\,\phi(x)\, dx. \tag{4.5}$$

Finally, using a partition of unity subordinate to the cover $U_0 \cup \bigcup_{i\in I} U_i$, from (4.5) and the STEP 1, we have the claim. □

We are finally in position to prove the main theorem.

Proof of Theorem 4.1. Let us consider the characteristic function of Ω, χ_Ω and approximate it using a standard mollifier with $\chi_\Omega^\varepsilon = \rho_\varepsilon * \chi_\Omega \in C_0^\infty(\Omega_{2\varepsilon})$. We explicitly observe that $\nabla\chi_\Omega^\varepsilon = 0$ in $\Omega_{-2\varepsilon}$ since $\chi_\Omega^\varepsilon(x) = 1$ for any $x \in \Omega_{-2\varepsilon}$. We can then compute

$$\begin{aligned}\lim_{\varepsilon\to 0}\langle -\operatorname{div}(f(k)\nabla d), \chi_\Omega^\varepsilon\rangle &= \lim_{\varepsilon\to 0}\int_{(\partial\Omega)_\varepsilon} f(k)\nabla d\cdot\nabla\chi_\Omega^\varepsilon\, dx\\ &= \lim_{\varepsilon\to 0}\int_{(\partial\Omega)_\delta} f(k)\nabla d\cdot dD(\chi_\Omega^\varepsilon)\\ &= \int_{(\partial\Omega)_\delta} f(k)\nabla d\cdot dD(\chi_\Omega)\\ &= \int_{\partial\Omega} f(k)\, d\mathcal{H}^{n-1},\end{aligned} \tag{4.6}$$

where $(A)_\delta$ denotes the δ-neighborhood of A. To conclude the proof it is sufficient to observe that from (4.2) we also have

$$\begin{aligned}&\lim_{\varepsilon\to 0}\langle -\operatorname{div}(f(k)\nabla d), \chi_\Omega^\varepsilon\rangle\\ &= \lim_{\varepsilon\to 0}\int_{\Omega_\varepsilon}\left(f(k)k - f'(k)|D^2 d|^2\right) dx\\ &\quad + \int_{\Sigma_0}\frac{|\nu_1(x)-\nu_2(x)|}{\sqrt{2}}[f(k^1(x)) + f(k^2(x))]\, d\mathcal{H}^{n-1}(x)\\ &= \int_\Omega\left(f(k(x))k(x) - f'(k(x))|D^2 d(x)|^2\right) dx\\ &\quad + \int_{\Sigma_0}\frac{|\nu_1(x)-\nu_2(x)|}{\sqrt{2}}[f(k^1(x)) + f(k^2(x))]\, d\mathcal{H}^{n-1}(x).\end{aligned}$$

□

5 The anisotropic case

In this section we briefly describe how our approach can be also used in the anisotropic setting to provide a representation for boundary integrals involving anisotropic curvatures. We restrict ourself here to a formal discussion implicitly assuming all the regularity properties and the necessary assumptions needed to perform the calculations, without mentioning them.

We start recalling some useful notations (*cf.* for instance [4]). Let φ^0 be a one-homogeneous convex function that will be our anisotropy and let φ be its dual function. The anisotropic distance function of a point x from a set C is defined by

$$d_\varphi(x, C) = \inf_{y \in C} \varphi(y - x).$$

For a given domain Ω we consider the signed anisotropic distance function from $\partial\Omega$, defined as

$$d_\varphi(x) = d_\varphi(x, \Omega) - d_\varphi(x, \mathbb{R}^n \setminus \Omega).$$

We denote by n_φ the Cahn-Hoffman vector and k_φ the anisotropic mean curvature. More precisely, defined

$$T_{\varphi^0} := \frac{1}{2}\nabla\big((\varphi^0)^2\big),$$

we set

$$n^*_\varphi = \frac{n(x)}{\varphi^0(n(x))} \ , \quad n_\varphi = T_{\varphi^0}(n^*_\varphi).$$

We have, near $\partial\Omega$,

$$n^*_\varphi = \nabla d_\varphi \ , \quad \langle n^*_\varphi, n_\varphi\rangle = 1 \ , \quad k_\varphi = -\operatorname{div} n_\varphi.$$

As in the isotropic case, a formal application of the Gauss theorem leads to

$$\begin{aligned}\int_{\partial\Omega} f(k_\varphi)\varphi^0(n)\, d\mathcal{H}^{n-1} &= \int_{\partial\Omega} f(k_\varphi)\varphi^0(n)\langle n^*_\varphi, n_\varphi\rangle\, d\mathcal{H}^{n-1} \\ &= \int_{\partial\Omega} f(k_\varphi)\, n_\varphi \cdot n\, d\mathcal{H}^{n-1} \approx -\int_\Omega \operatorname{div}\big(f(k_\varphi)\, n_\varphi\big).\end{aligned}$$

The main issue, as in the previous case, is to give a representation of the measure $-\operatorname{div}\big(f(k_\varphi)\, n_\varphi\big)$. Let us observe that, outside of the singular set of d_φ we have

$$\operatorname{div}\big(f(k_\varphi)\, n_\varphi\big) = f'(k_\varphi)\nabla k_\varphi \cdot n_\varphi - f(k_\varphi)k_\varphi\,;$$

moreover, the term $\nabla k_\varphi \cdot n_\varphi$ can be expressed in terms of the squared norm of the φ-Weingarten operator. To this aim for instance we can consider a proper test function ϕ and compute

$$\begin{aligned}\int_\Omega (n_\varphi \cdot \nabla)\,\mathrm{div}\, n_\varphi \phi &= -\int_\Omega \partial_j (n_\varphi^i \phi) \partial_i (n_\varphi^j)\\ &= -\int_\Omega \partial_j (n_\varphi^i) \partial_i (n_\varphi^j) \phi - \int_\Omega n_\varphi^i \partial_i (n_\varphi^j) \partial_j \phi\\ &= \int_\Omega \|\nabla n_\varphi\|^2 \phi,\end{aligned}$$

where ∇n_φ denotes the Jacobian matrix of n_φ and $\|\nabla n_\varphi\|^2$ denotes the square of its Euclidean norm, *i.e.* the sum of the squares of the components of the matrix ∇n_φ. In the last equality we have used the symmetry of the Jacobian of the Cahn-Hoffman vector and the Euler formula for homogeneous functions. We explicitly note that $\|\nabla n_\varphi\|^2$ is the squared norm of the anisotropic Weingarten operator that correspond to $\|D^2 d\|^2$ in the isotropic setting (*cf.* Lemma 3.2 and Remark 3.3 in [3]).

Proceeding similarly as in the previous section, using d_φ instead of d, we may derive the following representation formula

$$\int_{\partial\Omega} f(k_\phi)\varphi^0(n) d\mathcal{H}^{n-1} = \int_\Omega f'(k_\varphi)\|\nabla n_\varphi\|^2 dx - \int_\Omega f(k_\varphi) k_\varphi \, dx + \delta_\varphi^f(\Sigma_\varphi),$$

where

$$\delta_\varphi^f(\Sigma_\varphi) = \int_{\Sigma_0^\varphi} \big(f(k_\varphi^1(x)) n_\varphi(y_1) - f(k_\varphi^2(x)) n_\varphi(y_2) \big) \frac{n_\varphi^*(y_1) - n_\varphi^*(y_2)}{|n_\varphi^*(y_1) - n_\varphi^*(y_2)|} d\mathcal{H}^{n-1}.$$

References

[1] L. AMBROSIO, N. FUSCO and D. PALLARA, "Functions of Bounded Variation and Free Discontinuity Problems", Oxford Science Publications, 2000.

[2] P. AVILES and Y. GIGA, *The distance function and defect energy*, Proc. Roy. Soc. Edinburgh Sect. A **126** (1996), 923–938.

[3] G. BELLETTINI and I. FRAGALÀ, *Elliptic approximations of prescribed mean curvature surfaces in Finsler geometry*, Asymptotic Anal. **22** (2000), 87–111.

[4] G. BELLETTINI and L. MUGNAI, *Anisotropic geometric functionals and gradient flows*, In: "Nonlocal and Abstract Parabolic Equations and Their Application", Vol. 86, 2009, 21–43.

[5] F. DEMENGEL and R. TEMAM, *Convex functions of a measure and applications*, Indiana University Mathematics Journal **33** (1984), 673–709.

[6] H. FEDERER, "Geometric Measure Theory", Repr. of the 1969 ed., Berlin: Springer-Verlag, 1996.

[7] D. GILBARG and N. S. TRUDINGER, "Elliptic Partial Differential Equations of Second Order", Reprint of the 1998 ed., Berlin: Springer, 2001.

[8] C. GOFFMAN and J. SERRIN, *Sublinear functions of measures and variational integrals*, Duke Math. J. **31** (1964), 159–178.

[9] Y J HE, H. LI, H. MA and J. GE, *Compact embedded hypersurfaces with constant higher order anisotropic mean curvatures*, Indiana Univ. Math. J. **58** (2009), 853–868.

[10] E. HEINTZE and H. KARCHER, *A general comparison theorem with applications to volume estimates for submanifolds*, Ann. Sci. Éc. Norm. Supér. **11** (1978), 451–470.

[11] Y. LI and L. NIRENBERG, *The distance function to the boundary, finsler geometry, and the singular set of viscosity solutions of some hamilton-jacobi equations*, Comm. Pure Appl. Math. **58** (2005), 85–146.

[12] Y. LI and L. NIRENBERG, *Regularity of the distance function to the boundary*, Rend. Accad. Naz. Sci. XL Mem. Mat. Appl. **29** (2005), 257–264.

[13] S. MONTIEL and A. ROS, "Curves and Surfaces", Transl. from the Spanish by Sebastián Montiel. Transl. edited by Donald Babbitt, 2nd ed., Providence, RI: American Mathematical Society (AMS). Madrid: Real Sociedad Matemática Espanola, 2009.

[14] G. PSARADAKIS, *L^1 Hardy Inequalities with Weights*, Journal of Geometric Analysis, online 02-2012, doi=10.1007/s12220-012-9302-8.

[15] L. SCHWARTZ, "Théorie des distributions" (Distribution theory), (Théorie des distributions), Nouveau tirage, Paris: Hermann, 1998.

Boundary regularity for the Poisson equation in reifenberg-flat domains

Antoine Lemenant and Yannick Sire

Abstract. This paper is devoted to the investigation of the boundary regularity for the Poisson equation

$$\begin{cases} -\Delta u = f & \text{in } \Omega \\ u = 0 & \text{on } \partial\Omega \end{cases}$$

where f belongs to some $L^p(\Omega)$ and Ω is a Reifenberg-flat domain of $\mathbb{R}^N$. More precisely, we prove that given an exponent $\alpha \in (0, 1)$, there exists an $\varepsilon > 0$ such that the solution u to the previous system is locally Hölder continuous provided that Ω is (ε, r_0)-Reifenberg-flat. The proof is based on Alt-Caffarelli-Friedman's monotonicity formula and Morrey-Campanato theorem.

The goal of the present paper is to prove a boundary regularity result for the Poisson equation with homogeneous Dirichlet boundary conditions in non smooth domains. We consider the case of Reifenberg-flat domains as given by the following definition

Definition 0.1. Let ε, r_0 be two real numbers satisfying $0 < \varepsilon < 1/2$ and $r_0 > 0$. An (ε, r_0)-Reifenberg-flat domain $\Omega \subset \mathbb{R}^N$ is an open, bounded, and connected set satisfying the two following conditions:

(i) for every $x \in \partial\Omega$ and for any $r \leq r_0$, there exists a hyperplane $P(x, r)$ containing x which satisfies

$$\frac{1}{r} d_H(\partial\Omega \cap B(x, r), P(x, r) \cap B(x, r)) \leq \varepsilon.$$

(ii) for every $x \in \partial\Omega$, one of the connected component of

$$B(x, r_0) \cap \{x; d(x, P(x, r_0)) \geq 2\varepsilon r_0\}$$

is contained in Ω and the other one is contained in Ω^c.

Let $N \geq 2$. We consider the following problem in the (ε, r_0)-Reifenberg-flat domain $\Omega \subset \mathbb{R}^N$ for some $f \in L^q(\Omega)$,

$$(P1) \quad \begin{cases} -\Delta u = f & \text{in } \Omega \\ u = 0 & \text{on } \partial\Omega \end{cases}$$

The exact meaning and solvability of problem $(P1)$ is as follows: for any $f \in L^q(\Omega)$ with $2 \leq q \leq +\infty$, the application $v \mapsto \int_\Omega vf dx$ is a bounded Linear form on $W_0^{1,2}(\Omega)$, endowed with the scalar product $\int_\Omega \nabla u \cdot \nabla v \, dx$. Therefore, using Riesz representation Theorem we deduce the existence of a unique weak solution $u \in W_0^{1,2}(\Omega)$ for the problem $(P1)$.

Reifenberg-flat domains are less smooth than Lipschitz domains and it is well known that we cannot expect more regularity than Hölder for boundary regularity of the Poisson equation in Lipschitz domains (see [9], [18, Remark 17], or [10]).

Historically, Reifenberg-flat domains came into consideration because of their relationship with the regularity of the Poisson kernel and the harmonic measure, as shown in a series of famous and deep papers by Kenig and Toro (see for *e.g.* [12–14, 20]). In particular, they are Non Tangentially Accessible (in short NTA) domains as described in [11]. Notice that the Poisson kernel is defined as related to the solution of the equation $-\Delta u = 0$ in Ω with $u = f$ on $\partial\Omega$. In this paper we consider the equation $(P1)$ which is of different nature. However, our regularity result is again based on the monotonicity formula of Alt, Caffarelli and Friedman [1], which is known to be one of the key estimate in the study harmonic measure as well.

More recently, regularity of elliptic PDEs in Reifenberg-flat domains has been studied by Byun and Wang in [2–6] (see also the references therein). One of their main result regarding to equation of the type of $(P1)$ is the existence of a global $W^{1,p}(\Omega)$ bound on the solution. This fact will be used in Corollary 0.6 below.

The case of domains of $\mathbb{R}^N$ has been investigated by Caffarelli and Peral [7]. See also [10] for the case of Lipschitz domains. Some other type of elliptic problems in Reifenberg-flat domains can be found in [15–19].

The present paper is the first step towards a general boundary regularity theory for elliptic PDEs in divergence form on Reifenberg-flat domains. Our main result is the following.

Theorem 0.2. *Let* $1 \leq p, q, p_0 < +\infty$ *be some exponents satisfying* $\frac{1}{p} + \frac{1}{q} = \frac{1}{p_0}$ *and* $p_0 > \frac{N}{2}$. *Let* $\alpha > 0$ *be any given exponent such that*

$$\alpha < \frac{p_0 - \frac{N}{2}}{p_0}. \tag{0.1}$$

Then one can find an $\varepsilon = \varepsilon(N, \alpha)$ *such that the following holds. Let* $\Omega \subseteq \mathbb{R}^N$ *be an* (ε, r_0)*-Reifenberg-flat domain for some* $r_0 > 0$, *and let*

$u \in W_0^{1,2}(\Omega) \cap C^1(\Omega)$ be a solution for the problem $(P1)$ in Ω with $u \in L^p(\Omega)$ and $f \in L^q(\Omega)$. Then

$$u \in C^{0,\alpha}(B(x, r_0/12) \cap \overline{\Omega}) \quad \forall x \in \overline{\Omega}.$$

Moreover $\|u\|_{C^{0,\alpha}(B(x,r_0/12)\cap\overline{\Omega})} \leq C(N, r_0, \alpha, p_0, \|u\|_p, \|f\|_q)$.

Remark 0.3. To simplify the exposition, we decided throughout this paper to not allow $+\infty$ for the exponents p, q, p_0, but of course one could easily derive the corresponding statement by using the fact that $f \in L^p$ for all $p \geq 1$ in the case when $f \in L^\infty$ reducing to an application of Theorem 0.2.

Remark 0.4. In the proof of Theorem 0.2, we need to assume that $u \in C^1(\Omega)$. This appears for instance in the proof of the monotonicity theorem, in order to define $\frac{\partial u}{\partial \nu}$ on $\partial B(x, r) \cap \Omega$. Notice that by the classical regularity theory for elliptic equations, this occurs for instance as soon as $f \in L^q(\Omega)$ with $q > N$ because then f is Hölder continuous and therefore $u \in C^{2,\alpha}(\Omega)$.

Observe that in the statement of Theorem 0.2, some a priori L^p integrability on u is needed to get some Hölder regularity. More precisely, the key ingredient is that the product uf should be sufficiently integrable. In what follows we shall see at least two situations where we know that $u \in L^p$ for some p large enough, and consequently state two Corollaries where the integrability hypothesis is given on f only, without any a priori requierement on u.

First, we notice that when $u \in W_0^{1,2}(\Omega)$, the Sobolev inequality says that $u \in L^{2^*}(\Omega)$, with $2^* = \frac{2N}{N-2}$. Some simple computations shows that in this situation, u and f verify the statement of Theorem 0.2 provided that $2 \leq N \leq 5$, which leads to the following corollary.

Corollary 0.5. *Assume that $2 \leq N \leq 5$ and let $q \in I_N$ be given where*

$$I_N = \begin{cases} [2, +\infty) & \textit{if } N = 2 \\ \left(\dfrac{2N}{6-N}, +\infty\right) & \textit{for } 3 \leq N \leq 5. \end{cases}$$

Then for any $\alpha > 0$ verifying

$$\alpha < 1 - \frac{N}{2}\left(\frac{N-2}{2N} + \frac{1}{q}\right),$$

we can find an $\varepsilon = \varepsilon(N, \alpha)$ such that the following holds. Let $\Omega \subseteq \mathbb{R}^N$ be an (ε, r_0)-Reifenberg-flat domain for some $r_0 > 0$, let $f \in L^q(\Omega)$ and *let $u \in W_0^{1,2}(\Omega) \cap C^1(\Omega)$ be the unique solution for the problem $(P1)$ in Ω. Then*

$$u \in C^{0,\alpha}(B(x, r_0/12) \cap \overline{\Omega}) \quad \forall x \in \overline{\Omega}.$$

Moreover $\|u\|_{C^{0,\alpha}(B(x,r_0/12)\cap\overline{\Omega})} \leq C(N, r_0, \alpha, q, \|\nabla u\|_2, \|f\|_q)$.

Proof. Since $u \in W_0^{1,2}(\Omega)$, the Sobolev embedding says that $u \in L^p(\Omega)$, with $p = 2^* = \frac{2N}{N-2}$. And by assumption $f \in L^q$ for some $q \in I_N$ (notice that $q \geq 2$, which guarantees existence and uniqueness of the weak solution). We now try to apply Theorem 0.2 with those p and q. Let p_0 be defined by

$$\frac{1}{p} + \frac{1}{q} = \frac{1}{p_0}.$$

Then a simple computation yields that $p_0 > \frac{N}{2}$, provided that

$$q > \frac{2N}{6-N}.$$

This fixes the range of dimension $N \leq 5$ and notice that in this case $\frac{2N}{6-N} \geq 2$ except for $N = 2$, which justifies the definition of I_N. We then conclude by applying Theorem 0.2. □

In the proof of Corollary 0.5 we brutally used the Sobolev embedding on $W_0^{1,2}(\Omega)$ to obtain an L^p integrability on u. But under some natural hypothesis we can get more using a Theorem by Byun and Wang [2]. Precisely, if $f = \operatorname{div} F$ for some $F \in L^2$, then f lies in the dual space of $W_0^{1,2}(\Omega)$ which guarantees the existence and uniqueness of a weak solution u for $(P1)$, again by the Riesz representation Theorem. The theorem of Byun and Wang [2, Theorem 2.10] implies moreover that if $F \in L^r(\Omega)$ then $\nabla u \in L^r(\Omega)$ as well. But then the Sobolev inequality says that $u \in L^{r^*}$ which allows us to apply Theorem 0.2 for a larger range of dimensions and exponents. Of course this analysis is interesting only for $r \leq N$ because if $r > N$ we directly get some Hölder estimates by the classical Sobolev embedding. This leads to our second corollary.

Corollary 0.6. *Let $\frac{N}{3} < r \leq N$ and $q > \frac{rN}{3r-N}$ be given, so that moreover $q \geq 2$. Then for any $\alpha > 0$ satisfying*

$$\alpha < 1 - \frac{N}{2}\left(\frac{1}{r^*} + \frac{1}{q}\right), \qquad \textit{with } r^* := \frac{rN}{N-r},$$

(here $\frac{rN}{N-r}$ is understood as being $+\infty$ when $r = N$) we can find an $\varepsilon = \varepsilon(N, \alpha, |\Omega|, r)$ such that the following holds. Let $\Omega \subseteq \mathbb{R}^N$ be an (ε, r_0)-Reifenberg-flat domain for some $r_0 > 0$, let $f \in L^q(\Omega)$ and assume that

$f = \operatorname{div} F$ *for some* $F \in L^r(\Omega)$. *Let* $u \in W_0^{1,2}(\Omega) \cap C^1(\Omega)$ *be the unique weak solution for the problem* $(P1)$ *in* Ω. *Then*

$$u \in C^{0,\alpha}(B(x, r_0/12) \cap \overline{\Omega}) \quad \forall x \in \overline{\Omega}.$$

Moreover $\|u\|_{C^{0,\alpha}(B(x,r_0/12)\cap\overline{\Omega})} \leq C(N, r_0, \alpha, r, q, |\Omega|, \|f\|_q, \|F\|_r)$.

Proof. First we apply [2, Theorem 2.10] which provides the existence of a threshold $\varepsilon_0 = \varepsilon(N, |\Omega|, r)$ such that for any solution $u \in W_0^{1,2}(\Omega)$ of $(P1)$ with $f = \operatorname{div} F$ and $F \in L^r(\Omega)$, we have that $\nabla u \in L^r(\Omega)$, provided that Ω is (ε_0, r_0)-Reifenberg-flat. But then the Sobolev inequality implies that $u \in L^{r^*}$ with

$$r^* = \frac{rN}{N - r} \qquad \text{if } r < N,$$

and r^* is large as we want otherwise.

In order to apply Theorem 0.2 we define p_0 such that

$$\frac{1}{r^*} + \frac{1}{q} - \frac{1}{p_0},$$

and we only need to check that $p_0 > \frac{N}{2}$. This implies the following condition on r and q:

$$r > \frac{N}{3} \text{ and } q > \frac{rN}{3r - N},$$

as required in the statement of the Corollary. We finally conclude by applying Theorem 0.2. □

Our approach to prove Theorem 0.2 follows the one that was already used in [18] to control the energy of eigenfunctions near the boundary of Reifenberg-flat domains, and that we apply here to other PDE than the eigenvalue problem. The main ingredient in the proof is a variant of Alt-Caffarelli-Friedman's monotonicity formula [1, Lemma 5.1], to control the behavior of the Dirichlet energy in balls centered at the boundary, as well as in the interior of Ω. Then we conclude by Morey-Campanato Theorem.

ACKNOWLEDGEMENTS. The authors wishes to thank the anonymous referee for his or her careful reading of the paper and for providing to us precious comments and corrections.

1 The monotonicity Lemma

We begin with a technical Lemma which basically contains the justification of an integration by parts. The proof is exactly the same as the first step of [18, Lemma 15] but we decided to provide here the full details for the convenience of the reader.

Lemma 1.1. *Let $u \in W_0^{1,2}(\Omega) \cap C^1(\Omega)$ be a solution for the problem $(P1)$. Then for every $x_0 \in \partial\Omega$ and a.e. $r > 0$ we have*

$$\begin{aligned}\int_{B(x_0,r)\cap\Omega} |\nabla u|^2 |x|^{2-N} dx \leq r^{2-N} &\int_{\partial B(x_0,r)\cap\Omega} u\frac{\partial u}{\partial \nu} dS \\ &+ \frac{(N-2)}{2} r^{1-N} \int_{\partial B(x_0,r)\cap\Omega} u^2 dS \quad (1.1)\\ &+ \int_{B(x_0,r)\cap\Omega} uf|x|^{2-N} dx.\end{aligned}$$

Proof. Although (1.1) can be formally obtained through an integration by parts, the rigorous proof is a bit technical. In the sequel we use the notation $\Omega_r^+ := B(x_0, r) \cap \Omega$, and $S_r^+ := \partial B(x_0, r) \cap \Omega$. We find it convenient to define, for a given $\varepsilon > 0$, the regularized norm

$$|x|_\varepsilon := \sqrt{x_1^2 + x_2^2 + \cdots + x_N^2 + \varepsilon},$$

so that $|x|_\varepsilon$ is a C^∞ function. A direct computation shows that

$$\Delta(|x|_\varepsilon^{2-N}) = (2-N)N\frac{\varepsilon}{|x|_\varepsilon^{N+2}} \leq 0,$$

in other words $|x|_\varepsilon^{2-N}$ is superharmonic, and hopefully enough this goes in the right direction regarding to the next inequalities.

We use one more regularization thus we let $u_n \in C_c^\infty(\Omega)$ be a sequence of functions converging in $W^{1,2}(\mathbb{R}^N)$ to u. We now proceed as in the proof of Alt, Caffarelli and Friedman monotonicity formula [1]: by using the equality

$$\Delta(u_n^2) = 2|\nabla u_n|^2 + 2u_n\Delta u_n \quad (1.2)$$

we deduce that

$$2\int_{\Omega_r^+} |\nabla u_n|^2 |x|_\varepsilon^{2-N} = \int_{\Omega_r^+} \Delta(u_n^2)|x|_\varepsilon^{2-N} - 2\int_{\Omega_r^+} (u_n\Delta u_n)|x|_\varepsilon^{2-N}. \quad (1.3)$$

Since $\Delta(|x|_\varepsilon^{2-N}) \leq 0$, the Gauss-Green Formula yields

$$\int_{\Omega_r^+} \Delta(u_n^2)|x|_\varepsilon^{2-N} dx = \int_{\Omega_r^+} u_n^2 \Delta(|x|_\varepsilon^{2-N}) dx + I_{n,\varepsilon}(r) \leq I_{n,\varepsilon}(r), \quad (1.4)$$

where

$$I_{n,\varepsilon}(r) = (r^2+\varepsilon)^{\frac{2-N}{2}} \int_{\partial\Omega_r^+} 2u_n \frac{\partial u_n}{\partial \nu} dS + (N-2)\frac{r}{(r^2+\varepsilon)^{\frac{N}{2}}} \int_{\partial\Omega_r^+} u_n^2 dS.$$

In other words, (1.3) reads

$$2\int_{\Omega_r^+} |\nabla u_n|^2 |x|_\varepsilon^{2-N} dx \leq I_{n,\varepsilon}(r) - 2\int_{\Omega_r^+} (u_n \Delta u_n)|x|_\varepsilon^{2-N} dx. \quad (1.5)$$

We now want to pass to the limit, first as $n \to +\infty$, and then as $\varepsilon \to 0^+$. To tackle some technical problems, we first integrate over $\rho \in [r, r+\delta]$ and divide by δ, thus obtaining

$$\frac{2}{\delta}\int_r^{r+\delta} \left(\int_{\Omega_\rho^+} |\nabla u_n|^2 |x|_\varepsilon^{2-N} dx \right) d\rho \leq A_n - R_n, \quad (1.6)$$

where

$$A_n = \frac{1}{\delta}\int_r^{r+\delta} I_{n,\varepsilon}(\rho) d\rho$$

and

$$R_n = 2\frac{1}{\delta}\int_r^{r+\delta} \left(\int_{\Omega_\rho^+} (u_n \Delta u_n)|x|_\varepsilon^{2-N} dx \right) d\rho.$$

First, we investigate the limit of A_n as $n \to +\infty$: by applying the coarea formula, we rewrite A_n as

$$A_n = \frac{1}{\delta}\left(2 \int_{\Omega_{r+\delta}^+ \setminus \Omega_r^+} (|x|^2+\varepsilon)^{\frac{2-N}{2}} u_n \, \nabla u_n \cdot x \, dx + (N-2) \int_{\Omega_{r+\delta}^+ \setminus \Omega_r^+} \frac{|x|}{(|x|^2+\varepsilon)^{\frac{N}{2}}} u_n^2 dx \right).$$

Since u_n converges to u in $W^{1,2}(\mathbb{R}^N)$ when $n \to +\infty$, then by using again the coarea formula we get that

$$A_n \longrightarrow \frac{1}{\delta}\int_r^{r+\delta} I_\varepsilon(\rho) d\rho \qquad n \to +\infty,$$

where

$$I_\varepsilon(\rho) = (\rho^2+\varepsilon)^{\frac{2-N}{2}} \int_{\partial\Omega_\rho^+} 2u \frac{\partial u}{\partial \nu} dS + (N-2)\frac{\rho}{(\rho^2+\varepsilon)^{\frac{N}{2}}} \int_{\partial\Omega_\rho^+} u^2 dS.$$

Next, we investigate the limit of R_n as $n \to +\infty$. By using Fubini's Theorem, we can rewrite R_n as

$$R_n = \int_\Omega (u_n \Delta u_n) G(x) dx,$$

where

$$G(x) = |x|_\varepsilon^{2-N} \frac{1}{\delta} \int_r^{r+\delta} \mathbf{1}_{\Omega_\rho^+}(x) d\rho.$$

Since

$$\frac{1}{\delta} \int_r^{r+\delta} \mathbf{1}_{\Omega_\rho^+}(x) d\rho = \begin{cases} 1 & \text{if } x \in \Omega_r^+ \\ \dfrac{r+\delta-|x|}{\delta} & \text{if } x \in \Omega_{r+\delta}^+ \setminus \Omega_r^+ \\ 0 & \text{if } x \notin \Omega_{r+\delta}^+ \end{cases},$$

then G is Lipschitz continuous and hence by recalling $u_n \in C_c^\infty(\Omega)$ we get

$$\begin{aligned} \left| \int_\Omega \Big(u_n \Delta u_n - u \Delta u \Big) G dx \right| &\leq \left| \int_\Omega \Big(\Delta u_n - \Delta u \Big) u_n G dx \right| \\ &+ \left| \int_\Omega \Big(u_n - u \Big) \Delta u G dx \right| \\ &= \left| \int_\Omega \Big(\nabla u_n - \nabla u \Big) \Big(u_n \nabla G + G \nabla u_n \Big) dx \right| \\ &+ \left| \int_\Omega \nabla u \Big((\nabla u_n - \nabla u) G + (u_n - u) \nabla G \Big) dx \right| \\ &\leq \|\nabla u_n - \nabla u\|_{L^2(\Omega)} \Big(\|u_n\|_{L^2(\Omega)} \|\nabla G\|_{L^\infty(\Omega)} \\ &+ \|\nabla u_n\|_{L^2(\Omega)} \|G\|_{L^\infty(\Omega)} \Big) \\ &+ \|\nabla u\|_{L^2(\Omega)} \Big(\|\nabla u_n - \nabla u\|_{L^2(\Omega)} \|G\|_{L^\infty(\Omega)} \\ &+ \|u_n - u\|_{L^2(\Omega)} \|\nabla G\|_{L^\infty(\Omega)} \Big) \end{aligned}$$

and hence the expression at the first line converges to 0 as $n \to +\infty$.

By combining the previous observations and by recalling that u satisfies the equation in the problem $(P1)$ we infer that by passing to the limit $n \to \infty$ in (1.6) we get

$$\begin{aligned} \frac{2}{\delta} \int_r^{r+\delta} \left(\int_{\Omega_\rho^+} |\nabla u|^2 |x|_\varepsilon^{2-N} dx \right) dr &\leq \frac{1}{\delta} \int_r^{r+\delta} I_\varepsilon(\rho) d\rho \\ &+ \frac{2}{\delta} \int_r^{r+\delta} \left(\int_{\Omega_\rho^+} u f |x|_\varepsilon^{2-N} dx \right) d\rho. \end{aligned}$$

Finally, dividing by 2, by passing to the limit $\delta \to 0^+$, and then $\varepsilon \to 0^+$ we obtain (1.1). □

Next, we will need the following Lemma of Gronwall type.

Lemma 1.2. *Let $\gamma > 0$, $r_0 > 0$, $\psi : (0, r_0) \to \mathbb{R}$ be a continuous function and $\varphi : (0, r_0) \to \mathbb{R}$ be an absolutely continuous function that satisfies the following inequality for a.e. $r \in (0, r_0)$,*

$$\varphi(r) \leq \gamma r \varphi'(r) + \psi(r). \tag{1.7}$$

Then

$$r \mapsto \frac{\varphi(r)}{r^{\frac{1}{\gamma}}} + \frac{1}{\gamma} \int_0^r \frac{\psi(s)}{s^{1+\frac{1}{\gamma}}} ds$$

is a nondecreasing function on $(0, r_0)$.

Proof. We can assume that

$$\int_0^{r_0} \frac{\psi(s)}{s^{1+\frac{1}{\gamma}}} ds < +\infty,$$

otherwise the Lemma is trivial. Under our hypothesis, the function

$$F(r) := \frac{\varphi(r)}{r^{\frac{1}{\gamma}}} + \frac{1}{\gamma} \int_0^r \frac{\psi(s)}{s^{1+\frac{1}{\gamma}}} ds$$

is differentiable a.e. and absolutely continuous. A computation gives

$$\begin{aligned} F'(r) &= \frac{\varphi'(r) r^{\frac{1}{\gamma}} - \varphi(r) \frac{1}{\gamma} r^{\frac{1}{\gamma}-1}}{r^{\frac{2}{\gamma}}} + \frac{1}{\gamma} \frac{\psi(r)}{r^{1+\frac{1}{\gamma}}} \\ &= \frac{\varphi'(r) - \varphi(r) \frac{1}{\gamma} r^{-1}}{r^{\frac{1}{\gamma}}} + \frac{1}{\gamma} \frac{\psi(r)}{r^{1+\frac{1}{\gamma}}} \end{aligned}$$

thus (1.7) yields

$$r\gamma F'(r) = \frac{r\gamma\varphi'(r) - \varphi(r) + \psi(r)}{r^{\frac{1}{\gamma}}} \geq 0,$$

which implies that F is nondecresing. □

We now prove the monotonicity Lemma, which is inspired by Alt, Caffarelli and Friedman [1, Lemma 5.1.]. The following statement and its proof, is an easy variant of [18, Lemma 15], where the same estimate is performed on Dirichlet eigenfunctions of the Laplace operator. We decided to write the full details in order to enlighten the role of f in the inequalities.

Lemma 1.3. *Let $\Omega \subseteq \mathbb{R}^N$ be a bounded domain and let $u \in W^{1,2}(\Omega) \cap C^1(\Omega)$ be a solution for the problem $(P1)$. Given $x_0 \in \overline{\Omega}$ and a radius $r > 0$, we denote by $\Omega_r^+ := B(x_0, r) \cap \Omega$, by $S_r^+ := \partial B(x_0, r) \cap \Omega$ and by $\sigma(r)$ the first Dirichlet eigenvalue of the Laplace operator on the spherical domain S_r^+. If there are constants $r_0 > 0$ and $\sigma^* \in]0, N-1[$ such that*

$$\inf_{0<r<r_0} (r^2\sigma(r)) \geq \sigma^*, \tag{1.8}$$

then the function

$$r \mapsto \left(\frac{1}{r^\beta}\int_{\Omega_r^+} \frac{|\nabla u|^2}{|x-x_0|^{N-2}}\,dx\right) + \beta\int_0^r \frac{\psi(s)}{s^{1+\beta}}ds \tag{1.9}$$

is non decreasing on $]0, r_0[$, where $\beta \in]0, 2[$ is given by

$$\beta = \sqrt{(N-2)^2 + 4\sigma^*} - (N-2)$$

and

$$\psi(s) := \int_{\Omega_s^+} |uf||x-x_0|^{2-N}dx.$$

We also have the bound

$$\int_{\Omega_{r_0/2}^+} \frac{|\nabla u|^2}{|x-x_0|^{N-2}}\,dx \leq C(N, r_0, \beta)\|\nabla u\|^2_{L^2(\Omega)} + \psi(r_0). \tag{1.10}$$

Remark 1.4. Of course the Lemma is interesting only when

$$\int_0^{r_0} \frac{\psi(s)}{s^{1+\beta}}ds < +\infty. \tag{1.11}$$

This will be satisfied if u and f are in some L^p spaces with suitable exponents, as will be shown in Lemma 2.1.

Remark 1.5. Notice that it is not known in general whether $\nabla u \in L^p(\Omega)$ for some $p > 2$ and therefore it is not obvious to find a bound for the left hand side of (1.10).

Proof. We assume without losing generality that $x_0 = 0$ and to simplify notation we denote by B_r the ball $B(x, r)$.

By Lemma 1.1 we know that for a.e. $r > 0$,

$$\int_{\Omega_r^+} |\nabla u|^2|x|^{2-N}dx \leq r^{2-N}\int_{S_r^+} 2u\frac{\partial u}{\partial \nu}dS + \frac{(N-2)}{2}r^{1-N}\int_{S_r^+} u^2dS + \int_{\Omega_r^+} uf|x|^{2-N}dx. \tag{1.12}$$

Let us define

$$\psi(r) := \int_{\Omega_r^+} |uf||x|^{2-N} dx \tag{1.13}$$

and assume that (1.11) holds (otherwise there is nothing to prove).

Next, we point out that the definition of σ^* implies that

$$\int_{S_r^+} u^2 dS \leq \frac{1}{\sigma^*} r^2 \int_{S_r^+} |\nabla_\tau u|^2 dS \qquad r \in]0, r_0[, \tag{1.14}$$

where ∇_τ denotes the tangential gradient on the sphere. Also, let $\alpha > 0$ be a parameter that will be fixed later. Then by combining Cauchy-Schwarz inequality, (1.14) and the inequality $ab \leq \frac{\alpha}{2}a^2 + \frac{1}{2\alpha}b^2$, we get

$$\begin{aligned}
\left|\int_{S_r^+} u \frac{\partial u}{\partial \nu} dS\right| &\leq \Big(\int_{S_r^+} u^2 dS\Big)^{\frac{1}{2}} \Big(\int_{S_r^+} |\frac{\partial u}{\partial \nu}|^2 dS\Big)^{\frac{1}{2}} \\
&\leq \frac{r}{\sqrt{\sigma^*}} \Big(\int_{S_r^+} |\nabla_\tau u|^2 dS\Big)^{\frac{1}{2}} \Big(\int_{S_r^+} |\frac{\partial u}{\partial \nu}|^2 dS\Big)^{\frac{1}{2}} \\
&\leq \frac{r}{\sqrt{\sigma^*}} \Big(\frac{\alpha}{2} \int_{S_r^+} |\nabla_\tau u|^2 dS + \frac{1}{2\alpha} \int_{S_r^+} |\frac{\partial u}{\partial \nu}|^2 dS\Big).
\end{aligned} \tag{1.15}$$

Hence,

$$\begin{aligned}
& r^{2-N} \int_{S_r^+} 2u \frac{\partial u}{\partial \nu} dS + (N-2) r^{1-N} \int_{S_r^+} u^2 dS \\
&\leq r^{2-N} \frac{2r}{\sqrt{\sigma^*}} \Big[\frac{\alpha}{2} \int_{S_r^+} |\nabla_\tau u|^2 dS + \frac{1}{2\alpha} \int_{S_r^+} |\frac{\partial u}{\partial \nu}|^2 dS\Big] \\
&\quad + (N-2) r^{1-N} \frac{1}{\sigma^*} r^2 \int_{S_r^+} |\nabla_\tau u|^2 dS \\
&\leq r^{3-N} \Big[\Big(\frac{\alpha}{\sqrt{\sigma^*}} + \frac{N-2}{\sigma^*}\Big) \int_{S_r^+} |\nabla_\tau u|^2 dS + \frac{1}{\alpha\sqrt{\sigma^*}} \int_{S_r^+} |\frac{\partial u}{\partial \nu}|^2 dS\Big].
\end{aligned} \tag{1.16}$$

Next, we choose $\alpha > 0$ in such a way that

$$\frac{\alpha}{\sqrt{\sigma^*}} + \frac{N-2}{\sigma^*} = \frac{1}{\alpha\sqrt{\sigma^*}},$$

namely

$$\alpha = \frac{1}{2\sqrt{\sigma^*}} [\sqrt{(N-2)^2 + 4\sigma^*} - (N-2)].$$

Hence, by combining (1.12), (1.14) and (1.16) we finally get

$$\int_{\Omega_r^+} |\nabla u|^2 |x|^{2-N} dx \le r^{3-N} \gamma(N,\sigma^*) \int_{S_r^+} |\nabla u|^2 dS + \psi(r), \qquad (1.17)$$

where

$$\gamma(N,\sigma^*) = \left[\sqrt{(N-2)^2 + 4\sigma^*} - (N-2)\right]^{-1}.$$

Let us set

$$\varphi(r) := \int_{\Omega_r^+} |\nabla u|^2 |x|^{2-N} dx$$

and observe that

$$\varphi'(r) = r^{2-N} \int_{S_r^+} |\nabla u|^2 \, dS \quad a.e. \, r \in]0, r_0[\,,$$

hence (1.17) implies that

$$\varphi(r) \le \gamma r \varphi'(r) + \psi(r), \qquad (1.18)$$

with $\gamma = \gamma(N,\sigma^*)$.

But now Lemma 1.2 exactly says that the function

$$r \mapsto \frac{1}{r^\beta} \int_{\Omega_r^+} \frac{|\nabla u|^2}{|x - x_0|^{N-2}} dx + \beta \int_0^r \frac{\psi(s)}{s^{1+\beta}} ds \qquad (1.19)$$

is non decreasing on $(0, r_0)$, where $\beta \in (0,2)$ is given by

$$\beta = \frac{1}{\gamma} = \sqrt{(N-2)^2 + 4\sigma^*} - (N-2),$$

which proves the monotonicity result.

To finish the proof of the Lemma it remains to establish (1.10). For this purpose, we start by finding a radius $r_1 \in (r_0/2, r_0)$ such that $\int_{S_{r_1}^+} |\nabla u|^2 dS$ is less than average, which means

$$\int_{S_{r_1}^+} |\nabla u|^2 dS \le \frac{2}{r_0} \int_{\Omega_{r_0}^+} |\nabla u|^2 \, dx \le \frac{2}{r_0} \|\nabla u\|^2_{L^2(\Omega)}.$$

By combining (1.17) with the fact that $r_0/2 \le r_1 \le r_0$ we infer that

$$\begin{aligned} \int_{\Omega_{r_1}^+} |\nabla u|^2 |x|^{2-N} dx &\le C(N, r_0, \beta) \|\nabla u\|^2_{L^2(\Omega)} + \psi(r_1) \\ &\le C(N, r_0, \beta) \|\nabla u\|^2_{L^2(\Omega)} + \psi(r_0). \end{aligned} \qquad (1.20)$$

It follows that

$$\int_{\Omega^+_{r_0/2}} \frac{|\nabla u|^2}{|x|^{N-2}}\,dx \le \int_{\Omega^+_{r_1}} \frac{|\nabla u|^2}{|x|^{N-2}}\,dx$$
$$\le C(N, r_0, \beta)\|\nabla u\|^2_{L^2(\Omega)} + \psi(r_0),$$

and (1.10) is proved. □

2 An elementary computation

In order to apply Lemma 1.3, the first thing to check is that (1.11) holds. The purpose of the following Lemma is to prove that it is the case when u and f are in suitable L^p spaces.

Lemma 2.1. *Let $\Omega \subset \mathbb{R}^N$ be an arbitrary domain and let $p_0 > \frac{N}{2}$. Then for any $g \in L^{p_0}(\Omega)$, denoting*

$$\psi(r) := \int_{B(0,r)} |g||x|^{2-N}dx,$$

we have

$$|\psi(r)| \le C(N, p_0)\|g\|_{p_0} r^{\frac{2p_0-N}{p_0}}. \tag{2.1}$$

As a consequence

$$\int_0^{r_0} \frac{\psi(s)}{s^{1+\beta}}ds < +\infty$$

for any $\beta > 0$ satisfying

$$\beta < \frac{2p_0 - N}{p_0}. \tag{2.2}$$

Proof. First, we observe that $|x|^{2-N} \in L^m(B(0,r))$ for any m that satisfies

$$(N-2)m < N, \quad \text{which implies} \quad m < \frac{N}{N-2}.$$

Moreover a computation gives that under this condition,

$$\||x|^{2-N}\|^m_{L^m(B(0,r))} = \int_{B(0,r)} |x|^{(2-N)m} = C(N,m)r^{N-m(N-2)}. \tag{2.3}$$

Let us define m as being the conjugate exponent of p_0, namely

$$\frac{1}{m} := 1 - \frac{1}{p_0}.$$

Then the hypothesis $p_0 > \frac{N}{2}$ implies that $m < \frac{N}{N-2}$, and by use of Hölder inequality and by (2.3) we can estimate

$$|\psi(r)| \leq \|g\|_{p_0} \||x|^{2-N}\|_{L^m(B(0,r))} \leq C(N,m)\|g\|_{p_0} r^{\frac{N-m(N-2)}{m}},$$

or equivalently in terms of p_0,

$$|\psi(r)| \leq C(N, p_0)\|g\|_{p_0} r^{\frac{2p_0-N}{p_0}}. \tag{2.4}$$

The conclusion of the Lemma follows directly from (2.4). □

3 Interior estimate

We come now to our first application of Lemma 1.3, which is a decay estimate for the energy at interior points, for arbitrary domains.

Proposition 3.1. *Let $p, q, p_0 > 1$ be some exponents satisfying $\frac{1}{p} + \frac{1}{q} = \frac{1}{p_0}$ and $p_0 > \frac{N}{2}$. Let $\beta > 0$ be any given exponent such that*

$$\beta < \frac{2p_0 - N}{p_0}. \tag{3.1}$$

Let $\Omega \subseteq \mathbb{R}^N$ be any domain, $x_0 \in \Omega$, and let $u \in W_0^{1,2}(\Omega) \cap C^1(\Omega)$ be a solution for the problem $(P1)$ in Ω with $u \in L^p(\Omega)$ and $f \in L^q(\Omega)$. Then

$$\int_{B(x_0,r)\cap\Omega} |\nabla u|^2 dx \leq C r^{N-2+\beta} \|u\|_p \|f\|_q \quad \forall r \in (0, \mathrm{dist}(x_0, \partial\Omega)/2), \tag{3.2}$$

with $C = C(N, \beta, p_0, \mathrm{dist}(x_0, \partial\Omega))$.

Proof. We assume that $x_0 = 0$ and we define $r_1 := \mathrm{dist}(x_0, \partial\Omega)$ in such a way that $B(x, r)$ is totally contained inside Ω for $r \leq r_1$ and under the notation of Lemma 1.3, $\Omega_r^+ = B(0, r)$ and by $S_r^+ = \partial B(0, r)$ for any $r \leq r_1$. Under our hypothesis and in virtue of Lemma 2.1 that we apply with $g = uf$, we know that (1.11) holds for any $\beta > 0$ that satisfies

$$\beta < \frac{2p_0 - N}{p_0}. \tag{3.3}$$

Notice that $\frac{2p_0-N}{p_0} > 0$, because $p_0 > N/2$.

In the sequel we chose any exponent $\beta > 0$ satisfying (3.3), so that (1.11) holds and moreover Lemma 2.1 says that

$$\begin{aligned} \int_0^{r_1} \frac{\psi(s)}{s^{1+\beta}} ds &\leq C(N, p_0)\|u\|_p \|f\|_q \int_0^{r_1} s^{\frac{2p_0-N}{p_0} - 1 - \beta} ds \\ &\leq C(N, p_0, \beta)\|u\|_p \|f\|_q. \end{aligned} \tag{3.4}$$

We are now ready to prove (3.2), β still being a fixed exponent satisfying (3.3). We recall that the first eigenvalue of the spherical Dirichlet Laplacian on the unit sphere is equal to $N-1$, thus hypothesis (1.8) in the present context corresponds to

$$\inf_{0<r<r_1} (r^2\sigma(r)) = N-1, \tag{3.5}$$

so that (1.8) holds for any $\sigma^* < N-1$. Let us choose σ^* exactly equal to the one that satisfies

$$\beta = \sqrt{(N-2)^2 + 4\sigma^*} - (N-2).$$

one easily verifies that $\sigma^* < N-1$ because of (3.3).

As a consequence, we are in position to apply Lemma 1.3 which ensures that, if u is a solution for the problem $(P1)$, then the function in (1.9) is non decreasing. In particular, by monotonicity we know that for every $r \leq r_1/2$,

$$\begin{aligned} \frac{1}{r^{N-2+\beta}} \int_{\Omega_r^+} |\nabla u|^2\, dx &\leq \left(\frac{1}{r^\beta} \int_{\Omega_r^+} \frac{|\nabla u|^2}{|x-x_0|^{N-2}}\, dx \right) \\ &\quad + \beta \int_0^r \frac{\psi(s)}{s^{1+\beta}} ds \\ &\leq \left(\frac{1}{(r_1/2)^\beta} \int_{\Omega_{r_1/2}^+} \frac{|\nabla u|^2}{|x-x_0|^{N-2}}\, dx \right) \\ &\quad + \beta \int_0^{r_1/2} \frac{\psi(s)}{s^{1+\beta}} ds, \end{aligned} \tag{3.6}$$

and we conclude that for every $r \leq r_1/2$,

$$\int_{\Omega_r^+} |\nabla u|^2\, dx \leq K r^{N-2+\beta},$$

with

$$K = \left(\frac{1}{(r_1/2)^\beta} \int_{\Omega_{r_1/2}^+} \frac{|\nabla u|^2}{|x-x_0|^{N-2}}\, dx \right) + \beta \int_0^{r_1/2} \frac{\psi(s)}{s^{1+\beta}} ds.$$

Let us now provide an estimate on K. To estimate the first term in K we use (1.10) to write

$$\int_{\Omega_{r_1/2}^+} \frac{|\nabla u|^2}{|x-x_0|^{N-2}}\, dx \leq C(N, r_1, \beta) \|\nabla u\|^2_{L^2(\Omega)} + \psi(r_1) \tag{3.7}$$

Then we use (2.1) to estimate

$$\psi(r_1) \leq C(N, p_0, r_1) \|u\|_p \|f\|_q,$$

and from the equation satisfied by u we get

$$\|\nabla u\|_{L^2(\Omega)}^2 = \int_\Omega u f dx \leq \|u\|_p \|f\|_q$$

so that in total we have

$$\int_{\Omega^+_{r_1/2}} \frac{|\nabla u|^2}{|x - x_0|^{N-2}} dx \leq C(N, r_0, \beta, p_0)\|u\|_p \|f\|_q.$$

Finally, the last estimate together with (3.4) yields

$$K \leq C(N, r_1, \beta, p_0)\|u\|_p \|f\|_q,$$

and this ends the proof of the proposition. □

4 Boundary estimate

We now use Lemma 1.3 again to provide an estimate on the energy at boundary points, this time for Reifenberg-flat domains.

Proposition 4.1. *Let $p, q, p_0 > 1$ be some exponents satisfying $\frac{1}{p} + \frac{1}{q} = \frac{1}{p_0}$ and $p_0 > \frac{N}{2}$. Let $\beta > 0$ be any given exponent such that*

$$\beta < \frac{2p_0 - N}{p_0}. \tag{4.1}$$

Then one can find an $\varepsilon = \varepsilon(N, \beta)$ such that the following holds. Let $\Omega \subseteq \mathbb{R}^N$ be any (ε, r_0)-Reifenberg-flat domain for some $r_0 > 0$, let $x_0 \in \partial\Omega$ and let $u \in W^{1,2}_0(\Omega) \cap C^1(\Omega)$ be a solution for the problem $(P1)$ in Ω with $u \in L^p(\Omega)$ and $f \in L^q(\Omega)$. Then

$$\int_{B(x_0,r)\cap\Omega} |\nabla u|^2 dx \leq C r^{N-2+\beta} \|u\|_p \|f\|_q \quad \forall\, r \in (0, r_0/2), \tag{4.2}$$

with $C = C(N, r_0, \beta, p_0)$.

Proof. As before we assume that $x_0 = 0$ and we denote by $\Omega^+_r := B(0, r) \cap \Omega$ and by $S^+_r := \partial B(0, r) \cap \Omega$. To obtain the decay estimate on $\int_{\Omega^+_r} |\nabla u|^2 dx$ we will follow the proof of Proposition 3.1: the main difference is that for boundary points, (3.5) does not hold. This is where Reifenberg-flatness will play a role.

Let $\beta > 0$, be an exponent satisfying (4.1), so that invoquing Lemma 2.1 we have

$$\int_0^{r_0} \frac{\psi(s)}{s^{1+\beta}} ds \leq C(N, p_0, \beta)\|u\|_p \|f\|_q < +\infty. \tag{4.3}$$

Next, we recall that the first eigenvalue of the spherical Dirichlet Laplacian on a half sphere is equal to $N-1$ (as for the total sphere). For $t \in (-1,1)$, let S_t be the spherical cap $S_t := \partial B(0,1) \cap \{x_N > t\}$ so that $t=0$ corresponds to a half sphere. Let $\lambda_1(S_t)$ be the first Dirichlet eigenvalue in S_t. In particular, $t \mapsto \lambda_1(S_t)$ is continuous and monotone in t (the continuity follows for instance by considering $u \circ \phi_t$ as a competitor in the Rayleigh quotient where u is a given eigenvector in one domain S_{t_0} and ϕ_t is a diffeomorphism from S_t to S_{t_0} so that $u \circ \phi_t$ converges strongly in $W^{1,2}$ to u when $t \to t_0$). Therefore, since $\lambda_1(S_t) \to 0$ as $t \downarrow -1$, there is $t^*(\beta) < 0$ such that

$$\beta = \sqrt{(N-2)^2 + 4\lambda_1(S_{t^*})} - (N-2).$$

By applying the definition of Reifenerg-flat domain, we infer that, if $\varepsilon < t^*(\beta)/2$, then $\partial B(0,r) \cap \Omega$ is contained in a spherical cap homothetic to S_{t^*} (after possibly a rotation) for every $r \leq r_0$. Since the eigenvalues scale of by factor r^2 when the domain expands of a factor $1/r$, by the monotonicity property of the eigenvalues with respect to domains inclusion, we have

$$\inf_{r<r_0} r^2 \lambda_1(\partial B(0,r) \cap \Omega) \geq \lambda_1(S_{t^*}) = \frac{\beta}{2}\Big(\frac{\beta}{2} + N - 2\Big). \tag{4.4}$$

As a consequence, we are in position to apply the monotonicity Lemma (Lemma 1.3) which ensures that, if u is a solution for the problem $(P1)$ and $x_0 \in \partial\Omega$, then the function in (1.9) is non decreasing. We then conclude as in the proof of Proposition 3.1, *i.e.* by monotonicity we know that for every $r \leq r_0 < 1$,

$$\begin{aligned} \frac{1}{r^{N-2+\beta}} \int_{\Omega_r^+} |\nabla u|^2 \, dx &\leq \left(\frac{1}{r^\beta} \int_{\Omega_r^+} \frac{|\nabla u|^2}{|x-x_0|^{N-2}} \, dx \right) \\ &\quad + \beta \int_0^r \frac{\psi(s)}{s^{1+\beta}} ds \\ &\leq \left(\frac{1}{(r_0/2)^\beta} \int_{\Omega_{r_0/2}^+} \frac{|\nabla u|^2}{|x-x_0|^{N-2}} \, dx \right) \\ &\quad + \beta \int_0^{r_0/2} \frac{\psi(s)}{s^{1+\beta}} ds, \end{aligned} \tag{4.5}$$

hence for every $r \leq r_0/2$,

$$\int_{\Omega_r^+} |\nabla u|^2 \, dx \leq K r^{N-2+\beta},$$

with,

$$K = \left(\frac{1}{(r_0/2)^\beta}\int_{\Omega^+_{r_0/2}} \frac{|\nabla u|^2}{|x-x_0|^{N-2}}\,dx\right) + \beta\int_0^{r_0/2}\frac{\psi(s)}{s^{1+\beta}}ds.$$

Then we estimate K exactly as in the end of the proof of Proposition 3.1, using (1.10), (2.1) and (4.3) to bound

$$K \leq C(N, r_0, \beta, p_0)\|u\|_p\|f\|_q,$$

and this ends the proof of the proposition. □

5 Global decay result

Gathering together Proposition 3.1 and Proposition 4.1 we deduce the following global result.

Proposition 5.1. *Let $p, q, p_0 > 1$ be some exponents satisfying $\frac{1}{p}+\frac{1}{q} = \frac{1}{p_0}$ and $p_0 > \frac{N}{2}$. Let $\beta > 0$ be any given exponent such that*

$$\beta < \frac{2p_0 - N}{p_0}. \tag{5.1}$$

Then one can find an $\varepsilon = \varepsilon(N, \beta)$ such that the following holds. Let $\Omega \subseteq \mathbb{R}^N$ be any (ε, r_0)-Reifenberg-flat domain for some $r_0 > 0$, and let $u \in W^{1,2}_0(\Omega)\cap C^1(\Omega)$ be a solution for the problem $(P1)$ in Ω with $u \in L^p(\Omega)$ and $f \in L^q(\Omega)$. Then

$$\int_{B(x,r)\cap\Omega} |\nabla u|^2 dx \leq C r^{N-2+\beta}\|u\|_p\|f\|_q \quad \forall x \in \overline{\Omega},\ \forall\, r \in (0, r_0/6), \tag{5.2}$$

with $C = C(N, r_0, \beta, p_0)$.

Proof. By Proposition 3.1 and Proposition 4.1, we already know that (5.2) holds true for every $x \in \partial\Omega$, or for points x such that $\operatorname{dist}(x, \partial\Omega) \geq r_0/3$. It remains to consider balls centered at points $x \in \Omega$ verifying

$$\operatorname{dist}(x, \partial\Omega) \leq r_0/3.$$

Let x be such a point. Then Proposition 3.1 directly says that (5.2) holds for every radius r such that $0 < r \leq \operatorname{dist}(x, \partial\Omega)/2$, and it remains to extend this for the radii r in the range

$$\operatorname{dist}(x, \partial\Omega)/2 \leq r \leq r_0/6. \tag{5.3}$$

For this purpose, let $y \in \partial\Omega$ be such that

$$\mathrm{dist}(x, \partial\Omega) = \|x - y\| \leq r_0/3.$$

Denoting $d(x) := \mathrm{dist}(x, \partial\Omega)$ we observe that for the r that satisfies (5.3) we have

$$B(x, r) \subseteq B(y, r + d(x)) \subseteq B(y, 3r). \tag{5.4}$$

Then since $y \in \partial\Omega$, Proposition 4.1 says that

$$\int_{B(y,r)\cap\Omega} |\nabla u|^2 dx \leq C r^{N-2+\beta} \|u\|_p \|f\|_q \quad \forall\, r \in (0, r_0/2), \tag{5.5}$$

so that (5.2) follows, up to change C with $3^{N-2+\beta}C$. □

6 Conclusion and main result

The classical results on Campanato Spaces can be found for instance in [8]. We define the space

$$\mathcal{L}^{p,\lambda}(\Omega) := \left\{ u \in L^p(\Omega) \ ; \ \sup_{x,\rho} \left(\rho^{-\lambda} \int_{B(x,r)\cap\Omega} |u - u_{x,r}|^p dx \right) < +\infty \right\}$$

where the supremum is taken over all $x \in \Omega$ and all $\rho \leq \mathrm{diam}(\Omega)$, and where $u_{x,r}$ means the average of u on the ball $B(x, r)$. A proof of the next result can be found in [8, Theorem 3.1.].

Theorem 6.1 (Campanato). *If $N < \lambda \leq N + p$ then*

$$\mathcal{L}^{p,\lambda}(\Omega) \simeq C^{0,\alpha}(\overline{\Omega}), \qquad \textit{with } \alpha = \frac{\lambda - N}{p}.$$

(here $\simeq$ means that the identity map is bi-continuous).

We can now prove our main result.

Proof of Theorem 0.2. Considering u as a function of $W^{1,2}(\mathbb{R}^N)$ by setting 0 outside Ω, and applying Proposition 5.1 with $\beta = 2\alpha$, we obtain that

$$\int_{B(x,r)} |\nabla u|^2 dx \leq C r^{N-2+\beta} \|u\|_p \|f\|_q \quad \forall x \in \overline{\Omega},\ \forall\, r \in (0, r_0/6), \tag{6.1}$$

with $C = C(N, r_0, \beta, p_0)$. Recalling now the classical Poincaré inequality in a ball $B(x,r)$

$$\int_{B(x,r)} |u - u_{x,r}|dx \leq C(N)r^{1+\frac{N}{2}} \left(\int_{B(x,r)} |\nabla u|^2 dx\right)^{\frac{1}{2}},$$

we get

$$\int_{B(x,r)} |u - u_{x,r}|dx \leq Cr^{N+\frac{\beta}{2}} \quad \forall x \in \overline{\Omega},\ \forall\, r \in (0, r_0/6), \tag{6.2}$$

with $C = C(N, r_0, \beta, p_0, \|u\|_p, \|f\|_q)$. But this implies that

$$u \in \mathcal{L}^{1,N+\frac{\beta}{2}}(B(x, r_0/12) \cap \overline{\Omega}) \quad \forall x \in \overline{\Omega}.$$

Moreover $\frac{\beta}{2} < \frac{p_0 - \frac{N}{2}}{p_0} \leq 1$ and hence Theorem 6.1 says that

$$u \in C^{0,\alpha}(B(x, r_0/12) \cap \overline{\Omega})$$

with $\alpha = \frac{\beta}{2}$, and the norm is controlled by $C = C(N, r_0, \beta, p_0, \|u\|_p, \|f\|_q)$.

□

References

[1] H. W. ALT, L. A. CAFFARELLI and A. FRIEDMAN, *Variational problems with two phases and their free boundaries*, Trans. Amer. Math. Soc. (2) **282** (1984), 431–461.

[2] S. BYUN and L. WANG, *Elliptic equations with BMO nonlinearity in Reifenberg domains*, Adv. Math. (6) **219** (2008), 1937–1971.

[3] S. BYUN and L. WANG, *Gradient estimates for elliptic systems in non-smooth domains*, Math. Ann. (3) **341** (2008), 629–650.

[4] S. BYUN, L. WANG and S. ZHOU, *Nonlinear elliptic equations with BMO coefficients in Reifenberg domains*, J. Funct. Anal. (1) **250** (2007), 167–196.

[5] S.-S. BYUN and L. WANG, *Fourth-order parabolic equations with weak BMO coefficients in Reifenberg domains*, J. Differential Equations (11) **245** (2008), 3217–3252.

[6] S.-S. BYUN and L. WANG, *Elliptic equations with measurable coefficients in Reifenberg domains*, Adv. Math. (5) **225** (2010), 2648–2673.

[7] L. A. CAFFARELLI and I. PERAL, *On $W^{1,p}$ estimates for elliptic equations in divergence form*, Comm. Pure Appl. Math. (1) 51 (1998), 1–21.

[8] M. GIAQUINTA, "Introduction to Regularity Theory for Non-linear Elliptic Systems", Lectures in Mathematics ETH Zürich. Birkhäuser Verlag, Basel, 1993.

[9] P. GRISVARD, "Singularities in Boundary Value Problems", Vol. 22 of *Recherches en Mathématiques Appliquées [Research in Applied Mathematics]*, Masson, Paris, 1992.

[10] D. JERISON and C. E. KENIG, *The inhomogeneous Dirichlet problem in Lipschitz domains*, J. Funct. Anal. (1) **130**, 161–219.

[11] D. S. JERISON and C. E. KENIG, *Boundary behavior of harmonic functions in nontangentially accessible domains*, Adv. in Math. (1) **46** (1982), 80–147.

[12] C. KENIG and T. TORO, *Harmonic measure on locally flat domains*, Duke Math. J. (3) **87** (1997), 509–551.

[13] C. KENIG and T. TORO, *Free boundary regularity for harmonic measures and Poisson kernels*, Ann. of Math. (2), (2) **150** (1999), 369–454.

[14] C. KENIG and T. TORO, *Poisson kernel characterization of Reifenberg flat chord arc domains*, Ann. Sci. École Norm. Sup. (4), (3) **36** (2003), 323–401.

[15] A. LEMENANT, *Energy improvement for energy minimizing functions in the complement of generalized Reifenberg-flat sets* Ann. Sc. Norm. Super. Pisa Cl. Sci. (5), (2) **9** (2010), 351–384.

[16] A. LEMENANT and E. MILAKIS, *Quantitative stability for the first Dirichlet eigenvalue in Reifenberg flat domains in $\mathbb{R}^N$*, J. Math. Anal. Appl. (2) **364** (2010), 522–533.

[17] A. LEMENANT and E. MILAKIS, *A stability result for nonlinear Neumann problems in Reifenberg flat domains in $\mathbb{R}^N$*, Publ. Mat. (2) **55** (2011), 413–432.

[18] A. LEMENANT, E. MILAKIS and L. V. SPINOLO, *Spectral stability estimates for the Dirichlet and Neumann Laplacian in rough domains*, J. Funct. Anal. (9) **264** (2013), 2097–2135.

[19] E. MILAKIS and T. TORO, *Divergence form operators in Reifenberg flat domains*, Math. Z. (1) **264** (2010), 15–41.

[20] T. TORO, *Geometry of measures: harmonic analysis meets geometric measure theory*, In: "Handbook of Geometric Analysis", No. 1, Vol. 7 of Adv. Lect. Math. (ALM), Int. Press, Somerville, MA, 2008, 449–465.

Limiting models in condensed matter Physics and gradient flows of 1-homogeneous functionals

Matteo Novaga and Giandomenico Orlandi

Abstract. We survey some recent results on variational and evolution problems concerning a certain class of convex 1-homogeneous functionals for vector-valued maps related to models in phase transitions (Hele-Shaw), superconductivity (Ginzburg-Landau) and superfluidity (Gross-Pitaevskii). Minimizers and gradient flows of such functionals may be characterized as solutions of suitable non-local vectorial generalizations of the classical obstacle problem.

1 Introduction

In this note we survey some results obtained in [2,6,8] on variational and evolution problems concerning a certain class of convex 1-homogeneous functionals for vector-valued maps related to models in phase transitions (Hele-Shaw), superconductivity (Ginzburg-Landau) and superfluidity (Gross-Pitaevskii). It has been recognized that minimizers and gradient flows of such functionals may be characterized as solutions of suitable vectorial generalizations of obstacle-type problems, where the constraints are non-local in nature.

This motivated us to investigate the general structure and qualitative properties of such solutions in analogy with classical obstacle problems, trying in particular to characterize situations where the non-local constraint is saturated, as well as qualitative properties of the corresponding coincidence sets, and interpreting their physical meaning according to the considered model.

The functionals we are interested in have the general form

$$
\begin{aligned}
J(u) &= \int_\Omega |A(x) \cdot Du)| \\
&= \sup \left\{ \int_\Omega u \cdot \operatorname{div}(A(x)^t \cdot \xi)\, dx \,:\, \xi \in C_c^1(\Omega; \mathbb{R}^{kq}),\ |\xi| \le 1 \right\},
\end{aligned}
\tag{1.1}
$$

where $\Omega \subset \mathbb{R}^n$, $u \in L^1(\Omega; \mathbb{R}^k)$ and $A \in C^1(\Omega, \mathbb{R}^{qn})$. The functional J is convex and 1-homogeneous, and in case $u \in BV(\Omega; \mathbb{R}^k)$, *i.e.* $Du = \nu|Du|$ is a $\mathbb{R}^{nk}$-valued measure with total variation $|Du|$ and $|\nu| = 1$ $|Du|$-a.e., the functional $J(u)$ represents the total variation of the measure $A(x) \cdot \nu|Du| \llcorner \Omega$.

1.1 Some examples

A first important example is given by the weighted Total Variation functional (see *e.g.* [4, 10–12, 15])

$$\begin{aligned} J_0(u) &= \int_\Omega \rho(x)|Du| \\ &= \sup\left\{\int_\Omega u \cdot \operatorname{div}(\rho(x)\xi)dx : \xi \in C_c^1(\Omega; \mathbb{R}^n), |\xi| \le 1\right\}, \end{aligned} \tag{1.2}$$

defined for $u \in L^1(\Omega)$ and a regular nonnegative weight $\rho(x) \ge 0$. Further examples are given, for $u \in L^1(\Omega; \mathbb{R}^n)$ a vector field, by

$$J(u) = \int_\Omega |\nabla \cdot u|,$$

i.e. the total variation of the divergence of u, and, in case $n = 3$, by

$$J(u) = \int_\Omega |\nabla \times u|,$$

i.e. the total variation of the curl of u. More generally, if $u(x) \in \Lambda^k\mathbb{R}^n$ for $x \in \Omega$, *i.e.* u is a k-differential form, we may consider

$$\begin{aligned} J_k(u) &= \int_\Omega \rho(x)|du| \\ &= \sup\left\{\int_\Omega u \cdot d^*(\rho(x)\xi)dx : \xi \in C_c^1(\Omega; \Lambda^{k+1}\mathbb{R}^n), |\xi| \le 1\right\}, \end{aligned} \tag{1.3}$$

so that J_{n-1} corresponds to the (weighted) total variation of the divergence, and J_1 to the (weighted) total variation of the curl in case $n = 3$, and J_0 yields again the weighted TV functional.

1.2 Gradient flows

Given an initial datum $u_0 \in L^2(\Omega; \mathbb{R}^k)$, the L^2-gradient flow for $J(u)$ is defined by the differential inclusion

$$u_t \in -\partial J(u) \qquad t \in [0, +\infty), \tag{1.4}$$

where ∂J denotes the subgradient of the convex functional J. The general theory of [9] guarantees the existence of a global weak solution

$u \in H^1([0,+\infty), L^2(\Omega;\mathbb{R}^k))$ of (1.4). The (weighted) Moreau-Yosida regularization of the convex functional J, which yields a discrete approximation scheme for (1.4) is given by

$$I(u) = J(u) + \lambda \int_\Omega |u-u_0|^2 g(x)\, dx\,, \qquad (1.5)$$

with $u_0 : \Omega \to \mathbb{R}^k$ square integrable with respect to the measure $g(x)\,dx$, where $g(x) \geq 0$ is a nonnegative regular weight function and $\lambda > 0$ a parameter. For $n = 2$, $k = 1$ and $f(x) = g(x) \equiv 1$, (1.5) corresponds to the Rudin-Osher-Fatemi Total Variation based denoising model [21]. Anisotropic versions of TV functionals and applications to active contour and edge detection have been studied in [12]. Related homogeneous functionals for the description of landsliding have been also studied in [10, 15].

1.3 Formulation for differential forms

For u a k-differential form on Ω, define

$$I_k(u) = J_k(u) + \lambda \int_\Omega |u-u_0|^2 g(x)\, dx\,,$$

with u_0 a k-form in $L^2(g(x)\,dx)$. Functionals like I_k enter in the description of some phenomena in condensed matter Physics such as superconductivity and superfluidity (*e.g.* Bose-Einstein condensation). For instance, in case $n = 3$, $\lambda = 1$, $u_0(x) = \omega(xdy - ydx)$, $\Omega = \{\rho(x) \geq 0\}$ and $g(x) = \rho(x)$, the functional I_1 arises as a reduced model describing, in suitably scaled units, the vortex density distribution in a trapped Bose-Einstein condensate rotating around the z-axis with an angular velocity $\omega > 0$. Here u corresponds to a superfluid current density, so that its exterior differential (corresponding to the curl) describes the vorticity of the superfluid. The derivation of this model as a limiting description of the Gross-Pitaevsky energy functional in certain asymptotic regimes has been rigorously proved in [5, 6] through a Γ-convergence analysis valid for general Ginzburg-Landau type models. In axisymmetric domains (see [2]) the superfluid current can be expressed, in cylindrical coordinates, by $u = v(r,z)d\theta$, so that the situation is actually described by a weighted TV regularization functional $I_0(v)$, defined on $\tilde{\Omega} = \Omega \cap \{\theta = \text{cost.}\}$ with $g(r,z) = r^{-1}\rho(r,z)$.

Finally, the functional I_{n-1}, corresponding to the discretization of the L^2-gradient flow $u_t = -\partial J_{n-1}(u)$, has been studied in [8], where some rigorous connection with a (weak formulation of the) Hele-Shaw model in phase transitions has been established.

In the following sections we will analyze some properties of J_k and I_k focusing mainly on the cases $k = 1$ and $k = n - 1$.

2 Gradient flow of J_k

In order to analyze the gradient flow (1.4) for the functional J_k one is led to consider first the approximating scheme (1.5) given, fixing $\epsilon > 0$, by the minimum problem (we consider for simplicity uniform densities $\rho = g = 1$)

$$\min I_k(u) = J_k(u) + \int_\Omega \frac{1}{2\epsilon}|u - f|^2\, dx, \tag{2.1}$$

for a given k-form $f \in L^2(\Omega, \Lambda^k(\mathbb{R}^n))$. The Euler-Lagrange equation corresponding to (2.1) is

$$\frac{f-u}{\epsilon} \in \partial J_k(u)\,.$$

Notice that since J_k is convex positively 1-homogeneous we have $\eta \in \partial J_k(u)$ if and only if

$$J_k(u) = \langle \eta, u\rangle \text{ and } \langle \eta, w\rangle \le J_k(w)\ \forall\, \eta \in \partial J_k(u),\ \ \forall\, w \in L^2(\Omega; \Lambda^k\mathbb{R}^n). \tag{2.2}$$

In particular, the convex conjugate function J_k^* corresponds to the indicatrix function $J_k^*(\eta) = 0$ if $\sup\langle \eta, w\rangle \le J(w)$, and $J_k^*(\eta) = +\infty$ otherwise. An element $\eta \in \partial J_k(u)$ may be represented as $\eta = d^*v$ (in the distributional sense), for a $(k+1)$-form v such that $|v| \le 1$ a.e. in Ω and $(*v)_T = *(v_N) = 0$ on $\partial\Omega$, where d^* correspond to the adjoint operator to d with respect to the Hodge star duality, and ω_T (resp. ω_N) denotes the tangential (resp. normal) component of a form ω on $\partial\Omega$: in particular we have $(*\omega)_T = *(\omega_N)$, and the boundary condition $(*v)_T = 0$ corresponds to the vanishing of the normal component of v on $\partial\Omega$. Moreover, the condition $J_k(u) = \langle \eta, u\rangle = \int_\Omega v \cdot du$ implies that $v = du/|du|$ (and in particular $|v| = 1$) whenever $du \neq 0$. We may hence write the Euler-Lagrange equation (2.1) as

$$\frac{f-u}{\epsilon} = d^*v \text{ in } \Omega \qquad \text{and} \qquad (*v)_T = 0 \text{ on } \partial\Omega\,. \tag{2.3}$$

Accordingly, the gradient flow (1.4) corresponds to the differential system

$$u_t = d^*v \text{ in } \Omega \qquad \text{and} \qquad (*v)_T = 0 \text{ on } \partial\Omega \tag{2.4}$$

under the constraints $|v| \le 1$ and $J_k(u) = \int_\Omega v \cdot du$ valid for any time $t > 0$.

2.1 Dual formulation

Equation (2.3) is equivalent to

$$u \in \partial J_k^* \left(\frac{f-u}{\epsilon} \right), \tag{2.5}$$

where

$$J_k^*(\eta) := \sup_{w \in L^2(\Omega, \Lambda^k \mathbb{R}^n)} \int_\Omega \eta \cdot w \, dx - J_k(w) = \begin{cases} 0 & \text{if } \|\eta\|_* \le 1 \\ +\infty & \text{otherwise} \end{cases}$$

and

$$\|\eta\|_* = \sup \left\{ \int_\Omega \eta \cdot w \, dx \, : \, J_k(w) \le 1 \right\}.$$

Note that

$$J_k(w) + J_k^*(\eta) \ge \int_\Omega w \cdot \eta \, dx$$

for all w, η. The equality holds iff $\int_\Omega \eta \cdot w \, dx = J_k(w)$, and in such case we have $\|\eta\|_* \le 1$.

Letting u be a minimizer of (2.1) and $\eta = (f-u)/\epsilon$ we get from (2.5)

$$\eta - \frac{f}{\epsilon} + \frac{1}{\epsilon} \partial J_k^*(\eta) \ni 0 \tag{2.6}$$

which shows that η is the unique minimizer of

$$\min_\eta J_k^*(\eta) + \frac{\epsilon}{2} \int_\Omega \left| \eta - \frac{f}{\epsilon} \right|^2 dx = \frac{\epsilon}{2} \min_{\|\eta\|_* \le 1} \int_\Omega \left| \eta - \frac{f}{\epsilon} \right|^2 dx. \tag{2.7}$$

This corresponds to the dual problem of (2.1), which can be interpreted as the L^2-projection of f on the convex set $\{\|\eta\|_* \le \epsilon\}$. In particular, we deduce the existence of a critical threshold ϵ_c below which minimizers are necessarily trivial (see also [18] for the same result in the case of the Total Variation model corresponding to J_0). This can be summarized in the following

Proposition 2.1. *The k-form* $u \equiv 0$ *is a minimizer of* (2.1) *if and only if*

$$\epsilon \ge \epsilon_c := \|f\|_* . \tag{2.8}$$

2.2 Non local obstacle-type problems

Note that $\|\eta\|_* < \infty$ implies that

$$\int_\Omega \eta w = 0$$

for all w such that $dw = 0$. By Hodge decomposition, this implies that $\eta = d^* g$ for some $(k+1)$-form g, with $g_N = 0$ on $\partial\Omega$. It follows that

$$\begin{aligned}\|\eta\|_* &= \sup_{\int_\Omega |dw|\le 1} \int_\Omega d^* g \cdot w\, dx \\ &= \sup_{\int_\Omega |dw|\le 1} \int_\Omega g \cdot dw\, dx + \int_{\partial\Omega} w \wedge *g_N \\ &= \sup_{\int_\Omega |dw|\le 1} \int_\Omega g \cdot dw\, dx.\end{aligned} \tag{2.9}$$

We then get the following characterization of the norm $\|\cdot\|_*$:

$$\|\eta\|_* = \inf_{\substack{d^* g = \eta \\ g_N|_{\partial\Omega}=0}} \|g\|_{L^\infty(\Omega, \Lambda^{k+1}(\mathbb{R}^n))}. \tag{2.10}$$

We may hence view the norm $\|\cdot\|_*$ related to the dual convex function J_k^* as a non-local L^∞ norm and, accordingly, the dual problem (2.7) can be interpreted as a non-local vector-valued version of the classical obstacle problem, which reduces to the classical one in the case $k = n-1$, since in (2.9) we may choose n-forms dw such that the functions $*dw$ approximate a Dirac mass located where the essential supremum of $|g|$ is attained.

Let us briefly deduce (2.10). Indeed, it is immediate to show the $\le$ inequality. On the other hand, by the Hahn-Banach Theorem, there exists a differential form $g' \in L^\infty(\Omega; \Lambda^{k+1}(\mathbb{R}^n))$, with $d^* g' = d^* g = \eta$ (in the distributional sense) such that

$$\|\eta\|_* = \sup_{\int_\Omega |dw|\le 1} \int_\Omega g \cdot dw\, dx = \sup_{\int_\Omega |\psi|\le 1} \int_\Omega g' \cdot \psi\, dx = \|g'\|_{L^\infty(\Omega; \Lambda^{k+1}(\mathbb{R}^n))}.$$

Fix now φ_0 such that $d^*\varphi_0 = \eta$. We can write $g = \varphi_0 + d^*\psi$, so that (2.10) becomes

$$\|\eta\|_* = \min_{\psi:\, (\varphi_0 + d^*\psi)\cdot\nu_A = 0} \|\varphi_0 + d^*\psi\|_{L^\infty(A)}. \tag{2.11}$$

The Euler-Lagrange equation of (2.11) is a kind of generalization of the *infinity Laplacian equation* (see *e.g.* [7,20])

$$\mathrm{d}_\infty(\varphi_0 + d^*\psi) = 0.$$

Indeed when $k = n - 2$, by duality, problem (2.11) becomes

$$\min_{\psi \in W_0^{1,\infty}(\Omega)} \|\nabla\psi + \varphi_0\|_{L^\infty(\Omega)}, \tag{2.12}$$

whose corresponding Euler-Lagrange equation is

$$\langle(\nabla^2\psi + \nabla\varphi_0)(\nabla\psi + \varphi_0), (\nabla\psi + \varphi_0)\rangle = 0, \tag{2.13}$$

which is a non-homogeneous ∞-Laplacian equation reminiscent of the Aronsson problem [1]. For such class of equations it is not known if there are conditions on φ_0 guaranteeing uniqueness of solutions.

2.3 Some properties of the gradient flow of J_{n-1}

When $k = n - 1$ we can identify by duality a square integrable $(n-1)$-form in Ω with a vector field $u \in L^2(\Omega, \mathbb{R}^N)$, so that J_{n-1} is equivalent to the functional

$$J_{n-1}(u) := \int_\Omega |\nabla \cdot u|\, dx \tag{2.14}$$

that is, the total mass of $\nabla \cdot u$ as a measure.

The gradient flow of J_{n-1} is actualy equivalent to a constrained variational problem for a function w such that $\Delta w = \nabla \cdot u$. Consider indeed the formulation

$$\begin{cases} u_t &= \nabla v \\ u(0) &= u_0 \end{cases} \tag{2.15}$$

where v satisfies $|v| \le 1$ and

$$J_{N-1}(u) + \int_\Omega u \cdot \nabla v = 0.$$

It is well-known that the solution of (2.15) is unique and that $-\nabla v(t) = \partial^0 J_{N-1}(u(t))$ is the right-derivative of $u(t)$ at any $t \ge 0$ [9]. Given the solution $(u(t), v(t))$ of (2.15), we let $w(t) := \int_0^t v(s)\, ds$, which takes its values in $[-t, t]$. It holds $u(t) = u_0 + \nabla w(t)$, and the function $w(t)$ solves the following obstacle problem (see [8]):

$$\min\left\{\frac{1}{2}\int_\Omega |u_0 + \nabla w|^2\, dx \,:\, w \in H_0^1(\Omega)\,, |w| \le t \text{ a.e.}\right\}. \tag{2.16}$$

Observe that in case we additionally have $\nabla \cdot u_0 \geq \alpha > 0$, this obstacle problem is known to be an equivalent formulation of the Hele-Shaw flow [14, 16] (see also [17] for a viscosity formulation). Therefore, it turns out that the flow of J_{n-1} provides a (unique) global weak solution to the Hele-Shaw problem, under suitable regularity assumptions on the initial datum u_0. Moreover this formulation allows to consider quite general initial data u_0, for which for instance $\nabla \cdot u_0$ may change sign, or be a measure. Further regularity properties of the function $w(t)$ and the evolution law of the contact set are deduced via comparison principles (see [8] for details).

3 The functional I_1

Let us turn to analyze the functional I_1 which expresses a Moreau-Yosida regularization of the functional $J_1(u) = \int_\Omega \rho |du|$, *i.e.* the total (weighted) mass of the exterior differential of a square integrable 1-form u. In the case $N = 3$ (and $\rho = 1$) it corresponds to the total mass of the curl of a vector field $u \in L^2(\Omega; \mathbb{R}^N)$ as a measure. This type of functional arises as a reduced description of vortex density in superfluids and respectively superconductors corresponding to asymptotic regimes of the three dimensional Gross-Pitaevskii model of Bose-Einstein condensates (resp. the Ginzburg-Landau model for 3-d superconductivity)

3.1 Asymptotics for the Gross-Pitaevskii model

Consider a Bose-Einstein condensate with mass m confined in a domain $\Omega \subset \mathbb{R}^3$ by a smooth trapping nonnegative potential $0 \leq a \in C^\infty(\mathbb{R}^3)$, $a(x) \to +\infty$ as $|x| \to +\infty$, and subjected to a forcing Φ_ϵ that in general depends on a scaling parameter ϵ. In the model case corresponding to a rotation around the z-axis, one has $\Phi_\epsilon := \frac{1}{2} c_\epsilon (x_1 dx^2 - x_2 dx^1)$, and $a(x)$ grows at least quadratically in $|x|$. The Gross-Pitaevskii functional in the ϵ-scaling regime reads

$$G_\epsilon(u) := \int_{\mathbb{R}^3} \frac{1}{2}|\nabla u|^2 - \Phi_\epsilon \cdot ju + \frac{1}{\epsilon^2}\left(\frac{|u|^4}{4} + a(x)\frac{|u|^2}{2}\right)$$

where u is a complex-valued wave function whose modulus describes the superfluid density, $\Phi_\epsilon = |\log \epsilon| \Phi$ for some fixed Φ and the $j(u) = \frac{i}{2}(u d\bar{u} - \bar{u} du)$ measures the superfluid current. A stable condensate may be described by a (local or global) minimizer of G_ϵ in the function space

$$H^1_a(\mathbb{R}^3; \mathbb{C}) := H^1_a := \text{completion of } C^\infty_c(\mathbb{R}^3; \mathbb{C}) \text{ with respect to } \|\cdot\|_a, \tag{3.1}$$

where the norm $\|\cdot\|_a$ is defined by $\|u\|_a^2 := \int_{\mathbb{R}^3} |du|^2 + (1+a)|u|^2$. Define also

$$H^1_{a,m}(\mathbb{R}^3;\mathbb{C}) := H^1_{a,m} := \{u \in H^1_a \ : \ \int |u|^2 = m\}.$$

In order to study the behavior of minimizers of G_ϵ in $H^1_{a,m}$ it is convenient to rewrite the energy as follows: define

$$\rho(x) := (\lambda - a(x))^+, \ \ w(x) := (\lambda - a(x))^-, \ \ \text{for } \lambda \text{ such that } \int_{\mathbb{R}^3} \rho\, dx = m. \tag{3.2}$$

The last condition clearly determines λ uniquely. The function ρ is called the *Thomas-Fermi density* in the physics literature, and gives to the leading-order the condensate density, in the limit $\epsilon \to 0$. Since $\int \lambda |u|^2 = \lambda m$ for all $u \in H^1_{a,m}$, it follows that u minimizes G_ϵ in $H^1_{a,m}$ if and only if u minimizes

$$\mathcal{G}_\epsilon(u) := \int_{\mathbb{R}^3} \frac{1}{2}|\nabla u|^2 - \Phi_\epsilon \cdot ju + \frac{1}{4\epsilon^2}(\rho - |u|^2)^2 + \frac{w}{2\epsilon^2}|u|^2 \tag{3.3}$$

in $H^1_{a,m}$. We will henceforth write the Gross-Pitaevskii functional in the more convenient form (3.3), and define $\Omega = \{x \in \mathbb{R}^3 : \rho(x) > 0\}$. The following convergence result is a consequence of the analysis in [5,6].

Proposition 3.1. *Assume that $\Phi_\epsilon = |\log\epsilon|\Phi$, with $\Phi \in L^4_{\text{loc}}(\Lambda^1\mathbb{R}^3)$ and that $|\Phi(x)|^2 \leq Ca(x)$ outside some compact set K. Assume that u_ϵ minimizes $\mathcal{G}_\epsilon$ in $H^1_{a,m}$. Then*

$$|u_\epsilon| \to \rho \quad \textit{in } L^4(\mathbb{R}^3)$$

for ρ defined in (3.2), *and there exists $j_0 \in L^{4/3}(\Lambda^1\Omega)$ such that*

$$|\log\epsilon|^{-1} ju_\epsilon \rightharpoonup j_0 \ \textit{weakly in } L^{4/3}(\mathbb{R}^3).$$

Moreover, $j_0 = \rho v_0$, where v_0 is the unique minimizer of

$$\mathcal{G}(v) := \int_\Omega \rho\left(\frac{|v|^2}{2} - v\cdot\Phi + \frac{1}{2}|dv|\right). \tag{3.4}$$

in the space

$$L^2_\rho(\Lambda^1\Omega) := \left\{v \in L^1_{\text{loc}}(\Lambda^1\Omega) \ : \ \int_\Omega \rho|v|^2\, dx < \infty\right\}. \tag{3.5}$$

(We set $\mathcal{G}(v) = +\infty$ if dv is not a Radon measure or if ρ is not $|dv|$-integrable.)

This convergence results for the Gross-Pitaevskii functional parallel those obtained in [5] for the Ginzburg-Landau functional. The functional $\mathcal{G}$ of the limiting variational problem corresponds to I_1, and the 1-form v measures the asymptotic superfluid current and hence its curl, corresponding to dv, gives a measure of the leading-order vortex density of the superfluid. From the analysis of $\mathcal{G}$ (compare with the analysis of I_1 leading to Proposition 2.1) one may thus characterize when minimizers of the limiting problem are vortex-free to the leading order, and the magnitude of the critical threshold of the forcing (resp. the first critical magnetic field in the superconductivity case) above which there is vortex nucleation, by obtaining a description of minimizers of $\mathcal{G}$ as solutions of a nonlocal vector-valued obstacle problem of the type of (2.7).

Let us now state more precisely a necessary and sufficient condition on Φ and ρ for minimizers of the limiting functional $\mathcal{G}$ to be vortex-free, by which we mean that $dv_0 = 0$ in Ω. Denoting by

$$(v, w)_\rho := \int_\Omega \rho\, v \cdot w\, dx, \qquad \|v\|_\rho := (v, v)_\rho^{1/2},$$

respectively the inner product and norm on the Hilbert space $L^2_\rho(\Lambda^1\Omega)$, let P_ρ denote the orthogonal projection with respect to the L^2_ρ inner product onto $(\ker d)_\rho$, where

$$(\ker d)_\rho := L^2_\rho\text{-closure of } \{\varphi \in C^\infty(\Lambda^1\Omega) : d\varphi = 0,\ \|\varphi\|_\rho < \infty\}. \quad (3.6)$$

We will also write $P_\rho^\perp$ for the complementary orthogonal projection. Note that if $w \in \text{Image}(P_\rho^\perp) = (\ker d)_\rho^\perp$, then $\int(\rho w) \cdot \varphi = 0$ for all $\varphi \in (\ker d)_\rho \supset \ker d$. Thus $\rho w \in (\ker d)^\perp$, and so it follows from the standard (unweighted) Hodge decomposition that

$$\forall w \in (\ker d)_\rho^\perp,\ \exists \beta \in H^1_N(\Lambda^2\Omega) \text{ such that } w = \frac{d^*\beta}{\rho} \text{ and } \int_\Omega \frac{|d^*\beta|^2}{\rho} = \|w\|^2_\rho. \quad (3.7)$$

Thus if $\Phi \in L^2_\rho$, there exists $\beta_\Phi \in H^1_N$ such that $d^*\beta_\Phi \in L^2_\rho$ and

$$\Phi = P_\rho \Phi + \frac{d^*\beta_\Phi}{\rho}. \quad (3.8)$$

We have the followig result (see [6]):

Theorem 3.2. *Suppose that Ω is a bounded, open subset of $\mathbb{R}^3$ and that $0 \le \rho \in C^1(\Omega)$ and $\Phi \in L^4_{loc}(\Lambda^1\mathbb{R}^3) \cap L^2_\rho(\Lambda^1\Omega)$ are given. Let $\beta_\Phi \in H^1_N(\Lambda^2\Omega)$ be such that $P_\rho^\perp\Phi = \frac{d^*\beta_\Phi}{\rho}$, and let β_0 minimize the functional*

$$\beta \mapsto \frac{1}{2}\int_\Omega \frac{|d^*\beta|^2}{\rho} \quad (3.9)$$

in the space

$$\left\{\beta \in H^1_N(\Lambda^2\Omega) \ : \ \frac{d^*\beta}{\rho} \in L^2_\rho(\Lambda^1\Omega), \ \ \|\beta - \beta_\Phi\|_{\rho*} \le \frac{1}{2}\right\}, \qquad (3.10)$$

where

$$\|\beta\|_{\rho*} := \sup\left\{\int_\Omega \beta \cdot dw : \ w \in C^\infty(\Lambda^1\bar{\Omega}), \ \int_\Omega \rho|dw| \le 1\right\}. \qquad (3.11)$$

Then $v_0 = P_\rho\Phi + \frac{d^*\beta_0}{\rho}$ *is the unique minimizer of* $\mathcal{G}(\cdot)$ *in* $L^2_\rho(\Lambda^1\Omega)$. *Moreover,*

$$\int_\Omega (\beta_\Phi - \beta_0)\cdot dv_0 = \frac{1}{2}\int_\Omega \rho|dv_0|\,. \qquad (3.12)$$

Finally, $dv_0 = 0$ *if and only if* $\|\beta_\Phi\|_{\rho*} \le \frac{1}{2}$.

Note that (3.12) states that the action of the vorticity distribution dv_0 on the potential $\beta_0 - \beta_\Phi$ is the largest possible given the constraint (3.10).

Observe that the form of the constraint in the limiting dual variational problem depends on the dimension. In particular, in 2d, as in 3d, it is the case that if v_0 minimizes $\mathcal{G}$, then $dv_0 = d(\frac{d^*\beta_0}{\rho})$, where the potential β_0 minimizes the functional (3.9) subject to the constraint (3.10). The difference is that in 2d, the potentials β are 2-forms on $\mathbb{R}^2$, and so they can be identified with functions. Since it is not hard to check that $\{d\omega : \int_\Omega \rho|d\omega| \le 1\}$ is weakly dense in the set of signed measures μ such that $\int_\Omega \rho d|\mu| \le 1\}$, the 2d constrained problem reduces to minimizing (3.9) in the set

$$\left\{\beta \in H^1(\Lambda^2\Omega) : \|\frac{1}{\rho}(\beta - \beta_\Phi)\|_{L^\infty} \le \frac{1}{2}\right\}. \qquad (3.13)$$

This is a classical (weighted) 2-sided obstacle problem, and for many Φ's, using the maximum principle, it in fact reduces to a one-sided obstacle problem. Thus we can view the problem in Theorem 3.2 as a nonlocal, vector-valued analog of the classical obstacle problem.

3.2 Rotational symmetry and weighted TV minimization

In the presence of rotational symmetry, the functional $\mathcal{G}$ reduces to a simpler 2-dimensional model corresponding to the weighted Total Variation minimization functional I_0. More precisely, assume that there exist $\tilde{\Omega} \subset [0,\infty) \times \mathbb{R}$, $\tilde{\rho} : \tilde{\Omega} \to (0,\infty)$ and $\varphi : \tilde{\Omega} \to \mathbb{R}$ such that $\Omega = \{(r\cos\alpha, r\sin\alpha, z) : (r,z) \in \tilde{\Omega}, \alpha \in \mathbb{R}\}$, $\rho(r\cos\alpha, r\sin\alpha, z) = \tilde{\rho}(r,z)$ and $\Phi(r\cos\alpha, r\sin\alpha, z) = \varphi(r,z)d\theta$.

Then it is easy to see that the unique minimizer v_0 of $\mathcal{G}$ is given in cylindrical coordinates by $v_0 = w_0(r,z)d\theta$, where w_0 minimizes the functional (of the type I_0)

$$\mathcal{G}^{red}(w) := \frac{1}{2}\int_{\tilde{\Omega}} \tilde{\rho}\left(|\nabla w| + \frac{(w-\varphi)^2}{r}\right) dr\, dz \tag{3.14}$$

in the space of functions $w : \tilde{\Omega} \to \mathbb{R}$ such that $\int_{\tilde{\Omega}} \frac{\tilde{\rho}}{r} w^2\, dr\, dz < \infty$.

One can use duality to rewrite the problem of minimizing $\mathcal{G}^{red}$ as a constrained variational problem. For instance, one can verify that v_0 minimizes $\mathcal{G}^{red}$ if and only if it minimizes the functional

$$w \mapsto \int_{\tilde{\Omega}} \frac{\tilde{\rho}}{r} w^2\, dr\, dz \tag{3.15}$$

subject to the constraint

$$\int_{\tilde{\Omega}} \frac{\tilde{\rho}}{r}(\varphi - w)\zeta\, dr\, dz \le \frac{1}{2}\int_{\tilde{\Omega}} \tilde{\rho}|\nabla\zeta| \qquad \text{for all } \zeta \in C^\infty(\tilde{\Omega}). \tag{3.16}$$

For the velocity field represented by the 1-form $v = w(r,z)d\theta$, the associated vorticity 2-form is $dv = \partial_r w\, dr \wedge d\theta + \partial_z w\, dz \wedge d\theta$. The vorticity vector field, that is, the vector field dual to dv, is then $\frac{1}{r}(\partial_r w\, \hat{e}_z - \partial_z w\, \hat{e}_r)$, where $\hat{e}_z$ and $\hat{e}_r$ denote unit vectors in the (upward) vertical and (outward) radial directions respectively. It is natural to interpret integral curves of this vector field as "vortex curves". Since the vorticity vector field has no $\hat{e}_\theta$ component and is always tangent to the level surfaces of w, we conclude that vortex curves have the form "θ = constant, w = constant", at least for regular values of w. Thus in the reduced 2d model, we interpret level sets of a minimizer w_0, or more precisely sets of the form $\partial\{(r,z) : w_0(r,z) > t\}$, as representing vortex curves. For similar reasons, one should think to the "vorticity measure" as being given by $\nabla^\perp w_0$, rather than ∇w_0.

3.3 Contact curves and vortex curves

It is interesting to ask whether one can define a useful analog of the "contact set" (as normally defined for classical obstacle problems) for the variational problems with nonlocal constraints formulated in Theorem 3.2. We address this question first for Bose-Einstein condensates in the presence of rotational symmetry, as discussed above. Thus, we assume that $w_0 : \tilde{\Omega} \to \mathbb{R}$ minimizes the functional (3.15) subject to the constraint (3.16). Starting from (3.16), an approximation argument shows that if E

is a set of locally finite perimeter in $\tilde{\Omega}$, then

$$\int \frac{\tilde{\rho}}{r}(\varphi - w_0)\chi_E \, dr \, dz \le \frac{1}{2}\int \tilde{\rho}|\nabla \chi_E|, \tag{3.17}$$

where χ_E denotes the characteristic function of E. We say that ∂E is a *contact curve* if equality holds in (3.17) (where ∂E should be understood as the 1-dimensional set that carries $|\nabla \chi_E|$).

Lemma 3.3. *For a.e. t, $\partial\{w_0 > t\}$ is a contact curve.*

It is natural to interpret $\partial\{w_0 > t\}$ as a "vortex curve", so the Lemma states, heuristically, that every vortex curve for w_0 is also a contact curve.

Proof. By using rotational symmetry to rewrite (3.12) in the (r, z) variables, or by using the fact that $0 = \frac{d}{dt}\mathcal{G}^{red}(e^t w_0)\big|_{t=0}$, we find that

$$\frac{1}{2}\int \tilde{\rho}|\nabla w_0| + \int \frac{\tilde{\rho}}{r}(w_0 - \varphi)w_0 \, dr \, dz = 0.$$

Using the coarea formula, we then get

$$\int_{-\infty}^{\infty}\left(\frac{1}{2}\int \tilde{\rho}|\nabla \chi_{\{w_0>t\}}| + \int \frac{\tilde{\rho}}{r}(w_0 - \varphi)\chi_{\{w_0>t\}} \, dr \, dz\right) dt = 0. \tag{3.18}$$

It follows from (3.17) that

$$\frac{1}{2}\int \tilde{\rho}|\nabla \chi_{\{w_0>t\}}| + \int \frac{\tilde{\rho}}{r}(w_0 - \varphi)\chi_{\{w_0>t\}} \, dr \, dz \ge 0$$

for every t, and then (3.18) implies that the equality holds for a.e. t. □

It is probably not true that every contact curve for the minimizer w_0 is also a vortex curve, in the generality that we consider here, due to the possibility of degenerate (nonlocal) obstacles, as in the classical obstacle problem. One might hope, however, that the vortex curves and contact curves coincide under reasonable physical assumptions (for example $\Phi = r^2 d\theta$, corresponding to a rotation of a condensate around the z axis).

In the work [2] we have investigated (also numerically) further properties of the contact set: in particular we prove that vortex curves are smooth, of finite length, and meet orthogonally the boundary of $\tilde{\Omega}$ (see [19], and compare also with [13] in the superconductivity case). Moreover, the level set corresponding to $\sup w_0$ is necessarily flat, hence the union of vortex curves forms a proper subset of $\tilde{\Omega}$. In the peculiar case of $\tilde{\Omega}$ being an ellipsoid with suitable eccentricity, following [3] one can also prove that the level set corresponding to $\inf w_0$ is also flat, whence one deduces the existence of a vortex-free zone around the rotation axis.

The situation is more complicated for Bose-Einstein condensates in a general domain $\Omega \subset \mathbb{R}^3$ without rotational symmetry, since in this case the analogs of vortex curves and contact curves may not in fact be curves and do not in general admit a very easy concrete characterization. Abstractly, they may be described as follows: if we write $\mathcal{Z}$ to denote the closure (in the sense of distributions) of

$$\{d\alpha : \alpha \in L^2(\Lambda^1\Omega), \int_\Omega \rho|d\alpha| \leq 1\},$$

then one can think of the set $\mathrm{extr}\mathcal{Z}$ of extreme points of (the convex set) $\mathcal{Z}$ as analogous to the objects — distributional boundaries of sets of finite weighted perimeter — used above to describe vortex and contact curves. Indeed, by the arguments in Remark 3 of [22] and general convexity considerations, one can show that $\mathrm{extr}\mathcal{Z}$ is a nonempty Borel subset of a suitable metric space, and for any T in the vector space generated by $\mathcal{Z}$ (that is, the space $\cup_{\lambda>0}\lambda\mathcal{Z}$), there is a measure μ_T on $\mathrm{extr}\mathcal{Z}$ such that

$$T = \int_{\mathrm{extr}\mathcal{Z}} \omega \, d\mu_T(\omega) \tag{3.19}$$

and

$$\int_\Omega \rho\, d|T| = \int_{\mathrm{extr}\mathcal{Z}} \left(\int_\Omega \rho\, d|\omega| \right) d\mu_T(\omega). \tag{3.20}$$

We remark that in the closely related situation of divergence-free vector fields on $\mathbb{R}^n$, a concrete characterization of elements of the analog of $\mathrm{extr}\mathcal{Z}$ as "elementary solenoids" is established in [22].

With this notation, an analog of Lemma 3.3 is

Lemma 3.4. *Let β_0 be the minimizer of the constrained variational problem* (3.9), (3.10)*), so that $v_0 = P_\rho\Phi + \frac{d^*\beta_0}{\rho}$ is the minimizer of $\mathcal{G}(\cdot)$. Then*

$$\int_\Omega (\beta_\Phi - \beta_0) \cdot d\omega \leq \frac{1}{2} \int_\Omega \rho d|\omega|. \tag{3.21}$$

for every $\omega \in \mathcal{Z}$. We say that $\omega \in$ extr$\mathcal{Z}$ *is a "generalized contact curve" if the above condition holds with equality.*

Furthermore, let μ_{dv_0} denote a measure on extr$\mathcal{Z}$ satisfying (3.19), (3.20) *(with T replaced by dv_0). Then μ_{dv_0} a.e. ω is a generalized contact curve.*

The proof is exactly as in Lemma 3.3, except for the fact that (3.19), (3.20) are used instead of the coarea formula. Then (3.21) follows immediately from the fact that β_0 satisfies (3.10), and the last assertion is a consequence of (3.12).

It would presumably be possible to adapt the results of [22] to the closely related situations considered here, in order to obtain concrete descriptions of extr$\mathcal{Z}$, although we are not sure that this would add much insight. It would also be interesting to know whether, if we consider the model case of uniform rotation about the z axis (for Bose-Einstein) or a constant applied magnetic field (for Ginzburg-Landau), the complexities sketched above do not in fact occur, and the vortex curves and contact curves for minimizers can in fact be identified with curves of finite length; this seems likely to be the case.

References

[1] G. ARONSSON, *Minimization problem for the functional* $\sup_x F(x, f(x), f'(x))$, Ark. Mat. **6** (1965), 33–53.

[2] P. ATHAVALE, R. L. JERRARD, M. NOVAGA and G. ORLANDI, *Weighted TV minimization and applications to vortex density models,* preprint, 2013.

[3] A. AFTALION and R. L. JERRARD, *Shape of vortices for a rotating Bose-Einstein condensate*, Physical Review A (2) **66** (2002), 023611/1–023611/7.

[4] F. ALTER, V. CASELLES and A. CHAMBOLLE, *A characterization of convex calibrables sets in* $\mathbb{R}^N$, Math. Annalen **332** (2005), 329–366.

[5] S. BALDO, R. L. JERRARD, G. ORLANDI and H. M. SONER, *Convergence of Ginzburg-Landau functionals in 3-d superconductivity*, Archive Rat. Mech. Analysis (3) **205** (2012), 699–752.

[6] S. BALDO, R. L. JERRARD, G. ORLANDI and H. M. SONER, *Vortex density models for superconductivity and superfluidity*, Comm. Math. Phys. (1) **318** (2013), 131–171.

[7] E. N. BARRON, L. C. EVANS and R. JENSEN, *The infinity Laplacian, Aronsson's equation and their generalizations*, Trans. Amer. Math. Soc. (1) **360** (2008), 77–101.

[8] A. BRIANI, A. CHAMBOLLE, M. NOVAGA and G. ORLANDI, *On the gradient flow of a one-homogeneous functional*, Confluentes Mathematici (4) **3**, 617–635.

[9] H. BRÉZIS, "Opérateurs maximaux monotones et semi-groupes de contractions dans les espaces de Hilbert", North-Holland, Amsterdam, 1973.

[10] G. CARLIER and M. COMTE, *On a weighted total variation minimization problem*, J. Funct. Anal. (1) **250** (2007), 214–226.

[11] V. CASELLES, A. CHAMBOLLE and M. NOVAGA, *Total Variation in imaging*, In: "Handbook of Mathematical Methods in Imaging", Springer, 2011, 1016–1057.

[12] V. CASELLES, G. FACCIOLO and E. MEINHARDT, *Anisotropic Cheeger Sets and Applications*, SIAM J. Imaging Sciences, (4) **2** (2009), 1211–1254.

[13] D. CHIRON, *Boundary problems for the Ginzburg-Landau equation*, Commun. Contemp. Math. **7** (2005), 597–648.

[14] C. M. ELLIOTT and V. JANOVSKÝ, *A variational inequality approach to Hele-Shaw flow with a moving boundary*, Proc. Roy. Soc. Edinburgh Sect. A **88** (1981), 93–107.

[15] I. IONESCU and T. LACHAND-ROBERT, *Generalized Cheeger sets related to landslides*, Calc. Var. Partial Differential Equations (2) **23** (2005), 227–249.

[16] B. GUSTAFSSON, *Applications of Variational inequalities to a moving boundary problem for Hele Shaw flows*, Siam J. Math. Anal. **16** (1985), 279–300.

[17] C. I. KIM and A. MELLET, *Homogenization of a Hele-Shaw problem in periodic and random media*, Arch. Rat. Mech. Anal. **194** (2009), 507–530.

[18] Y. MEYER, "Oscillating Patterns in Image Processing and Nonlinear Evolution Equations", University Lecture Series, 22. American Mathematical Society, Providence, RI, 2001.

[19] A. MONTERO and B. STEPHENS, *On the geometry of Gross-Pitaevskii vortex curves for generic data*, Proceedigs of the A.M.S., to appear, 2012.

[20] Y. PERES, O. SCHRAMM, S. SHEFFIELD and D. WILSON, *Tug-of-war and the infinity Laplacian*, J. Amer. Math. Soc. (1) **22** (2009), 167–210.

[21] L. RUDIN, S. J. OSHER and E. FATEMI, *Nonlinear total variation based noise removal algorithms*, Physica D **60** (1992), 259–268.

[22] S. K. SMIRNOV, *Decomposition of solenoidal vector charges into elementary solenoids, and the structure of normal one-dimensional flows*, Rossiĭskaya Akademiya Nauk. Algebra i Analiz **5** (1993), 206–238.

Maximally localized Wannier functions: existence and exponential localization

Adriano Pisante

Abstract. We describe recent results proved in [32] in collaboration with G.Panati, concerning a periodic Schrödinger operator and the maximally localized (composite) Wannier functions corresponding to a relevant family of its Bloch bands. More precisely, we discuss the minimization problem for the associated localization functional introduced in [22] and we review some rigorous results about the existence and exponential localization of its minimizers, in dimension $d \leq 3$. The proof combines ideas and methods from the Calculus of Variations and the regularity theory for harmonic maps between Riemannian manifolds.

1 Introduction

Many transport properties of electrons in crystalline solids are understood by the analysis of Schrödinger operators in the form

$$H = -\Delta + V_\Gamma(x) \qquad \text{acting in } L^2(\mathbb{R}^d) \tag{1.1}$$

where the function $V_\Gamma : \mathbb{R}^d \to \mathbb{R}$ is periodic with respect to a Bravais lattice $\Gamma \simeq \mathbb{Z}^d$. The function V_Γ represents (in Rydberg units) the electrostatic potential experienced by a test electron and generated by the ionic cores of the crystalline solid and, in a mean-field approximation, by the remaining electrons (see *e.g.* [5]).

A crucial problem in solid state physics is the construction of an orthonormal basis, canonically associated to the operator H, consisting of functions which are exponentially localized in space. The relevance comes both from computational and from theoretical issues (see *e.g.* [11], [13,26], [18,34]).

A convenient basis has been proposed by Wannier [41], and *Wannier functions* are nowadays a fundamental tool in solid state physics. The problem of proving the existence of exponentially localized Wannier functions was raised in 1959 by W. Kohn [19], who solved it in dimension $d = 1$ in the case of a single isolated Bloch band for a centrosymmetric potential. The latter condition has been later removed by J. des

Cloizeaux [8]. In higher dimension, the problem has been solved, always in the case of a single isolated Bloch band in [7,8] for centrosymmetric potentials and finally under general hypotheses in [27].

However, in dimension $d > 1$ the Bloch bands of crystalline solids are not, in general, isolated. Thus the interesting problem, in view of real applications, concerns the case of *multiband systems*, and in this context the more general notion of *composite Wannier functions* is relevant [1,7]. The existence of exponentially localized composite Wannier functions has been proved in [28] in dimension $d = 1$. As for $d > 1$, this problem remained unsolved until recently [3,31], where the first existence result was obtained.

To circumvent such conceptual difficulty, and in view of the application to numerical simulations, the solid-state physics community preferred to introduce the alternative notion of *maximally localized Wannier functions* [22]. The latter are defined as the minimizers of a suitable localization functional, known as the Marzari-Vanderbilt (MV) functional. In [22] it is also conjectured that the minimizers, whenever they exist, are exponentially localized. While such an approach provided excellent results in computational physics [23], a mathematical analysis of the MV functional was still missing until [32].

In [32] it is shown that minimizers of the MV functional do exist for $d \leq 3$ in a suitable function space and, more relevantly, they are exponentially localized. More precisely, let $m \geq 1$ be the number of Bloch bands corresponding to the composite Wannier functions; then the result is proved in three cases: if $m = 1$ for any $d \geq 1$, if $1 \leq d \leq 2$ for any $m \geq 1$, and if $d = 3$ under the constraint $2 \leq m \leq 3$. The proofs are dimension-dependent. In the first two cases, exponential localization holds true for any stationary point of the MV functional, *i.e.* for any solution of the corresponding Euler-Lagrange equations. On the other hand, in the three dimensional case the minimality property enters in a crucial way and the limitations on m arises from the methods exploited in the proof and they will be discussed in detail below.

Mathematically, the MV functional can be rewritten, after Bloch-Floquet transform and by exploiting the time-reversal symmetry of the operator (1.1) (*i.e.* the fact that H commutes with the complex-conjugation operator), as a perturbation $\tilde{F}_{\mathrm{MV}}$ of the Dirichlet energy for maps from $\mathbb{T}_d^*$ to $\mathcal{U}(m)$, where $\mathbb{T}_d^* \simeq \mathbb{R}^d/(2\pi\mathbb{Z})^d$ is a d-dimensional flat torus and $\mathcal{U}(m) \subset \mathrm{M}_m(\mathbb{C})$ is the unitary group. The exponential localization of a minimizer of the MV functional is related to the analyticity of the corresponding minimizer of $\tilde{F}_{\mathrm{MV}}$ via a Paley-Wiener type theorem. The existence of a minimizer of the latter functional follows essentially from the direct method of calculus of variations.

In [32], analyticity of the minimizers of $\tilde{F}_{\mathrm{MV}}$ is obtained by adapting ideas and methods from the regularity theory for harmonic maps (see [6, 20, 32] and references therein). The crucial step is to prove that any minimizer of $\tilde{F}_{\mathrm{MV}}$ is continuous. In the two dimensional case, this fact is a consequence of the hidden structure of the nonlinear terms in the Euler-Lagrange equation for the $\tilde{F}_{\mathrm{MV}}$ functional. In the three dimensional case, the continuity follows instead from the deeper fact that minimizers at smaller and smaller scales look like minimizing harmonic maps from $\mathbb{T}_d^*$ to $\mathcal{U}(m)$. Thus, in the latter case the key technical point is to show constancy of the tangent maps as in the important paper [38] and that is precisely the point where in the present proof the constraint on m comes into play. As a consequence, under the limitation mentione above, continuity of the minimizers of $\tilde{F}_{\mathrm{MV}}$ holds and, in turn, analytic regularity.

2 Wannier functions and Bloch bundles

In this section we recall, following [30], that, in view of their invariance under Γ-translations, Schrödinger operators in the form (1.1) can be decomposed as a direct integral of simpler *fiber Hamiltonian* operators by the (modified) Bloch-Floquet transform. Our definitions, in contrast with the physics literature, implies that the fiber Hamiltonian (2.10) has a k-independent domain. As it will be clear below, this fact is very convenient from the mathematical viewpoint.

The lattice Γ, corresponding to the Bravais lattice in physics, is described as

$$\Gamma = \left\{\gamma \in \mathbb{R}^d : \gamma = \textstyle\sum_{j=1}^{d} n_j \, \gamma_j \ \text{ for some } n_j \in \mathbb{Z}\right\},$$

where $\{\gamma_1, \ldots, \gamma_d\}$ are fixed linearly independent vectors in $\mathbb{R}^d$. The dual lattice, w.r.t. the ordinary inner product, is $\{\Gamma^* := \{k\} \in \mathbb{R}^d : k \cdot \gamma \in 2\pi\mathbb{Z}$ for all $\gamma \in \Gamma\}$. To fix the notation, we denote by Y the centered fundamental domain of Γ, namely

$$Y = \left\{x \in \mathbb{R}^d : x = \textstyle\sum_{j=1}^{d} \alpha_j \, \gamma_j \ \text{ for } \alpha_j \in [-\frac{1}{2}, \frac{1}{2}]\right\}.$$

Analogously, we define the centered fundamental domain Y^* of Γ^* by setting

$$Y^* = \left\{k \in \mathbb{R}^d : k = \textstyle\sum_{j=1}^{d} k'_j \, \gamma_j^* \ \text{ for } k'_j \in [-\frac{1}{2}, \frac{1}{2}]\right\},$$

where $\{\gamma_j^*\}$ is the dual basis to $\{\gamma_j\}$, *i.e.* $\gamma_j^* \cdot \gamma_i = 2\pi\delta_{j,i}$. When the opposite faces of Y^* are identified, one obtains the torus $\mathbb{T}_d^* := \mathbb{R}^d/\Gamma^*$, equipped with the flat Riemannian metric induced by $\mathbb{R}^d$.

For $\psi \in \mathcal{S}(\mathbb{R}^d)$, one defines the modified Bloch-Floquet transform as

$$(\tilde{\mathcal{U}}_{\mathrm{BF}}\psi)(k,y) := \frac{1}{|Y^*|^{\frac{1}{2}}} \sum_{\gamma\in\Gamma} \mathrm{e}^{-\mathrm{i}k\cdot(y+\gamma)}\, \psi(y+\gamma), \qquad y\in\mathbb{R}^d,\ k\in\mathbb{R}^d. \tag{2.1}$$

One immediately reads the periodicity properties

$$\begin{aligned} \big(\tilde{\mathcal{U}}_{\mathrm{BF}}\psi\big)(k, y+\gamma) &= \big(\tilde{\mathcal{U}}_{\mathrm{BF}}\psi\big)(k,y) && \text{for all } \gamma\in\Gamma\,, \\ \big(\tilde{\mathcal{U}}_{\mathrm{BF}}\psi\big)(k+\lambda, y) &= \mathrm{e}^{-\mathrm{i}\lambda\cdot y}\,\big(\tilde{\mathcal{U}}_{\mathrm{BF}}\psi\big)(k,y) && \text{for all } \lambda\in\Gamma^*\,. \end{aligned} \tag{2.2}$$

For any fixed $k \in \mathbb{R}^d$, $\big(\tilde{\mathcal{U}}_{\mathrm{BF}}\psi\big)(k,\cdot)$ is a Γ-periodic function and can thus be regarded as an element of $\mathcal{H}_{\mathrm{f}} := L^2(\mathbb{T}^d)$, $\mathbb{T}^d$ being the flat torus $\mathbb{R}^d/\Gamma$. On the other hand, the second equation in (2.2) can be read as a pseudoperiodicity property, involving a unitary representation of the group Γ^*, given by

$$\tau:\Gamma^*\to\mathcal{U}(\mathcal{H}_{\mathrm{f}})\,,\quad \lambda\mapsto\tau(\lambda)\,,\quad \big(\tau(\lambda)\varphi\big)(y) = \mathrm{e}^{\mathrm{i}\lambda\cdot y}\varphi(y). \tag{2.3}$$

Following [30], it is convenient to introduce the Hilbert space

$$\mathcal{H}_\tau := \Big\{\varphi\in L^2_{\mathrm{loc}}(\mathbb{R}^d,\mathcal{H}_{\mathrm{f}}) : \varphi(k-\lambda) = \tau(\lambda)\,\varphi(k)\ \forall\lambda\in\Gamma^*, \text{ for a.e. } k\in\mathbb{R}^d\Big\}, \tag{2.4}$$

equipped with the inner product

$$\langle\varphi,\psi\rangle_{\mathcal{H}_\tau} = \int_{Y^*} dk\,\langle\varphi(k),\psi(k)\rangle_{\mathcal{H}_{\mathrm{f}}}\,.$$

Obviously, there is a natural isomorphism between $\mathcal{H}_\tau$ and $L^2(Y^*,\mathcal{H}_{\mathrm{f}})$ given by restriction from $\mathbb{R}^d$ to Y^*. The map defined by (2.1) extends to a unitary operator

$$\tilde{\mathcal{U}}_{\mathrm{BF}} : L^2(\mathbb{R}^d) \longrightarrow \mathcal{H}_\tau \simeq \int_{Y^*}^{\oplus} \mathcal{H}_{\mathrm{f}}\,dk,$$

with inverse given by

$$\left(\tilde{\mathcal{U}}_{\mathrm{BF}}^{-1}\varphi\right)(x) = \frac{1}{|Y^*|^{\frac{1}{2}}}\int_{Y^*} dk\ \mathrm{e}^{\mathrm{i}k\cdot x}\varphi(k,[x]),$$

where $[\,\cdot\,]$ refers to the a.e. unique decomposition $x = \gamma_x + [x]$, with $\gamma_x\in\Gamma$ and $[x]\in Y$.
Finally, from the definition (2.1) one easily checks that

$$\tilde{\mathcal{U}}_{\mathrm{BF}}(-\mathrm{i}X_j\psi) = \partial_{k_j}\varphi\,, \tag{2.5}$$

whenever the functions are sufficiently regular (here X_j is the multiplication operator by x_j). As a consequence, one gets

$$\begin{aligned} \psi \in W^{m,2}(\mathbb{R}^d),\ m \in \mathbb{N} &\Leftrightarrow \tilde{\mathcal{U}}_{\mathrm{BF}}\, \psi \in L^2(Y^*, W^{m,2}(\mathbb{T}^d)), \\ \langle x \rangle^m\, \psi \in L^2(\mathbb{R}^d),\ m \in \mathbb{N} &\Leftrightarrow \tilde{\mathcal{U}}_{\mathrm{BF}}\, \psi \in \mathcal{H}_\tau \cap W^{m,2}_{\mathrm{loc}}(\mathbb{R}^d, L^2(\mathbb{T}^d)), \end{aligned} \tag{2.6}$$

where, as usual, $\langle x \rangle = (1 + |x^2|)^{1/2}$.
Taking the previous equivalence into account one concludes that $\varphi \in \mathcal{H}_\tau$ is in $C^\infty(\mathbb{R}^d, \mathcal{H}_{\mathrm{f}})$, if and only if the corresponding function ψ decreases faster than the inverse of any polynomial, *i.e.* $P(X_1, \dots, X_d)\psi$ is in $L^2(\mathbb{R}^d)$ for any polynomial P.

In order to discuss exponentially decaying functions in analogy with the Paley-Wiener theorem, for any $\beta > 0$ we define

$$\Omega_\alpha = \left\{ \kappa \in \mathbb{C}^d : |\mathrm{Im}(\kappa_j)| < \beta \quad \forall j \in \{1, \dots, d\} \right\} \tag{2.7}$$

and we consider the space of complex-analytic functions

$$\mathcal{H}^{\mathbb{C}}_{\tau,\beta} = \left\{ \Phi \in L^2_{\mathrm{loc}}(\Omega_\beta, \mathcal{H}_{\mathrm{f}}) : \Phi(z - \lambda) = \tau(\lambda)\Phi(z) \text{ for all } \lambda \in \Gamma^*, z \in \Omega_\beta \right\}.$$

Let ϕ be the restriction to $\mathbb{R}^d$ of a function $\Phi \in \mathcal{H}^{\mathbb{C}}_{\tau,\beta}$ analytic in the strip Ω_β. Assume that

$$\int_{Y^*} \|\Phi(k + ih)\|^2_{\mathcal{H}_{\mathrm{f}}}\, dk \le C \qquad \forall h \text{ with } |h_j| < \beta$$

with a constant C uniform in h. Then, it is not hard to show that the function $\psi := \tilde{\mathcal{U}}_{\mathrm{BF}}^{-1}\, \phi$ satisfies

$$\int_{\mathbb{R}^d} \mathrm{e}^{2\beta|x|}\, |\psi(x)|^2 dx < +\infty. \tag{2.8}$$

A function $\psi \in L^2(\mathbb{R}^d)$ satisfying (2.8) for some $\beta > 0$ is said to be ***exponentially localized***, while a function $\psi \in L^2(\mathbb{R}^d)$ such that $P(X_1, ..., X_d)\psi$ is in $L^2(\mathbb{R}^d)$ for any polynomial P is said ***almost-exponentially localized***.

Now, to assure that $H = -\Delta + V_\Gamma$ is self-adjoint in $L^2(\mathbb{R}^d)$ on the domain $W^{2,2}(\mathbb{R}^d)$, we make the following Kato-type assumption on the Γ-periodic potential [33, Theorem XIII.96]:

$$\begin{aligned} &V_\Gamma \in L^2_{\mathrm{loc}}(\mathbb{R}^d) \text{ for } d \le 3, \\ &V_\Gamma \in L^p_{\mathrm{loc}}(\mathbb{R}^d) \text{ with } p > d/2 \text{ for } d \ge 4. \end{aligned} \tag{2.9}$$

Under the previous assumption on V_Γ, the transformed Hamiltonian is a fibered operator over Y^*, *i.e.*

$$\tilde{\mathcal{U}}_{\mathrm{BF}}\, H\, \tilde{\mathcal{U}}_{\mathrm{BF}}^{-1} = \int_{Y^*}^{\oplus} dk\, H(k)$$

with fiber operator

$$H(k) = \big(-\mathrm{i}\nabla_y + k\big)^2 + V_\Gamma(y)\,, \qquad k \in \mathbb{R}^d, \tag{2.10}$$

acting on the k-independent domain $\mathcal{D}_0 = W^{2,2}(\mathbb{T}^d) \subset L^2(\mathbb{T}^d)$. Each fiber operator $H(k)$ is self-adjoint, has compact resolvent and thus pure point spectrum accumulating at infinity. The eigenvalues are labeled increasingly, *i.e.* $E_0(k) \le E_1(k) \le E_2(k) \le \ldots$, and repeated according to their multiplicity. Since the fiber Hamiltonians are τ-covariant, see [30], in the sense that

$$H(k+\lambda) = \tau(\lambda)^{-1}\, H(k)\, \tau(\lambda), \qquad \forall \lambda \in \Gamma^*, \tag{2.11}$$

the eigenvalues are Γ^*-periodic, *i.e.* $E_n(k+\lambda) = E_n(k)$ for all $\lambda \in \Gamma^*$.

We denote by $\sigma_*(k)$ the set $\{E_i(k) : n \le i \le n+m-1\}$, $k \in Y^*$, corresponding to a physically relevant family of m Bloch bands, and we assume the following *gap condition*:

$$\inf_{k \in \mathbb{T}_d^*} \mathrm{dist}\big(\sigma_*(k), \sigma(H(k)) \setminus \sigma_*(k)\big) > 0. \tag{2.12}$$

It is possible to see that, up to complexifying the parameter, the family $\{H(\kappa)\}_{\kappa\in\mathbb{C}^d}$ is an entire analytic family of type (A) [33, Chapter XII]. Hence $\{H(\kappa)\}_{\kappa\in\mathbb{C}^d}$ is an entire analytic family in the sense of Kato [33, Theorem XII.9]. As a consequence, under the assumption (2.12), it is possible to deduce the following property of the family of spectral projectors $\{P_*(k)\}_{k\in\mathbb{R}^d}$ corresponding to $\sigma_*(k)$:

(P_1) the map $k \mapsto P_*(k)$ is smooth from $\mathbb{R}^d$ to $\mathcal{B}(\mathcal{H}_{\mathrm{f}})$ (equipped with the operator norm);

($\tilde{\mathrm{P}}_1$) the map $k \mapsto P_*(k)$ extends to a $\mathcal{B}(\mathcal{H}_{\mathrm{f}})$-valued analytic function on the domain Ω_α for some $\alpha > 0$;

(P_2) the map $k \mapsto P_*(k)$ is τ-covariant, *i.e.*

$$P_*(k+\lambda) = \tau(\lambda)^{-1}\, P_*(k)\, \tau(\lambda) \qquad \forall k \in \mathbb{R}^d, \quad \forall \lambda \in \Gamma^*;$$

($\tilde{\mathrm{P}}_2$) the map $\kappa \mapsto P_*(\kappa)$ is τ-covariant, *i.e.*

$$P_*(\kappa+\lambda) = \tau(\lambda)^{-1}\, P_*(\kappa)\, \tau(\lambda) \qquad \forall \kappa \in \Omega_\alpha, \quad \forall \lambda \in \Gamma^*;$$

(P_3) there exists an antiunitary operator[1] C acting on $\mathcal{H}_f$ such that

$$P_*(-k) = C\, P_*(k)\, C^{-1} \qquad \text{and} \qquad C^2 = 1.$$

While properties (P_1) and (P_2) are a consequence of the fact that the operator H commutes with the lattice translations, jointly with the gap condition (2.12), property (P_3) follows from the fact that the operator (1.1) is real, and corresponds to the time-reversal symmetry of the physical system (see [31] for details).

We now consider a family σ_* of m Bloch bands satisfying (2.12). Since the eigenvalues in the family σ_* generically intersect each other and the eigenprojectors of $H(k)$ corresponding to single eigenvalues are not smooth at the intersection point, the natural notion for functions spanning Ran $P_*(k)$ at any k turns out to be the following one [1,7].

Definition 2.1. Let $\{P_*(k)\}_{k\in\mathbb{R}^d} \subset \mathcal{B}(\mathcal{H}_f)$ be a family of orthogonal projectors satisfying (P_1) and (P_2), with $\dim P_*(k) \equiv m < +\infty$. A function $\chi \in \mathcal{H}_\tau$ is called a **quasi-Bloch function** (for the family $\{P_*(k)\}$) if

$$P_*(k)\chi(k,\cdot) = \chi(k,\cdot) \quad \text{and} \quad \chi(k,\cdot) \neq 0 \qquad \forall k \in Y^*. \tag{2.13}$$

A **Bloch frame** (for the family $\{P_*(k)\}$) is a set $\{\chi_a\}_{a=1,\dots,m}$ of quasi-Bloch functions such that $\{\chi_1(k),\dots,\chi_m(k)\}$ is an orthonormal basis of Ran $P_*(k)$ at (almost-)every $k \in Y^*$.

A Bloch frame is fixed only up to a k-dependent unitary matrix (*Bloch gauge freedom*) $U(k) \in \mathcal{U}(m)$, *i.e.* if $\{\chi_a\}_{a=1,\dots,m}$ is a Bloch frame then the functions $\widetilde{\chi}_a(k) = \sum_{b=1}^m \chi_b(k) U_{b,a}(k)$ also define a Bloch frame.

Definition 2.2. The **composite Wannier functions** corresponding to a Bloch frame $\{\chi_a\}_{a=1,\dots,m}$ are the functions

$$w_a(x) := \left(\tilde{\mathcal{U}}_{\mathrm{BF}}^{-1}\chi_a\right)(x), \qquad a \in \{1,\dots,m\}.$$

We emphasize that, for V_Γ satisfying (2.9), the composite Wannier functions are actually in $W^{2,2}(\mathbb{R}^d)$ and, in general, one cannot expect better regularity properties. For example, if V_Γ has a Coulomb singularity, the Wannier functions are not smooth, as it happens for the eigenfunctions of the hydrogen atom.

[1] By *antiunitary* operator we mean a surjective antilinear operator $C : \mathcal{H} \to \mathcal{H}$, such that $\langle C\varphi, C\psi\rangle_{\mathcal{H}} = \langle \psi, \varphi\rangle_{\mathcal{H}}$ for any $\varphi, \psi \in \mathcal{H}$.

We note also that, the existence of an exponentially (resp. almost-exponentially) localized composite Wannier functions is equivalent to the existence of an analytic (resp. smooth) Bloch frame. Property ($\tilde{\mathrm{P}}_1$) (resp. (P_1)) assures such existence locally around a given point. However, as already noticed [7, 28], there might be a topological obstruction to obtaining a global analytic (resp. smooth) Bloch frame, in view of the competition between the regularity and the τ-equivariance (remember that the Bloch frame must be in $\mathcal{H}_\tau$ by definition).

To clarify this topological obstruction, following [31] it is convenient to introduce the concept of **Bloch bundle**. To be precise, to a family of orthogonal projectors $\{P_*(k)\}_{k\in\mathbb{R}^d}$ satisfying (P_1) and (P_2) is canonically associated a Hermitian smooth vector bundle $\mathcal{E}_*$ over $\mathbb{T}_d^*$, called the *Bloch bundle*. If ($\tilde{\mathrm{P}}_1$) and ($\tilde{\mathrm{P}}_2$) are also satisfied, then $\mathcal{E}_*$ is the restriction to $\mathbb{T}_d^* = \mathbb{R}^d/\Gamma^*$ of a holomorphic Hermitian vector bundle $\tilde{\mathcal{E}}_*$ over Ω_α/Γ^*.

The idea of the construction is to consider $\sqcup_{k\in\mathbb{R}^d}\operatorname{Ran} P_*(k)$ as a subbundle of the trivial bundle $\mathbb{R}^d \times \mathcal{H}_{\mathrm{f}}$ over $\mathbb{R}^d$ and to use the τ-equivariance of the projectors to obtain a (quotient) vector bundle $\mathcal{E}_*$ over the quotient space $\mathbb{T}_d^* = \mathbb{R}^d/\Gamma^*$.

To construct $\mathcal{E}_*$, one firstly introduces on the set $\mathbb{R}^d \times \mathcal{H}_{\mathrm{f}}$ the equivalence relation $\sim_\tau$, where

$$(k,\varphi) \sim_\tau (k',\varphi') \Leftrightarrow (k',\varphi') = (k+\lambda\,,\ \tau(\lambda)^{-1}\varphi) \quad \text{for some } \lambda \in \Gamma^*.$$

The equivalence class with representative (k,φ) is denoted by $[k,\varphi]$. Then the total space $\mathcal{E}_*$ of the vector bundle is defined by

$$\mathcal{E}_* := \left\{[k,\varphi] \in (\mathbb{R}^d \times \mathcal{H}_{\mathrm{f}})/\!\sim_\tau : \quad \varphi \in \operatorname{Ran} P_*(k)\right\}.$$

The projection to the base space $\pi : \mathcal{E}_* \to \mathbb{T}_d^*$ is $\pi[k,\varphi] = \mu(k)$, where μ is the projection modulo Γ^*. One checks that $\mathcal{E}_* \xrightarrow{\pi} \mathbb{T}_d^*$ is a smooth complex vector bundle with typical fiber $\mathbb{C}^m$. In particular, the local triviality follows from (P_1) and the use of the Kato-Nagy's formula (see [16, Sec. I.6.8]). In addition, the vector bundle $\mathcal{E}_*$ carries a natural Hermitian structure from $\mathcal{H}_{\mathrm{f}}$.

As for the analytic case, one introduces an equivalence relation $\sim_\tau$ over $\Omega_\alpha \times \mathcal{H}_{\mathrm{f}}$ as above. Then the total space of $\tilde{\mathcal{E}}_*$ is defined by

$$\tilde{\mathcal{E}}_* = \{[\kappa,\varphi] \in (\Omega_\alpha \times \mathcal{H}_{\mathrm{f}})/\!\sim_\tau : \quad \varphi \in \operatorname{Ran} P_*(\kappa)\}$$

and local triviality follows again from Kato-Nagy's formula. Since $\kappa \mapsto P_*(\kappa)$ extends $k \mapsto P_*(k)$ by ($\tilde{\mathrm{P}}_1$), and both the maps are τ-covariant, $\mathcal{E}_* \to \mathbb{T}_d^*$ is clearly a restriction of $\tilde{\mathcal{E}}_* \to \Omega_\alpha/\Gamma^*$. The Hermitian structure over the vector bundle $\tilde{\mathcal{E}}_*$ is defined as in the smooth case.

The vector bundle $\mathcal{E}_*$ is equipped with a natural $\mathfrak{u}(m)$-connection (*Berry connection*), induced by the trivial connection on the trivial (infinite dimensional) vector bundle $(\mathbb{R}^d \times \mathcal{H}_{\mathrm{f}})/\!\sim_\tau \to \mathbb{T}_d^*$.

The Bloch bundle encodes the geometrical obstruction to the existence of a global smooth (resp. analytic) Bloch frame. Indeed, under the assumptions (P_1) (resp. $(\tilde{\mathrm{P}}_1)$) and (P_2) (resp. $(\tilde{\mathrm{P}}_2)$), with $\dim P_*(k) \equiv m < +\infty$ on the projectors, the following equivalence is implicit in [31]:

(A) **existence of a regular Bloch frame:** there exists a Bloch frame $\{\chi_a\}_{a=1}^m$ such that each χ_a is in $C^\infty(\mathbb{R}^d, \mathcal{H}_{\mathrm{f}})$ (resp. each χ_a is the restriction to $\mathbb{R}^d$ of a function $\tilde{\chi}_a \in \mathcal{H}_{\tau,\alpha}^{\mathbb{C}}$ analytic on Ω_α);

(B) **triviality of the Bloch bundle:** the vector bundle $\mathcal{E}_*$, associated to the family $\{P_*(k)\}_{k\in\mathbb{R}^d}$, is trivial in the category of smooth Hermitian vector bundles over $\mathbb{T}_d^*$ (resp. the vector bundle $\tilde{\mathcal{E}}_*$ is trivial in the category of holomorphic Hermitian vector bundles over Ω_α/Γ^*).

The key fact here is that, as a consequence of [31], under the assumption (P_3) of time-reversal symmetry the Bloch bundle is always trivial in low dimension (see [31], Theorem 2). Thus, up to a Gram-Schmidt orthonormalization procedure one has the following existence result for a Bloch frame.

Theorem 2.3. *Let $\{P_*(k)\}_{k\in\mathbb{R}^d} \subset \mathcal{B}(\mathcal{H}_{\mathrm{f}})$ be a family of orthogonal projectors satisfying properties $(\tilde{\mathrm{P}}_1)$, $(\tilde{\mathrm{P}}_2)$ and (P_3). Assume $d \le 3$ and $m \ge 1$, or $d \ge 1$ and $m = 1$. Then there exists a Bloch frame $\{\chi_a\}_{a=1,\dots,m}$ such that each χ_a is real-analytic, i.e. $\chi_a \in C^\omega(\mathbb{R}^d, \mathcal{H}_{\mathrm{f}}) \cap \mathcal{H}_\tau$.*

3 The Marzari-Vanderbilt localization functional

The long-lasting uncertainty about the existence of exponentially localized composite Wannier functions in three dimensions, settled only recently [3, 31], and the need of an approach suitable for numerical simulations, lead the solid state physics community to explore new paths. In an important paper [22], Marzari and Vanderbilt introduced the following concept.

For a system of L^2-normalized composite Wannier functions $w = \{w_1, \dots, w_m\} \subset L^2(\mathbb{R}^d)$ the **Marzari-Vanderbilt localization functional** is

$$F_{MV}(w) = \sum_{a=1}^{m} \int_{\mathbb{R}^d} |x|^2 |w_a(x)|^2 dx - \sum_{a=1}^{m} \sum_{j=1}^{d} \left(\int_{\mathbb{R}^d} x_j |w_a(x)|^2 dx \right)^2 . \quad (3.1)$$

We emphasize that the above definition of $F_{\mathrm{MV}}(w)$ includes the crucial constraint that the corresponding Bloch functions $\varphi_a(k,\cdot) = (\mathcal{U}_{\mathrm{BF}} w_a)(k,\cdot)$, for $a \in \{1, \dots, m\}$, are a Bloch frame, *i.e.* $\{\varphi_1(k, \cdot), \dots, \varphi_m(k, \cdot)\}$ is an orthonormal set in $\mathcal{H}_f$ for each $k \in Y^*$ and

$$\mathrm{Span}_{\mathbb{C}} \{\varphi_1(k, \cdot), \dots, \varphi_m(k, \cdot)\} = P_*(k)(\mathcal{H}_f), \quad \forall k \in Y^*. \tag{3.2}$$

As already discussed inthe previous section, the latter condition actually implies $w_a \in W^{2,2}(\mathbb{R}^d) = \mathcal{D}(H)$.

Definition 3.1. Let $\{P_*(k)\}_{k\in\mathbb{R}^d} \subset \mathcal{B}(\mathcal{H}_\mathrm{f})$ be a family of projectors satisfying properties $(\tilde{\mathrm{P}}_1)$ and (P_2), with $\dim P_*(k) \equiv m < +\infty$. A system of **maximally localized composite Wannier functions** is a global minimizer $w = \{w_1, \dots, w_m\}$ of the localization functional F_{MV} in the space $\mathcal{W}^m := (\mathcal{D}(H) \cap \mathcal{D}(X))^m$, under the constraint that $\{\varphi_1, \dots, \varphi_m\}$, for $\varphi_a = \tilde{\mathcal{U}}_{\mathrm{BF}} w_a$, is a Bloch frame.

A natural problem, raised in [22], is the following.

Problem. Let $\{P_*(k)\}_{k\in\mathbb{R}^d} \subset \mathcal{B}(\mathcal{H}_\mathrm{f})$ be a family of projectors satisfying properties $(\tilde{\mathrm{P}}_1)$ and (P_2), with $\dim P_*(k) \equiv m < +\infty$.

(MV_1) **(Existence)** prove that there exists a system of maximally localized composite Wannier functions;

(MV_2) **(Localization)** prove that any maximally localized composite Wannier function is exponentially localized, in the sense that there exists $\beta > 0$ such that (2.8) holds.

Since the (modified) Bloch-Floquet transform $\tilde{\mathcal{U}}_{\mathrm{BF}} : L^2(\mathbb{R}^d) \to L^2(Y^*; \mathcal{H}_\mathrm{f})$ is an isometry and it satisfies $(\tilde{\mathcal{U}}_{BF} X_j g)(k, y) = i \frac{\partial}{\partial k_j} (\tilde{\mathcal{U}}_{BF} g)(k, y)$, the functional (3.1) can be rewritten in terms of the Bloch frame $\varphi = \{\varphi_1, \dots, \varphi_m\}$ as

$$\tilde{F}_{MV}(\varphi) = \sum_{a=1}^{m} \sum_{j=1}^{d} \left\{ \int_{Y^*} dk \int_{\mathbb{T}^d} \left| \frac{\partial \varphi_a}{\partial k_j} \right|^2 dy - \left(\int_{Y^*} dk \int_{\mathbb{T}^d} \overline{\varphi_a}\, i \frac{\partial \varphi_a}{\partial k_j} dy \right)^2 \right\}. \tag{3.3}$$

Correspondingly, in view of (2.6), the space $\mathcal{W} = \mathcal{D}(H) \cap \mathcal{D}(X) = \mathcal{D}(H) \cap \mathcal{D}(\langle X \rangle)$ is mapped by the Bloch-Floquet transform into

$$\mathcal{H}_\tau \cap L^2_{\mathrm{loc}}(\mathbb{R}^d, W^{2,2}(\mathbb{T}^d)) \cap W^{1,2}_{\mathrm{loc}}(\mathbb{R}^d, L^2(\mathbb{T}^d)) =: \widetilde{\mathcal{W}}.$$

Hereafter, to solve problem (MV_1) for any d, we will make the following

Assumption 1: there exists a Bloch frame $\chi = \{\chi_1, \dots, \chi_m\} \subset \mathcal{H}_\tau$ such that $\chi_a \in \widetilde{\mathcal{W}}$ for every $a \in \{1, \dots, m\}$.

In the case most relevant to us, *i.e.* Schrödinger operators with real potential, for $d \leq 3$ the previous assumption is automatically satisfied, since

Theorem 2.3 provide the existence of a Bloch frame which is even real-analytic. While this extra regularity is unessential for problem (MV_1), it will be crucial when dealing with (MV_2). Notice that, under the previous assumption, the set of admissible Bloch frames in Definition 3.1 is not empty, so problem (MV_1) makes sense.

By using the previous Bloch frame, we can lift the functional (3.3) to $W^{1,2}$-maps from $\mathbb{T}_d^*$ to the unitary group $\mathcal{U}(m)$, *i.e.* to Γ^*-periodic maps from $\mathbb{R}^d$ to $\mathcal{U}(m)$. Indeed, given any map $U \in W^{1,2}(\mathbb{T}_d^*, \mathcal{U}(m))$ one defines a Bloch frame $\varphi = \{\varphi_1, \ldots, \varphi_m\} \subset \tilde{\mathcal{W}}$ by setting $\varphi = \chi \cdot U$, *i.e.* $\varphi_a(k,\cdot) = \sum_b \chi_b(k,\cdot)\, U_{b,a}(k)$. Vice versa, if $\varphi = \{\varphi_1, \ldots, \varphi_m\} \subset \tilde{\mathcal{W}}$ is a Bloch frame, then pointwise $\varphi_a(k,\cdot) = \sum_b \chi_b(k,\cdot)\, U_{b,a}(k)$ with $U_{b,a}(k) = \langle \chi_b(k)\,,\, \varphi_a(k)\rangle$, hence $U \in W^{1,2}(\mathbb{T}_d^*, \mathcal{U}(m))$.
For the given reference frame χ, the functional (3.3) in terms of the gauge U becomes

$$\begin{aligned}\tilde{F}_{MV}(U;\chi) = &\sum_{j=1}^{d} \int_{\mathbb{T}_d^*} \left[\mathrm{tr}\left(\frac{\partial U^*}{\partial k_j}(k) \frac{\partial U}{\partial k_j}(k) \right) + m \sum_{a=1}^{m} \left\| \frac{\partial \chi_a(k,\cdot)}{\partial k_j} \right\|_{\mathcal{H}_f}^2 \right] dk \\ &+ \sum_{j=1}^{d} \int_{\mathbb{T}_d^*} \mathrm{tr}\left[\left(U(k) \frac{\partial U^*}{\partial k_j}(k) - \frac{\partial U}{\partial k_j}(k) U^*(k) \right) A_j(k) \right] dk \\ &+ \sum_{a=1}^{m} \sum_{j=1}^{d} \left(\int_{\mathbb{T}_d^*} \left[U^*(k) \left(\frac{\partial U}{\partial k_j}(k) + A_j(k) U(k) \right) \right]_{aa} dk \right)^2 .\end{aligned} \tag{3.4}$$

Here the matrix coefficients $A_j \in L^2(\mathbb{T}_d^*; \mathfrak{u}(m))$ are given by the formula

$$\left[A_j(k)\right]_{cb} = \left\langle \chi_c(k,\cdot)\,,\, \frac{\partial \chi_b(k,\cdot)}{\partial k_j} \right\rangle_{\mathcal{H}_f} \tag{3.5}$$

When χ is real-analytic, the functions $A_j \in C^\omega(\mathbb{T}_d^*; \mathfrak{u}(m))$ represent the antihermitian connection 1-form induced on the (sub)bundle $\mathcal{E}_*$ by the trivial connection on the bundle $\mathbb{R}^d \times \mathcal{H}_f$.
Moreover,

$$\begin{aligned}&\inf\left\{ F_{MV}(w) : \begin{array}{c} \{w_1, \ldots, w_m\} \subset \mathcal{W} \\ \tilde{\mathcal{U}}_{\mathrm{BF}}\, w \text{ is a Bloch frame} \end{array} \right\} \\ &\quad = \inf\left\{ \tilde{F}_{MV}(U;\chi) : U \in W^{1,2}(\mathbb{T}_d^*; \mathcal{U}(m)) \right\}\end{aligned} \tag{3.6}$$

Therefore, problem (MV_1) is equivalent to showing that the r.h.s. of (3.6) is attained. Analogously, in view of the Paley-Wiener type theorem mentioned in the previous section, problem (MV_2) corresponds to show that any minimizer of $\tilde{F}_{MV}(\cdot;\chi)$ is real-analytic, provided that χ is also real-analytic.

Here a couple of remarks are in order. First, when reformulating problem (MV_1) in terms of the functional $\tilde{F}_{MV}(U;\chi)$ the regularity of the reference frame χ plays no essential role, provided χ is in $\tilde{\mathcal{W}}$, since one can always pass from a given reference frame to a new one without affecting the existence of minima. This observation justifies the fact that a minimizer (whenever it exists) can be evaluated starting from any reference Bloch frame $\chi \subset \tilde{\mathcal{W}}$, even a discontinuous one, as it happens in numerical simulations.

Secondly, to compute the infimum of $\tilde{F}_{MV}(U;\chi)$ it is sufficient to consider smooth change of gauges. More precisely, for $d \leq 3$ and for any fixed Bloch frame $\chi \subset \tilde{\mathcal{W}}$ one has

$$\begin{aligned}&\inf\left\{\tilde{F}_{MV}(U;\chi) : U \in W^{1,2}(\mathbb{T}_d^*;\mathcal{U}(m))\right\}\\&\quad=\inf\left\{\tilde{F}_{MV}(U;\chi) : U \in C^{\infty}(\mathbb{T}_d^*;\mathcal{U}(m))\right\}.\end{aligned}$$

As a consequence, when computing the above infimum numerically, one can let U vary in any set S such that $C^{\infty}(\mathbb{T}_d^*;\mathcal{U}(m)) \subset S \subset W^{1,2}(\mathbb{T}_d^*;\mathcal{U}(m))$ (since for $d \leq 3$ there is strong density of smooth maps, see *e.g.* [12, Theorem 1.3 and Section 5]).

Now we claim that the right hand side in (3.6) is attained. The proof is a simple modification of the direct method in the calculus of variations, in order to handle a natural invariance of the functional (3.1). Indeed, if $\{w_1,\ldots,w_m\}$ are composite Wannier functions satisfying (3.2) and $\{\gamma_1,\ldots,\gamma_m\} \subset \Gamma$, then

$$F_{MV}(w_1,\ldots,w_m) = F_{MV}(\tilde{w}_1,\ldots,\tilde{w}_m), \quad \tilde{w}_a(x) = w_a(x+\gamma_a), \quad 1 \leq a \leq m. \tag{3.7}$$

Moving to Bloch functions, we have

$$\tilde{\varphi}_a(k,\cdot) \equiv \left(\tilde{\mathcal{U}}_{BF}\,\tilde{w}_a\right)(k,\cdot) = e^{ik\cdot\gamma_a}\left(\tilde{\mathcal{U}}_{BF}\,w_a\right)(k,\cdot) = e^{ik\cdot\gamma_a}\varphi_a(k,\cdot),$$

so that $\{\tilde{\varphi}_1(k,\cdot),\ldots,\tilde{\varphi}_m(k,\cdot)\}$ is still orthonormal in $\mathcal{H}_f$ and (3.2) holds. Correspondingly, the functional (3.4) has the invariance

$$\tilde{F}_{MV}(U;\chi) = \tilde{F}_{MV}(\tilde{U};\chi), \quad \tilde{U}(k) = \mathrm{diag}\left(e^{ik\cdot\gamma_a}\right)U(k). \tag{3.8}$$

Thus, it is possible to see that the functional (3.4) is nonnegative, weakly lower semicontinuous and coercive at least "transversally to the orbit" given by the action of Γ on $W^{1,2}(\mathbb{T}_d^*;\mathcal{U}(m))$.

Theorem 3.2. *Let $\{P_*(k)\}_{k\in\mathbb{R}^d} \subset \mathcal{B}(\mathcal{H}_f)$ be a family of orthogonal projectors satisfying properties* ($\tilde{\text{P}}_1$) *and* (P_2)*, with* $\dim P_*(k) \equiv m$*. Assume that there exists a Bloch frame $\chi \subset \tilde{\mathcal{W}}$ (e.g. under the assumptions of Theorem 2.3). Then there exists $U \in W^{1,2}(\mathbb{T}_d^*; \mathcal{U}(m))$ which is a minimizer on $W^{1,2}(\mathbb{T}_d^*; \mathcal{U}(m))$ of the localization functional $\tilde{F}_{MV}(\cdot, \chi)$ defined by (3.4).*

Finally, in order to derive the Euler-Lagrange equations corresponding to the functional (3.4), we consider the unitary group $\mathcal{U}(m)$ as isometrically embedded into the space $M_m(\mathbb{C})$ with the standard real Euclidean product $\langle A\,,\,B\rangle = \operatorname{Re}\operatorname{tr}(A^*B)$. Recall that $\mathcal{U}(m)$ is a compact real Lie group, the induced metric is biinvariant, its Lie algebra is given by the real vector space of complex antihermitian matrices.

Let $\varphi \in C^\infty(\mathbb{T}_d^*; M_m(\mathbb{C}))$ and for $\varepsilon \neq 0$ fixed let $U(k) + \varepsilon\varphi(k)$ be a free variation of U in the direction φ. In a sufficiently small tubular neighborhood $\mathcal{O}$ of $\mathcal{U}(m)$ in $M_m(\mathbb{C})$ there is a well defined smooth nearest point projection map $\Pi : \mathcal{O} \to \mathcal{U}(m)$, so we can consider the induced variations

$$U_\varepsilon(k) := \Pi(U(k) + \varepsilon\varphi(k)) = U\left(\mathbb{I} + \varepsilon\frac{1}{2}\left[U^*\varphi - (U^*\varphi)^*\right]\right) + o(\varepsilon)\,. \tag{3.9}$$

Thus, simple calculations show that a given map $U \in W^{1,2}(\mathbb{T}_d^*; \mathcal{U}(m))$ satisfies $\dfrac{d}{d\varepsilon}\tilde{F}_{MV}(U_\varepsilon; \chi)|_{\varepsilon=0} = 0$ if and only if U is a weak solution to the Euler-Lagrange equations

$$-\Delta U + \sum_{j=1}^{d}\frac{\partial U}{\partial k_j}U^{-1}\frac{\partial U}{\partial k_j} + \sum_{j=1}^{d}\left[\frac{\partial U}{\partial k_j}U^{-1}A_jU - \frac{\partial A_j}{\partial k_j}U - A_j\frac{\partial U}{\partial k_j}\right] + \tag{3.10}$$

$$+\sum_{j=1}^{d}\left[-\left(\frac{\partial U}{\partial k_j} + A_jU\right)G^j + UG^jU^{-1}\left(\frac{\partial U}{\partial k_j} + A_jU\right)\right] = 0\,.$$

Here the constant (purely imaginary) diagonal matrices $\{G^j\} \subset M_m(\mathbb{C})$ are defined as $G^j = \operatorname{diag}\left(\int_{\mathbb{T}_d^*} U^*(k)\left[\frac{\partial U}{\partial k_j}(k) + A_j(k)U(k)\right]dk\right)$, where $\left[\operatorname{diag} M\right]_{ab} = M_{ab}\delta_{ab}$.

4 Regularity of minimizers and exponential localization

The goal of this section is to review the regularity results for minimizers of $\tilde{F}(\cdot;\chi)$ and localization property for maximally localized Wannier functions establish in [32]. Roughly speaking we are going to explain why, possibly under suitable restrictions on m, a change of gauge $U \in W^{1,2}(\mathbb{T}_d^*; \mathcal{U}(m))$ which minimizes the functional (3.4) is real-analytic, whenever χ is real-analytic. In turn, via Paley-Wiener type theorem, this fact will always imply that maximally localized Wannier functions are exponentially localized. Since the details of the proofs are subtle and often very technical, we will just convey the main ideas behind the proofs, deferring to [32] for all the details.

To deal with the regularity of the minimizers, we make the following assumption which, as already noticed after stating the weaker Assumption 1, is automatically satisfied for $d \leq 3$ or $m = 1$ (compare Theorem 2.3).

Assumption 2: there exists a Bloch frame $\chi = \{\chi_1, \dots, \chi_m\} \subset \mathcal{H}_\tau$ such that $\chi_a \in C^\omega(\mathbb{R}^d, \mathcal{H}_{\mathrm{f}})$ for every $a \in \{1, \dots, m\}$.

The simplest case is when $d = 1$ and $m \geq 1$ is arbitrary. In this case the analytic regularity of any solution U to (3.10) is standard result in ODE theory.

The first nontrivial situation is the case $m = 1$, $d \geq 1$. Since $\mathcal{U}(1)$ is abelian the equation (3.10) reduces to

$$-\Delta U + \sum_{j=1}^{d} \frac{\partial U}{\partial k_j} U^{-1} \frac{\partial U}{\partial k_j} = \left(\sum_{j=1}^{d} \frac{\partial A_j}{\partial k_j} \right) U \,. \tag{4.1}$$

Recall that in any sufficiently small ball $B \subset \mathbb{T}_d^*$ we have $U(k) = e^{if(k)}$ for some $f \in W^{1,2}(B; \mathbb{R})$ (see [2]). Thus, equation (4.1) reads $\Delta f = i \sum_{j=1}^{d} \frac{\partial A_j}{\partial k_j} \in C^\omega(B; \mathbb{R})$, because the matrices in the Berry connection are antihermitian and real-analytic. Since the Laplacian is analytic-hypoelliptic we conclude $f \in C^\omega(B)$ and in turn $U \in C^\omega(\mathbb{T}_d^*; \mathcal{U}(1))$ since the ball B can be choosen arbitrarily.

When considering the case $n, m \geq 2$ the situation is much more difficult. The argument requires two steps:

(A) Continuity of minimizers of the functional (3.4) (difficult).

(B) Real-analiticity of continuous solutions to (3.10) (easy)

(B) The second step is considerably easier than the first one, since it essentially follows from classical regularity results for elliptic systems. Indeed, just exploiting continuity in a subtle way, *i.e.* by choosing appropriate test functions (see [15]), one can prove that for solution to (3.10)

$$U \in C^0 \cap W^{1,2} \text{ (weak solution)} \Longrightarrow U \in W^{2,2} \cap W^{1,4} \text{ (strong solution)}$$

Then, the regularity problem becomes essentially linear, in the sense that basic regularity results for the Laplacian and a standard bootstrap argument gives $U \in W^{2,p}$ for any $p > 2$. Then, $U \in C^{1,\alpha}$ for any $\alpha \in (0, 1)$ by Sobolev embedding, and another bootsprap argument gives $U \in C^{l,\alpha}$ for any $l \geq 1$ by Schauder theory, *i.e.* $U \in C^{\infty}$. Finally, real-analiticity follows from the classical results in [24].

(A) Concerning the first step we will argue in different ways following a *dimension-dependent* argument. As we will see, when $d = 2$ the continuity property will be true for *any* solution to (3.10) (and for any $m \geq 1$). On the contrary, when $d = 3$ continuity will follow as a consequence of *minimality* (and under further restriction on m).

First we are going to justify continuity of any weak solutions to (3.10) in the case $d = 2$. The argument here is only sketched, since, up to minor technical modifications, it is very similar to the proof of continuity for weakly harmonic maps from a two dimensional domain into spheres (see *e.g.* [20], Chapter III, Section 3.2 pag. 57-61).

We observe that if $U \in W^{1,2}$ is a weak solution to (3.10) then U is also a weak solution to

$$\Delta U = \sum_{j=1}^{2} \frac{\partial U}{\partial k_j} U^{-1} \frac{\partial U}{\partial k_j} + f \tag{4.2}$$

for a suitable $f \in L^2$ defined in an obvious way by (3.10). If we set $B^j = \frac{1}{2}\left(U^* \frac{\partial U}{\partial k_j} - \frac{\partial U^*}{\partial k_j} U\right)$ then $B^j \in L^2(\mathbb{T}_2^*; \mathfrak{u}(m))$ for $j \in \{1, 2\}$, where $\mathfrak{u}(m)$ is the Lie algebra of antihermitian matrices. Now we modify B_j and f as follows

$$\tilde{B} = B - \nabla\Delta^{-1} \operatorname{div} B\,, \qquad \tilde{f} = f + \nabla U \cdot \nabla\Delta^{-1} \operatorname{div} B\,,$$

so that $\tilde{B}^j \in L^2(\mathbb{T}_2^*; \mathfrak{u}(m))$ for $j \in \{1, 2\}$ satisfy $\operatorname{div} \tilde{B} = 0$ in $\mathcal{D}'(\mathbb{T}_2^*)$ and $\tilde{f} \in L^p$ for some $p > 1$. Thus U is a weak solution to

$$\Delta U = \sum_{j=1}^{2} \frac{\partial U}{\partial k_j} \tilde{B}^j + \tilde{f} \tag{4.3}$$

and the right hand side of (4.3) is in the local Hardy space $\mathcal{H}^1_{\mathrm{loc}}(\mathbb{T}_2^*) \subset L^1(\mathbb{T}_2^*)$ in view of the *div-curl* lemma (see [9]). As a consequence, $U \in W^{2,1}(\mathbb{T}_2^*; \mathcal{U}(m))$ and in turn $U \in C^0(\mathbb{T}_2^*; \mathcal{U}(m))$ (since $\nabla U \in L^{2,1}$, because of the refined Sobolev embedding into Lorentz spaces).

Next, we analyze the case $d = 3$. As already mentioned, here *energy minimality* comes into play in a crucial way. This is not surprising and it is indeed a very well-known phenomenon in the regularity theory for harmonic maps. In that case everywhere discountinuous solutions of the Euler-Lagrange equations may exist, *e.g.* in the model case $U : \Omega \subset \mathbb{R}^3 \to S^2$, and partial regularity is possible only in some restricted class of solutions (see [20] and the references therein).

In order to prove real-analiticity of minimizers we try to prove a decay property for some scaling-invariant quantity, much in the spirit of the Morrey decay Lemma (see [24]). More precisely, the key point in [32] is to show that the L^2–deviation of U from its average $⨍ U$ has a power-like decay at small scales, *i.e.*

$$⨍_B \left| U - ⨍_B U \right|^2 \leq C|B|^{2\alpha/3}, \tag{4.4}$$

for some absolute $C > 0$ and $\alpha > 0$ and for any sufficiently small ball $B \subset \mathbb{T}_3^*$. Once this is true, a well known characterization of Holder spaces actually yields $U \in C^{0,\alpha}$ which is the desirered conclusion.

As a matter of fact it will be better first to prove that the l.h.s. of (4.4) is uniformly bounded, combining the Poincare inequality

$$⨍_B \left| U - ⨍_B U \right|^2 \leq C|B|^{2/3} ⨍_B |\nabla U|^2, \tag{4.5}$$

with some control on the energy of a minimizer U on small balls, and then to show that both quantities in (4.5) actually decay at smaller and smaller scales. However, to get a quantitative decay of the former quantity, it is technically convenient to rephrase that property in terms of the quantitative behaviour of the BMO norm of U at smaller and smaller scales (for an introduction to functions of Bounded Mean Oscillation we refer *e.g.* to [36]).

As a matter of fact, the continuity of minimizers of the functional (3.4) relies on the following four ingredients.

1) **(Perturbed) Monotonicity formula.** Arguing by inner variations, as for the monotonicity identity for harmonic maps (see *e.g.* [20,32,37]), one can prove that the r.h.s. in (4.5) is uniformly bounded and (asymptotically) nonincreasing at small scales.
2) **Strong compactness for scaled maps.** When analyzing U at small scales one considers $U_R(k) = U(k_0+Rk)$ to zoom U around any fixed point $k_0 \in \mathbb{T}_3^*$. Using the Luckhaus interpolation lemma (see *e.g.* [35], Chapter 2), it is possible to show that scaled maps are compact in $W^{1,2}_{loc}$ and converge, up to subsequences to a degree-zero homogeneous, locally energy minimizing harmonic map $U_0 : \mathbb{R}^3 \to \mathcal{U}(m)$ (a so-called *tangent map*).

3) **Liouville Theorem.** Constancy of tangent harmonic maps U_0 *should follow* from energy minimality. We will comment more precisely on that in the last section, since this is by far the most difficult point in the proof and at present it can be proved **only when** $m = 2$ **or** $m = 3$.
4) **Epsilon-regularity.** Much in the spirit of the seminal paper [37], it follows that if the r.h.s. in (4.5) is sufficiently small, then the l.h.s. (actually the related BMO norm) decays according to (4.4). This can be accomplished by adapting the iteration argument in BMO from [6] for partial regularity for stationary harmonic maps into spheres.

Combining the previous tools continuity follows. Indeed, fix $k_0 \in \mathbb{T}_3^*$ and for each $R = R_n \searrow 0$ we define $U_R(k) = U(k_0 + Rk)$. According to 1) such a sequence is bounded in $W^{1,2}_{\mathrm{loc}}(\mathbb{R}^3; \mathcal{U}(m))$ and, up to subsequences, it converges weakly to a degree-zero homogeneous map $U_0 \in W^{1,2}_{\mathrm{loc}}(\mathbb{R}^3; \mathcal{U}(m))$. According to 2), such convergence is strong and the limiting map U_0 is a degree-zero homogeneous local minimizer of the Dirichlet integral in $W^{1,2}_{\mathrm{loc}}(\mathbb{R}^3; \mathcal{U}(m))$. According to 3), at least when $2 \leq m \leq 3$ we have $U_0(k) \equiv const$, therefore $|B_R(k_0)|^{-1/3} \int_{B_{R_n}(k_0)} |\nabla U|^2 = \int_{B_1(0)} |\nabla U_{R_n}|^2 \to 0$ as $n \to \infty$, hence it can be made arbitrarily small at sufficiently small scale. Thus, in view of 4) continuity around k_0 follows, and $U \in C^0(\mathbb{T}_3^*; \mathcal{U}(m))$ since k_0 was arbitrary and U is actually Holder continuous.

To summarize, concerning regularity properties for minimizers of the localization functionals (3.3) and (3.4), the following theorem from [32] holds.

Theorem 4.1. *Let* $1 \leq d \leq 2$ *and* $m \geq 1$, *or* $d = 3$ *and* $1 \leq m \leq 3$, *or* $d \geq 4$ *and* $m = 1$. *Let* χ *be a real-analytic Bloch frame and* $\tilde{F}_{MV}(\cdot)$ *and* $\tilde{F}_{MV}(\cdot\,;\,\chi)$ *the functionals defined by (3.3) and (3.4) respectively. Then:*

(i) *any minimizer* $U \in W^{1,2}(\mathbb{T}_d^*; \mathcal{U}(m))$ *of* $\tilde{F}_{MV}(\cdot\,;\,\chi)$ *is real-analytic;*
(ii) *any minimizer* $\varphi = \{\varphi_1, \dots, \varphi_m\} \subset \tilde{\mathcal{W}}$ *of* $\tilde{F}_{MV}(\cdot)$ *is a real-analytic map from* $\mathbb{R}^d$ *to* $(\mathcal{H}_{\mathrm{f}})^m$.

Combining existence, regularity, and decay properties under the inverse (modified) Bloch-Floquet transform in the theorems presented in the previous and in the present section we finally arrive at the main result in [32]. The following theorem provides an affirmative answer to problems (MV_1) and (MV_2).

Theorem 4.2. *Let* σ_* *be a family of m Bloch bands for the operator* (1.1) *satisfying the gap condition* (2.12), *and let* $\{P_*(k)\}_{k\in\mathbb{R}^d}$ *be the corresponding family of spectral projectors. Assume* $1 \leq d \leq 2$ *and* $m \geq 1$, *or*

$d \geq 1$ and $m = 1$, or $d = 3$ and $1 \leq m \leq 3$. Then there exist composite Wannier functions $\{w_1, \dots, w_m\} \subset \mathcal{W}$ which minimize the localization functional (3.1) *under the constraint that the corresponding quasi-Bloch functions are an orthonormal basis for* Ran $P_*(k)$ *for each $k \in Y^*$. In addition, for any system of maximally localized composite Wannier function $w = \{w_1, \dots, w_m\}$ there exists $\beta > 0$ such that $e^{\beta|x|} w_a$ is in $L^2(\mathbb{R}^d)$ for every $a \in \{1, \dots, m\}$, i.e. the composite Wannier function w_a is exponentially localized.*

5 Harmonic maps into $\mathcal{U}(m)$ and Liouville theorem

In this final section we discuss the Lioville Theorem for minimizing (tangent) harmonic map into the unitary group $\mathcal{U}(m)$. We consider, for $U \in W^{1,2}_{\mathrm{loc}}(\mathbb{R}^3; \mathcal{U}(m))$ with $m \geq 2$, the energy functional

$$E(U;\Omega) = \int_\Omega \frac{1}{2} \sum_{j=1}^{3} \mathrm{tr}\left(\frac{\partial U^*}{\partial k_j} \frac{\partial U}{\partial k_j} \right) dk\,, \qquad \Omega \subset\subset \mathbb{R}^3. \tag{5.1}$$

We assume that U is a local minimizer of (5.1) in $\mathbb{R}^3$, *i.e.* that $E(U;\Omega) \leq E(W;\Omega)$ for any $\Omega \subset\subset \mathbb{R}^3$ and for any $W \in W^{1,2}_{\mathrm{loc}}(\mathbb{R}^3; \mathcal{U}(m))$ such that $\mathrm{supp}(U - W) \subset\subset \Omega$. Clearly, if $\Psi \in C^\infty_0(\Omega; \mathfrak{u}(m))$ and $\varepsilon \in \mathbb{R}$, then $U_\varepsilon(k) = U(k) \exp \varepsilon \Psi(k)$ is an admissible variation of U, hence local minimality gives

$$\left.\frac{d}{d\varepsilon}\right|_{\varepsilon=0} E(U_\varepsilon;\Omega) = 0\,, \qquad \left.\frac{d^2}{d\varepsilon^2}\right|_{\varepsilon=0} E(U_\varepsilon;\Omega) \geq 0\,. \tag{5.2}$$

Since the tangential variation Ψ can be chosen arbitrarily, the first condition in (5.2) easily implies that U is a weakly harmonic map, *i.e.* U is a weak solution to

$$-\Delta U + \sum_{j=1}^{3} \frac{\partial U}{\partial k_j} U^{-1} \frac{\partial U}{\partial k_j} = 0\,. \tag{5.3}$$

The aim here is to show that, when $m \geq 2$, any local minimizer $U \in W^{1,2}_{\mathrm{loc}}(\mathbb{R}^3; \mathcal{U}(m))$ which is degree-zero homogeneous (a minimizing *tangent map*), *i.e.*

$$U(k) = \omega\left(\frac{k}{|k|}\right) \quad \text{for some } \omega \in C^\infty(S^2; \mathcal{U}(m)),$$

is constant (here smothness away from the origin is a general consequence of weak harmonicity and degree-zero homogeneity). The argument we are going to describe here combines a stability inequality derived from (5.2) (see inequality (5.5) below) and a nontrivial quantization property for the energy of every harmonic map $\omega \in C^\infty(S^2; \mathcal{U}(m))$ proved in [40, Corollary 8].

The first important observation is that, up to a simple lifting trick, one can assume the target to be the special unitary group $\mathcal{SU}(m)$ (since for tangent maps the determinant must be constant). As a consequence, when $m = 2$ this Liouville property is already known and it follows from [38, Proposition 1], since $\mathcal{SU}(2) \equiv S^3(\sqrt{2})$. However, when the target is a sphere, the constants in the stability inequalities are uniformly bounded as the dimension of the target sphere increases (see [38, formula (*)]), which is not the case when the target is the (special) unitary group $\mathcal{SU}(m)$, $m \geq 3$, as we will see below. Thus, in contrast with [38] one has to use a different technique to get the constancy property of tangent maps and unfortunately at present the method works **only in the case** $m \leq 3$. In our opinion, if (5.5) cannot be improved, then it seems difficult to prove the Liouville property for any $m \geq 4$ using the so-called Bochner method as in the sphere-valued case (see [38] and [21]; see also [43, Chapter 5] and references therein) and for the general case some new idea is needed.

Assuming $\Omega \subset \mathbb{R}^3 \setminus \{0\}$, we know that U is a smooth harmonic map and we denote by V the variational vector field with compact support in Ω, *i.e.* $V(k) = \frac{d}{d\varepsilon}\big|_{\varepsilon=0} U_\varepsilon(k) = U_\varepsilon(k)\Psi(k)$ associated to the deformation $U_\varepsilon(k) = U(k)\exp\varepsilon\Psi(k)$ given by $\Psi \in C_0^\infty(\Omega; \mathfrak{u}(m))$. We regard $\mathcal{SU}(m) \subset M_m(\mathbb{C})$ as a Riemannian manifold with the metric induced by the embedding in $M_m(\mathbb{C})$, the latter being equipped with the Hilbert-Schmidt inner product

$\langle A, B\rangle = \operatorname{Re}\operatorname{tr}(A^*B)$. The second variation formula for the energy [20, Chapter 1] yields

$$\frac{d^2}{d\varepsilon^2}\bigg|_{\varepsilon=0} E(U_\varepsilon; \Omega) = \sum_j \int_\Omega \|\widetilde{\nabla}_{e_j} V\|^2 - R(dU(e_j), V, dU(e_j), V) \quad (5.4)$$

where $\{e_j\}$ is an orthonormal basis of $\mathbb{R}^3 \cong T\Omega$, $\widetilde{\nabla}$ is the pull-back via U of the Levi-Civita connection on $T\mathcal{SU}(m)$, R is the curvature (4, 0)-tensor on $T\mathcal{SU}(m)$, and $\|A\| = (\operatorname{tr}(A^*A))^{1/2}$ is the Hilbert-Schmidt norm in $M_m(\mathbb{C})$.

Now we focus on vector fields V which are obtained by orthogonal projection onto $T\mathcal{SU}(m)$ of a constant vector fields on $M_m(\mathbb{C})$ (localized in the domain by a fixed cut-off function $\eta \in C_0^\infty(\Omega)$) and we average over such constant vector fields the corresponding inequality given by (5.4). Similar ideas were already used in [14,38] and *e.g.* in [42]. In [32] a direct explicit computation gives the following stability inequality with a sharp constant

$$\int_{\Omega} \eta^2 \, \|U^{-1} dU\|^2 \le m(m^2-1) \int_{\Omega} |\nabla \eta|^2 \qquad \forall \eta \in C_0^\infty(\Omega)\,. \tag{5.5}$$

Since U is weakly harmonic and degree-zero homogeneous, then ω is also weakly harmonic, *i.e.* it is a critical point of the energy functional

$$\mathcal{E}(\omega) = \frac{1}{2} \int_{S^2} \left\|\omega^{-1} d\omega\right\|^2 d\mathrm{Vol} \tag{5.6}$$

defined on $W^{1,2}(S^2; \mathcal{U}(m))$ (hence ω is C^∞-smooth, see [20, Chapter 3]).

Now we take degree-zero homogeneity of U into account. By localizing the inequality (5.5) on S^2, *i.e.* by taking $\eta(k) = \rho(|k|)\psi(k/|k|)$ with $\rho \in C_0^\infty((0,\infty))$ and $\psi \in C^\infty(S^2)$ and optimizing in ρ (see [38, Lemma 1.3]), one obtains

$$\int_{S^2} |\psi|^2 \frac{1}{2} \left\|\omega^{-1} d\omega\right\|^2 d\mathrm{Vol} \le \frac{1}{2} m(m^2-1) \int_{S^2} \left(|\nabla \psi|^2 + \frac{1}{4}|\psi|^2 \right) d\mathrm{Vol}\,.$$

In view of the definition (5.6), setting $\psi \equiv 1$ in the previous inequality yields

$$\mathcal{E}(\omega) \le \frac{\pi}{2}\, m(m^2-1) \tag{5.7}$$

for any harmonic sphere ω corresponding to a minimizing tangent map U.

In order to conclude, we recall that a very precise description of the space of all the harmonic spheres $\omega \in C^\infty(S^2;\, \mathcal{U}(m))$ was given in [39], proving by algebraic techniques the so-called factorization into unitons. A similar factorization result was obtained in [40], using an energy induction argument.

Proposition 5.1 ([40], Corollary 7′). *Let $\omega : S^2 \to \mathcal{U}(m)$ be a nonconstant harmonic map. Then there exist a natural number $l \ge 1$ and a canonical factorization*

$$\omega = \omega_0 (p_1 - p_1^\perp) \cdots (p_l - p_l^\perp)\,, \quad 8l\pi \le \mathcal{E}(\omega)\,, \tag{5.8}$$

where $\omega_0 \in \mathcal{U}(m)$ and each p_j is the hermitian projection onto a subbundle of $S^2 \times \mathbb{C}^m$ holomorphic w.r.to the complex structure induced by the operator $\bar\partial + \bar\partial p_1 + \ldots \bar\partial p_{j-1}$.

Here we regard each projection p_j as a map in a complex Grassmannian $G_{k,m}(\mathbb{C})$, $k = \text{Rank}\, p_j$, and each factor $p_j - p_j^{\perp}$ in (5.8) as a corresponding map into $\mathcal{U}(m)$, through the isometric embedding $G_{k,m}(\mathbb{C}) \hookrightarrow \mathcal{U}(m)$ (the *Cartan embedding*) defined by assigning to each subspace a unitary operator corresponding the reflection w.r.to the subspace.

The main result in [40] shows that each factor in (5.8) changes the energy by an integer multiple of 8π. The consequence which will be relevant to us is the following quantization property for a critical point of (5.6).

Proposition 5.2 ([40], Corollary 8). *The energy $\mathcal{E}(\omega)$ of any harmonic map $\omega : S^2 \to \mathcal{U}(m)$ is an integer multiple of 8π. In particular, if $\omega \not\equiv const$ then $\mathcal{E}(\omega) \geq 8\pi$.*

Combining Proposition 5.2 and inequality (5.7) we see that $\mathcal{E}(\omega) = 0$, and hence ω is constant (and in turn U is constant) whenever $\frac{\pi}{2}\, m(m^2 - 1) < 8\pi$, *i.e.* for $m = 2$.

The case $m = 3$ requires additional care but the conclusion still holds. Indeed (5.7) implies that $\mathcal{E}(\omega) = 8\pi$, hence (5.8) contains just one factor, and with some extra work $\mathcal{E}(\omega) = 0$. Thus, we have the following result concerning the Liouville property.

Corollary 5.3. *Let $2 \leq m \leq 3$ and let $U(k) = \omega\left(\frac{k}{|k|}\right)$, $U \in W^{1,2}_{\text{loc}}(\mathbb{R}^3; \mathcal{U}(m))$ be a local minimizer of* (5.1). *Then U is constant.*

References

[1] E. I. Blount, *Formalism of Band Theory*, In: "Solid State Physics", Seitz, F., Turnbull, D. (eds.), Vol. 13, Academic Press, 1962, 305–373.

[2] F. Bethuel and X. M. Zheng, *Density of smooth functions between two manifolds in Sobolev spaces*, J. Funct. Anal. **80** (1988), 60–75.

[3] Ch. Brouder, G. Panati, M. Calandra, Ch. Mourougane and N. Marzari, *Exponential localization of Wannier functions in insulators*, Phys. Rev. Lett. **98** (2007), 046402.

[4] S. Campanato, *Proprietà di hölderianità di alcune classi di funzioni*, Ann. Scuola Norm. Sup. Cl. Sci. **17** (1963), 175–188.

[5] E. Cancs, A. Deleurence and M. Lewin, *A new approach to the modeling of local defects in crystals: the reduced Hartree-Fock case*, Commun. Math. Phys. **281** (2008), 129–177.

[6] S. Y. Chang, L. Wang and P. Yang, *Regularity of harmonic maps*, Comm. Pure Appl. Math. **52** (1999), 1099–1111.

[7] J. DES CLOIZEAUX, *Energy bands and projection operators in a crystal: Analytic and asymptotic properties*, Phys. Rev. **135** (1964), A685–A697.

[8] J. DES CLOIZEAUX, *Analytical properties of n-dimensional energy bands and Wannier functions*, Phys. Rev. **135** (1964), A698–A707.

[9] R. COIFMAN, P.-L. LIONS, Y. MEYER and S. SEMMES, *Compensated compactness and Hardy spaces*, J. Math. Pures Appl. **72** (1993), 247–286.

[10] B. A. DUBROVIN and S. P. NOVIKOV, *Ground state of a two-dimensional electron in a periodic magnetic field*, Zh. Eksp. Teor. Fiz. **79** (1980), 1006–1016. Translated in *Sov. Phys. JETP* **52**, Vol. 3 (1980) 511–516.

[11] S. GOEDECKER, *Linear scaling electronic structure methods*, Rev. Mod. Phys. **71** (1999), 1085–1111.

[12] F. HANG and F. H. LIN, *Topology of Sobolev mappings. II*, Acta Math. **191** (2003), 55–107.

[13] B. HELFFER and J. SJÖSTRAND, *Équation de Schrödinger avec champ magnetique et equation de Harper*. In: "Schrödinger operators", Lecture Notes in Physics 345, Springer, Berlin, 1989, 118–197.

[14] R. HOWARD and S. W. WEI, *Nonexistence of stable harmonic maps to and from certain homogeneous spaces and submanifolds of Euclidean space*, Trans. Amer. Math. Soc. **294** (1986), 319–331.

[15] J. JOST, "Riemannian Geometry and Geometric Analysis", Universitext, Springer, Berlin, 1995.

[16] T. KATO, "Perturbation Theory for Linear Operators", Springer, Berlin, 1966.

[17] S. KIEVELSEN, *Wannier functions in one-dimensional disordered systems: application to fractionally charged solitons*, Phys. Rev. B **26** (1982), 4269–4274.

[18] R. D. KING-SMITH and D. VANDERBILT, *Theory of polarization of crystalline solids*, Phys. Rev. B **47** (1993), 1651–1654.

[19] W. KOHN, *Analytic properties of bloch waves and Wannier functions*, Phys. Rev. **115** (1959), 809.

[20] F. H. LIN and C. Y. WANG, "The Analysis of Harmonic Maps and their Heat Flows", World Scientific, 2008.

[21] F. H. LIN and C. Y. WANG, *Stable stationary harmonic maps to spheres*, Acta Math. Sin. (Engl. Ser.) **22** (2006), 319–330.

[22] N. MARZARI and D. VANDERBILT, *Maximally localized generalized Wannier functions for composite energy bands*, Phys. Rev. B **56** (1997), 12847–12865.

[23] N. MARZARI, A. A. MOSTOFI, J. R. YATES, I. SOUZA and D. VANDERBILT, *Maximally localized Wannier functions: Theory and applications*, Rev. Mod. Phys. **84** (2012), 1419.

[24] C. B. MORREY, "Multiple Integrals in the Calculus of Variations" Grund. der math. Wissenschaften 130, Springer, New York, 1966.

[25] R. MOSER, "Partial Regularity for Harmonic Maps and Related Problems", World Scientific, 2005.

[26] Y.-S. LEE, M. B. NARDELLI and N. MARZARI, *Band structure and quantum conductance of nanostructures from maximally localized Wannier functions: the case of functionalized carbon nanotubes*, Phys. Rev. Lett. **95** (2005), 076804.

[27] G. NENCIU, *Existence of the exponentially localised Wannier functions*, Comm. Math. Phys. **91** (1983), 81–85.

[28] G. NENCIU, *Dynamics of band electrons in electric and magnetic fields: Rigorous justification of the effective Hamiltonians*, Rev. Mod. Phys. **63** (1991), 91–127.

[29] S. P. NOVIKOV, *Magnetic Bloch functions and vector bundles. Typical dispersion law and quantum numbers*, Sov. Math. Dokl. **23** (1981), 298–303.

[30] G. PANATI, H. SPOHN and S. TEUFEL, *Effective dynamics for Bloch electrons: Peierls substitution and beyond*, Comm. Math. Phys. **242** (2003), 547–578.

[31] G. PANATI, *Triviality of Bloch and Bloch-Dirac bundles*, Ann. Henri Poincaré **8** (2007), 995–1011.

[32] G. PANATI and A. PISANTE, *Bloch bundles, Marzari-Vanderbilt functional and maximally localized Wannier functions*, Commun. Math. Phys., to appear.

[33] M. REED and B. SIMON, "Methods of Modern Mathematical Physics", Vol. IV: "Analysis of Operators", Academic Press, New York, 1978.

[34] R. RESTA, *Theory of the electric polarization in crystals*, Ferroelectrics **136** (1992), 51–75.

[35] L. SIMON, "Theorems on Regularity and Singularity of Energy Minimizing Maps", Lectures in Mathematics ETH Zürich. Birkhäuser, Basel, 1996.

[36] E. STEIN, "Harmonic Analysis: Real-variable Methods, Orthogonality, Oscillatory Integrals", Princeton Mathematical Series, 43. Princeton University Press, Princeton, 1993.

[37] R. SCHOEN and K. UHLENBECK, *A regularity theory for harmonic maps*, J. Differential Geom. **17** (1982), 307–335.

[38] R. SCHOEN and K. UHLENBECK, *Regularity of minimizing harmonic maps into the sphere*, Invent. Math. **78** (1984), 89–100.

[39] K. Uhlenbeck, *Harmonic maps into Lie groups: classical solutions of the chiral model*, J. Differential Geom. **30** (1989), 1–50.

[40] G. Valli, *On the energy spectrum of harmonic 2-spheres in unitary groups*, Topology **27** (1988), 129–136.

[41] G. H. Wannier, *The structure of electronic excitation levels in insulating crystals*, Phys. Rev. **52** (1937), 191–197.

[42] W. Wei, *Liouville Theorems for stable harmonic maps into either strongly unstable, or δ−pinched manifolds*, Proceedings of Symposia in Pure Mathematic **44** (1986).

[43] Y. L. Xin, "Geometry of Harmonic Maps. Progress in Nonlinear Differential Equations and their Applications", Vol. 23, Birkhäuser, Boston, 1996.

Flows by powers of centro-affine curvature

Alina Stancu

Abstract. We consider compact, strictly convex, origin-symmetric, smooth hypersurfaces in $\mathbb{R}^{n+1}$ shrinking with speed given by powers of their centro-affine curvature. We show that, as long as the support function of the evolving convex bodies is bounded from both sides, the centro-affine curvature is also bounded above and below. We prove that the flow's singularity which appears when the support function goes to zero is a compact contained in a hyperplane of dimension $(n-1)$. This information is exploited in $\mathbb{R}^3$ to show that these flows shrink any admissible surface to a point and that, up to $SL(3)$ transformations, the rescaled images of the evolving surface converge, in the Hausdorff metric, to a ball.

1 Introduction

Let $\mathcal{K}^s_{\text{reg}}$ be the class of origin symmetric, strictly convex bodies in $\mathbb{R}^{n+1}$, $n \geq 2$, which have smooth boundaries with strictly positive Gauss curvature. By the volume of K, denoted $\text{Vol}(K)$, we will understand its Lebesgue measure as a compact convex set in $\mathbb{R}^{n+1}$, while the surface area, or later the affine surface area, will represent the volume of the boundary of K as a convex hypersurface with the appropriate metric. The space of linear transformations of $\mathbb{R}^{n+1}$ with determinant 1 will be denoted by $SL(n+1)$.

The preferred parametrization of ∂K is $X_K : \mathbb{S}^n \to \partial K$ which assigns to each unit vector u, the point on the boundary of K of outer normal u. This leads to a scalar representation of ∂K given by the so-called support function of K namely $h_K : \mathbb{S}^n \to (0, +\infty)$ with $h_K(u) = X(u) \cdot u$, in all unitary directions u, and, in fact, as ∂K is smooth,

$$X_K(u) = h_K(u)u + \bar{\nabla} h_K(u), \tag{1.1}$$

where $\bar{\nabla}$ is the connection on $\mathbb{S}^n$, and $\bar{g}$ the metric, inherited from $\mathbb{R}^{n+1}$. The Gauss curvature $\mathcal{K}$ of ∂K, viewed as a function on $\mathbb{S}^n$, is related to the support function of the convex body by

$$\frac{1}{\mathcal{K}} = \det(\bar{\nabla}^2 h + \bar{g}\, h). \tag{1.2}$$

To introduce the flow by centro-affine curvature, we need to define two $SL(n+1)$-invariants of K. We start first with the centro-affine curvature function of the strictly convex hypersurface ∂K, $K \in \mathcal{K}^s_{\text{reg}}$. This is defined by the formula

$$\mathcal{K}_0(u) = \frac{\mathcal{K}(u)}{(h_K(u))^{n+2}}, \quad \forall u \in \mathbb{S}^n, \tag{1.3}$$

which it is known that measures in a certain way the size of the centered osculating ellipsoid at the point of ∂K with outer normal vector u. (Note that $K \in \mathcal{K}^s_{\text{reg}}$ implies that K contains the origin of $\mathbb{R}^{n+1}$ in its interior.) Ellipsoids centered at the origin, and only them, have constant centro-affine curvature function.

Finally, the centro-affine normal vector field along ∂K is

$$\mathcal{N}_0(u) := -\mathcal{K}_0{}^{-\frac{1}{n+2}}(u) \left(\mathcal{K}^{\frac{1}{n+2}} u + \bar{\nabla} \left(\mathcal{K}^{\frac{1}{n+2}} \right) \right). \tag{1.4}$$

For more details, see [17] and, for some of the more classical framework, see [8].

Let $p \geq 1$ be some fixed real number. In [17], a class of curvature flows was defined on a set of convex bodies $\mathcal{K}_{\text{reg}}$ (including $\mathcal{K}^s_{\text{reg}}$) which, for $p \geq 1$, resumes to the following Cauchy problem

$$\begin{cases} \dfrac{\partial X(u,t)}{\partial t} = \mathcal{K}_0{}^{\frac{p}{n+p+1}}(u,t)\, \mathcal{N}_0(u,t) \\ X(u,0) = X_K(u), \end{cases} \tag{1.5}$$

for any fixed $K \in \mathcal{K}^s_{\text{reg}}$. A solution to the flow is a parametrized family denoted by $\{K_t\}_t$ and we will often identify the convex bodies with the convex hypersurfaces bounding them.

One can summarize the above evolution as the flow of strictly convex hypersurfaces by powers of the centro-affine curvature in the direction of the centro-affine normal. For $p = 1$, this flow is the affine normal flow studied by Sapiro-Tanenbaum in the plane, and by Andrews in all higher dimensions, [1, 14]. Thus the class of p-flows (1.5) introduces infinitely many $SL(n+1)$-invariant flows (flows which commute with the action of $SL(n+1)$ on $\mathbb{R}^{n+1}$) in addition to the previous, classical one. Note that centered ellipsoids, and only them, are homothetic solutions to this flows.

The first applications of this newer class of curvature flows is in within the L_p-Brunn-Minkowski theory of convex bodies introduced by Lutwak [11, 12] (which explains also the exponent of the $\mathcal{K}_0$ in the flow's definition), in connection with the classification of global $SL(n+1)$ invariants for convex bodies [10], isoperimetric type inequalities, partial differential equations, see [6, 17, 18].

The short time existence of solutions to the flow was established in [17] and is due to the strict parabolicity of the Cauchy problem on $\mathcal{K}^s_{\text{reg}}$. In [7], it was shown that, for a certain range of values p, the flow (1.5) shrinks any convex body in $\mathcal{K}^s_{\text{reg}}$ to a point and, if the shrinking bodies are rescaled to enclose constant volume, they converge in the C^∞ norm to ellipsoids, (see also [5] for equivalent results in the plane). More precisely, we proved:

Theorem 1.1 ([7]). *Let $1 \le p < \frac{n+1}{n-1}$ be a real number. Let $X_{K_0} : \mathbb{S}^n \to \mathbb{R}^{n+1}$ be a smooth, strictly convex embedding of $K_0 \in \mathcal{K}^s_{\text{reg}}$. Then there exists a unique solution $X : \mathbb{S}^n \times [0, T) \to \mathbb{R}^{n+1}$ of the flow* (1.5) *with initial data X_{K_0}. The rescaled convex bodies given by the embeddings $\left(\frac{\text{Vol}(\mathbb{S}^n)}{\text{Vol}(K_t)}\right)^{\frac{1}{n+1}} X_{K_t}$ converge in the C^∞ topology to the unit ball modulo $SL(n+1)$.*

In the previous theorem, the range of values for p is due to a technical lemma showing the existence of a lower bound of the Gauss curvature (independent on the initial conditions) as long as K_t stays in an annulus.

In this paper, we aim to improve the range of p for which we conclude the long term existence of the flow and its asymptotic to a ball modulo $SL(n+1)$. We succeed to extend p to $(1, \infty)$ in dimension 3 by using an idea inspired from [9], and by rendering the convergence to the sup-norm. Additionally, we show that in arbitrary dimension, the flow exists up until a time T when the convex bodies $K_t \subset \mathbb{R}^{n+1}$ collapse to a compact in an $(n-1)$-dimensional hyperplane. While it is possible that, even in dimension 4 and higher, the evolving convex bodies K_t shrink to a point for any initial shape in $\mathcal{K}^s_{\text{reg}}$, last step of the present proof works only in the lowest dimension. We will make a comment about a possible extension to higher dimensions in Section 3.

To study the flow (1.5), we pass to the equivalent scalar Cauchy problem

$$\begin{cases} \dfrac{\partial h(u,t)}{\partial t} = - h(u,t)\, \mathcal{K}_0^{\frac{p}{n+1+p}}(u,t) \\ h(u,0) = h_K(u). \end{cases} \tag{1.6}$$

Note also, from here, that the symmetry with respect to the origin of the initial data is preserved during the evolution and that the homothetic solutions are centered ellipsoids and only those.

The paper is structured as follows. In the next section, we will show that, as long as the support function is bounded from below (since the flow shrinks the bodies, the support function is also bounded from above), the Gauss curvature, hence the centro-affine curvature, is bounded from both sides as well, Proposition 2.1 and Proposition 2.3, respectively. We conclude that, if T denotes the maximal time of existence of the flow, the singularity which appears is due to $\lim_{t\to T} h_{\min}(t) = 0$. In section 4, we exploit the $SL(n+1)$-invariance of the flow, to show that, at time T, the evolving convex bodies in isotropic position collapse to a compact contained in an $(n-1)$-hyperplane. In low dimension, this conclusion is sufficiently rigid, so we can show, in the main theorem presented in the last section, that isotropic positions of the evolving surfaces have to collapse to a *circular* point under these flows.

2 Gauss curvature bounds

The following bounds on the Gauss curvature are similar to the ones obtained with Ivaki in [7]. However, here they depend on the initial conditions which is something we avoided in the latter paper for technical reasons mentioned above.

Given a convex body K, the inner radius of K is the radius of the largest ball contained in K; the outer radius of K is the radius of the smallest ball containing K.

Proposition 2.1 (Upper bound on the Gauss curvature). *For any smooth, strictly convex solution* $\{K_t\}_{[0,t_0]}$ *of the evolution equation* (1.6) *whose inner and outer radii* $r_\pm$ *are bounded* $0 < R_- \leq r_-(K_t) \leq r_+(K_t) \leq R_+ < +\infty$ *for* $t \in [0, t_0]$ *and some constants* $R_\pm$, *we have*

$$\mathcal{K}^{\frac{p}{n+p+1}}(t) \leq C_0(n, p, R_-, R_+), \quad \forall t \in [0, t_0],$$

where $C_0(n, p, R_-, R_+)$ *is a positive constant depending on* n, p, R_-, R_+ *and initial data.*

Proof. Let $\alpha := 1 - \frac{(n+2)p}{n+1+p}$, respectively, $\beta := -\frac{p}{n+1+p}$, and consider the function

$$\Psi = \frac{h^\alpha S_n^\beta}{h - R_-/2},$$

where the reciprocal of the Gauss curvature $\frac{1}{\mathcal{K}} =: S_n$ is denoted by the n-th symmetric polynomial in the radii of curvature as a function on the unit sphere $\mathbb{S}^n$. Using the maximum principle, we will show that Ψ is bounded from above by a function of n, p, R_-, R_+ and its initial value. At the point where the maximum of Ψ occurs, we have

$$0 = \bar{\nabla}_i \Psi = \bar{\nabla}_i \left(\frac{h^\alpha S_n^\beta}{h - R_-/2} \right) \quad \text{and} \quad \bar{\nabla}_i \bar{\nabla}_j \Psi \leq 0.$$

Hence, we obtain

$$\frac{\bar{\nabla}_i (h^\alpha S_n^\beta)}{h - R_-/2} = \frac{(h^\alpha S_n^\beta) \bar{\nabla}_i h}{(h - R_-/2)^2}$$

and, consequently,

$$\bar{\nabla}_i \bar{\nabla}_j \left(h^\alpha S_n^\beta\right) + \bar{g}_{ij} \left(h^\alpha S_n^\beta\right) \leq \frac{h^\alpha S_n^\beta \mathfrak{r}_{ij} - R_-/2 h^\alpha S_n^\beta \bar{g}_{ij}}{h - R_-/2}. \tag{2.1}$$

We calculate

$$\begin{aligned} \partial_t \Psi = & -\frac{\beta h^\alpha S_n^{\beta-1}}{h - R_-/2} (\dot{S}_n)_{ij} \left[\bar{\nabla}_i \bar{\nabla}_j \left(h^\alpha S_n^\beta\right) + \bar{g}_{ij} \left(h^\alpha S_n^\beta\right)\right] \\ & + \frac{S_n^\beta \partial_t h^\alpha}{h - R_-/2} + \frac{h^{2\alpha} S_n^{2\beta}}{(h - R_-/2)^2}, \end{aligned}$$

where $(\dot{S}_n)_{ij} := \dfrac{\partial S_n}{\partial \mathfrak{r}_{ij}}$ is the derivative of S_n with respect to the entry $\mathfrak{r}_{ij}$ of the radii of curvature matrix. By [13, Theorem 1, page 102], applied to the top symmetric polynomial S_n, we have that $\dot{S} := \left(\dfrac{\partial S}{\partial \tau_{ij}}\right)_{ij}$ is a positive definite bilinear form as long as ∂K_t has positive Gauss curvature at all points.

Using inequality (2.1), and denoting by $\mathcal{H}$ the mean curvature of the hypersurface ∂K_t, we conclude that, at the point where the maximum of Ψ is reached, we have

$$\partial_t \Psi \leq \Psi^2 \left(-n\beta - \alpha + 1 + \frac{\beta R_-}{2} \mathcal{H} \right). \tag{2.2}$$

By the arithmetic-geometric mean inequality, $\mathcal{H} \geq \dfrac{n}{S_n^{\frac{1}{n}}}$, thus

$$H \geq n \left(\frac{h - R_-/2}{h^\alpha S_n^\beta} \right)^{\frac{1}{n\beta}} \left(\frac{h^\alpha}{h - R_-/2} \right)^{\frac{1}{n\beta}} \geq n \Psi^{-\frac{1}{n\beta}} \left(\frac{R_+^\alpha}{R_+ - R_-/2} \right)^{\frac{1}{n\beta}}.$$

This enables us to rewrite the inequality (2.2) as

$$\begin{aligned}\partial_t \Psi &\leq \Psi^2 \left(-n\beta - \alpha + 1 + \frac{nR_-}{2} \beta \Psi^{-\frac{1}{n\beta}} \left(\frac{R_+^\alpha}{R_+/2} \right)^{\frac{1}{n\beta}} \right) \\ &= -\Psi^2 \left(C'(n, p, R_-, R_+) \Psi^{\frac{n+1+p}{np}} - C(n, p) \right),\end{aligned}$$

for some positive constants $C(n, p)$ and $C'(n, p, R_-, R_+)$. Hence, either

$$\Psi \leq (C(n, p)/C'(n, p, R_-, R_+))^{np/(n+1+p)},$$

or $\Psi_t \leq 0$, thus

$$\Psi \leq \max \{C(n, p, R_-, R_+), \Psi(0)\}$$

and the corresponding claim for $\mathcal{K}$ follows. □

Let K° denote the polar body of K with respect to the origin

$$K^\circ = \{y \in \mathbb{R}^{n+1} \mid x \cdot y \leq 1, \ \forall x \in K\}.$$

The evolution of the polar body is completely determined by the evolution of K under a centro-affine flow, or more generally under any geometric flow, and viceversa. In this respect, we recall:

Lemma 2.2 (The dual p-flow). *[17] Let $\{K_t\}_{[0,T)}$ be a smooth, strictly convex solution of the evolution equation (1.6). Then $\{K_t^\circ\}_{[0,T)}$ is a solution of the following evolution equation, the expanding p-flow (alternatively called the dual p-flow)*

$$\partial_t h^\circ = h^\circ \left(\frac{\mathcal{K}^\circ}{(h^\circ)^{n+2}} \right)^{-\frac{p}{n+1+p}}, \tag{2.3}$$

where h° denotes the support function of K_t°, and $\mathcal{K}^\circ$ the Gauss curvature of its boundary, as functions on $\mathbb{S}^n$.

Proposition 2.3 (Lower bound on the Gauss curvature). *Assume that $\{K_t\}_{[0,t_0]}$ is a smooth, strictly convex solution of equation (1.6) with constants $R_\pm$ bounding above and below its inner and outer radii $r_\pm$: $0 < R_- \leq r_-(K_t) \leq r_+(K_t) \leq R_+ < +\infty$ for $t \in [0, t_0]$. Then*

$$\mathcal{K}^{\frac{p}{n+p}}(t) \geq C_0^\circ(n, p, R_-, R_+), \quad \forall t \in [0, t_0],$$

where $C_0^\circ(n, p, R_-, R_+)$ is a positive constant depending on n, p, R_-, R_+ and initial data.

Proof. Recall the evolution equation (2.3) of the polar body of K_t and define $\alpha := 1 + \frac{(n+2)p}{n+1+p}$, respectively, $\beta := \frac{p}{n+1+p}$. Therefore the equation of the dual flow is $\partial_t h^\circ = (h^\circ)^\alpha S_n^{\circ\beta}$, where $S_n^{\circ\beta}$ is the same n-th symmetric polynomial as before but associated with the polar body. Since $R_- \le r_-(K_t) \le r_+(K_t) \le R_+$, we have

$$\frac{1}{R_+} \le r_-^\circ(K_t^\circ) \le r_+^\circ(K_t^\circ) \le \frac{1}{R_-}$$

for the corresponding inner/outer radii $r_\pm^\circ(K_t^\circ)$ of the polar body K_t°. Define

$$R_-^\circ := \frac{1}{R_+}, \quad R_+^\circ := \frac{1}{R_-}$$

and consider now the function

$$\Phi = \frac{(h^\circ)^\alpha S_n^{\circ\beta}}{2R_+^\circ - h^\circ}.$$

At the point where the minimum of Φ occurs,

$$0 = \bar\nabla_i \Phi = \bar\nabla_i \left(\frac{(h^\circ)^\alpha S_n^{\circ\beta}}{2R_+^\circ - h^\circ} \right) \quad \text{and} \quad \bar\nabla_i \bar\nabla_j \Phi \ge 0,$$

hence, we obtain

$$\frac{\bar\nabla_i((h^\circ)^\alpha S_n^{\circ\beta})}{2R_+^\circ - h^\circ} = -\frac{(h^\circ)^{2\alpha} S_n^{\circ 2\beta} \bar\nabla_i h^\circ}{(2R_+^\circ - h^\circ)^2}$$

and

$$\begin{aligned} \bar\nabla_i \bar\nabla_j &\left((h^\circ)^\alpha S_n^{\circ\beta}\right) + \bar g_{ij} \left((h^\circ)^\alpha S_n^{\circ\beta}\right) \\ &\ge \frac{-(h^\circ)^\alpha S_n^{\circ\beta} \mathfrak{r}_{ij}^\circ + 2R_+^\circ (h^\circ)^\alpha S_n^{\circ\beta} \bar g_{ij}}{2R_+^\circ - h^\circ}, \end{aligned} \tag{2.4}$$

where $\mathfrak{r}_{ij}^\circ := \bar\nabla_i \bar\nabla_j h^\circ + h^\circ \bar g_{ij}$.

Thus

$$\begin{aligned} \partial_t \Phi = {} & \frac{\beta (h^\circ)^\alpha S_n^{\circ\beta - 1}}{2R_+^\circ - h^\circ} (\dot S_n^\circ)_{ij} \left[\bar\nabla_i \bar\nabla_j \left((h^\circ)^\alpha S_n^{\circ\beta}\right) + \bar g_{ij} \left((h^\circ)^\alpha S_n^{\circ\beta}\right) \right] \\ & + \frac{S_n^{\circ\beta} \partial_t (h^\circ)^\alpha}{2R_+^\circ - h^\circ} + \frac{(h^\circ)^{2\alpha} S_n^{\circ 2\beta}}{(2R_+^\circ - h^\circ)^2}. \end{aligned}$$

and, using (2.4), we obtain

$$\partial_t \Phi \geq \Phi^2 \left(1 - n\beta - \alpha + 2\beta R_+^\circ \mathcal{H}^\circ\right). \tag{2.5}$$

We will now bound the mean curvature $\mathcal{H}^\circ$ of the polar from below by a negative power of Φ,

$$\mathcal{H}^\circ \geq n \left(\frac{2R_+^\circ - h^\circ}{(h^\circ)^\alpha S_n^{\circ\beta}}\right)^{\frac{1}{n\beta}} \left(\frac{(h^\circ)^\alpha}{2R_+^\circ - h^\circ}\right)^{\frac{1}{n\beta}} \geq n\Phi^{-\frac{1}{n\beta}} \left(\frac{R_-^{\circ\alpha}}{2R_+^\circ - R_-^\circ}\right)^{\frac{1}{n\beta}}.$$

Consequently, inequality (2.5) can be rewritten as follows

$$\begin{aligned}\partial_t \Phi &\geq \Phi^2 \left(1 - n\beta - \alpha + 2R_+^\circ n\beta \Phi^{-\frac{1}{n\beta}} \left(\frac{R_-^{\circ\alpha}}{2R_+^\circ - R_-^\circ}\right)^{\frac{1}{n\beta}}\right) \\ &= \Phi^2 \left(-C(n,p) + C'(n,p,R_-^\circ, R_+^\circ)\Phi^{-\frac{n+p}{np}}\right),\end{aligned}$$

for some positive constants $C(n,p)$ and $C'(n,p,R_-^\circ,R_+^\circ)$.
Hence

$$\partial_t \left(\frac{1}{\Phi}\right) \leq -C'(n,p,R_-^\circ,R_+^\circ)\left(\frac{1}{\Phi}\right)^{\frac{n+1+p}{np}} + C(n,p),$$

which implies, as in the previous lemma, that

$$\frac{1}{\Phi} \leq \max\left\{\left(\frac{C(n,p)}{C'(n,p,R_-^\circ,R_+^\circ)}\right)^{\frac{np}{n+1+p}}, \frac{1}{\Phi(0)}\right\}$$

or, equivalently, we have a bound from below for Φ.

Therefore, we have bounded from above the curvature $\mathcal{K}^\circ$ of the polar body in terms of n, p, R_-, R_+ and initial data. To complete the proof, we recall the following fact about the reciprocity of the centro-affine curvatures of convex bodies polar/dual to each other: for every $x \in \partial K$, there exists an $x^\circ \in \partial K^\circ$ such that

$$\left(\frac{\mathcal{K}}{h^{n+2}}\right)(x)\left(\frac{\mathcal{K}^\circ}{(h^\circ)^{n+2}}\right)(x^\circ) = 1,$$

where x and x° are related by $\langle x, x^\circ\rangle = 1$, with $\langle\, ,\rangle$ the inner product in $\mathbb{R}^{n+1}$. A proof of this identity can be found in [4]. By the above identity, an upper bound for $\mathcal{K}^\circ$ implies a lower bound for $\mathcal{K}$. Thus, the latter is now bounded from below by constants depending on n, p, R_-, R_+ and data at time zero. □

3 Analysis of the singularity

Consider now the usual 1-homogeneous extension of the support function of a convex body $K \subset \mathbb{R}^{n+1}$ which contains the origin, $h_K : \mathbb{S}^n \to \mathbb{R}_+$, to $H_K : \mathbb{R}^{n+1} \setminus \{0\} \to \mathbb{R}_+$ defined by $H_K(x) = |x|\, h\left(\frac{x}{|x|}\right)$, [15], where $|\,.\,|$ is the Euclidean norm in $\mathbb{R}^{n+1}$.

Lemma 3.1. *With the above notations,*

$$|\nabla H_K(x)| \leq \max_{u \in \mathbb{S}^n} h_K(u), \quad \forall x \in \mathbb{R}^{n+1} \setminus \{0\}. \tag{3.1}$$

Proof. By a direct computation, we can see that

$$|\nabla H_K(x)|^2 = h^2\left(\frac{x}{|x|}\right) + ||\bar{\nabla} h_K\left(\frac{x}{|x|}\right)||^2, \tag{3.2}$$

where $\bar{\nabla}$ and $||\,.\,||$ are the standard gradient, respectively, induced norm on $\mathbb{S}^n$. The right-hand side of (3.2) is the (Euclidean) norm of the position vector of the convex body K encountered earlier, namely $X_K : \mathbb{S}^n \to \mathbb{R}^{n+1}$, $X_K(u) = h_K(u)\, u + \bar{\nabla} h_K(u)$. The length of the position vector of K is maximal when $\bar{\nabla} h_K(u) = \mathbf{0}$, thus it equals the maximum of h_K on the sphere. □

Proposition 3.2. *Let $K \subset \mathbb{R}^{n+1}$ be a convex body with $\mathbf{0} \in \mathrm{Int}(K)$ and support function $h_K : \mathbb{S}^n \to \mathbb{R}_+$. Denote by $h_{\max} := \max_{u \in \mathbb{S}^n} h_K(u)$ and, similarly, by $h_{\min} := \min_{u \in \mathbb{S}^n} h_K(u)$.*

There exists a constant c_n depending only on the dimension n, and there exists a special linear transformation $L \in SL(n+1)$ such that

$$\frac{h_{\min} h_{\max}^n}{S(LK)^{\frac{n+1}{n}}} \geq c_n, \tag{3.3}$$

where $S(LK)$ stands for the surface area of the convex body $LK := L(K)$ (or, equivalently, the volume of the convex hypersurface bounding the body LK with the induced metric).

Proof. Let u_0 be an arbitrary fixed point on $\mathbb{S}^n$. By the convexity of the support function, and the previous lemma, for any $u \in \mathbb{S}^n$, we have $h(u) \leq h(u_0) + h_{\max}|u - u_0|$, where $|\ .\ |$ is the Euclidean distance in $\mathbb{R}^{n+1}$. Let $d\mu_{\mathbb{S}^n} = \sin^{n-1}\theta_1 \sin^{n-2}\theta_2 \ldots \sin\theta_{n-1}\, d\theta_1 ... d\theta_n$, where $\theta_n \in [0, 2\pi)$ and $\theta_i \in [0, \pi]$, $i = 1, \ldots, (n-1)$ are the coordinates are with respect to u_0, and let ω_m denote the volume of $\mathbb{S}^m$ as hypersurface in $\mathbb{R}^{m+1}$.

Then, as the reciprocal of the support function of K is the radial function of the polar K°, we have

$$\begin{aligned}
\mathrm{Vol}(K^\circ) &= \frac{1}{n+1}\int_{\mathbb{S}^n} \frac{1}{h^{n+1}(u)}\, d\mu_{\mathbb{S}^n}(u) \qquad (3.4)\\
&\geq \frac{1}{n+1}\int_{\mathbb{S}^n} \frac{1}{(h(u_0)+h_{\max} \mid u-u_0 \mid)^{n+1}}\, d\mu_{\mathbb{S}^n}(u)\\
&= \frac{1}{n+1}\int_0^{\pi} \frac{\omega_{n-1}\sin^{n-1}\theta}{(h(u_0)+h_{\max} 2\sin(\theta/2))^{n+1}}\, d\theta\\
&\geq \frac{1}{n+1}\int_0^{\pi/2} \frac{\omega_{n-1}\sin^{n-1}\theta}{(h(u_0)+h_{\max}\theta)^{n+1}}\, d\theta\\
&\geq \frac{\omega_{n-1}}{2^{n+1}(n+1)}\int_0^{\pi/3} \frac{\theta^{n-1}}{(h(u_0)+h_{\max}\theta)^{n+1}}\, d\theta\\
&= \frac{\omega_{n-1}}{2^{n+1}(n+1)\, h^n_{\max}}\int_{h(u_0)}^{h(u_0)+h_{\max}\pi/3} \frac{(\phi-h(u_0))^{n-1}}{\phi^{n+1}}\, d\phi\\
&= \frac{\omega_{n-1}}{2^{n+1}(n+1)\, h^n_{\max}}\int_0^{h_{\max}\pi/3} \frac{\psi^{n-1}}{(h(u_0)+\psi)^{n+1}}\, d\psi\\
&\geq \frac{\omega_{n-1}}{2^{n+1}(n+1)\, h^n_{\max}}\int_0^{h(u_0)\,\pi/3} \frac{\psi^{n-1}}{(h(u_0)+\psi)^{n+1}}\, d\psi\\
&= \frac{\omega_{n-1}}{2^{n+1}(n+1)\, h^n_{\max}\, h(u_0)}\int_0^{\pi/3} \frac{\xi^{n-1}}{(1+\xi)^{n+1}}\, d\xi\\
&= \frac{C_n}{h^n_{\max}\, h(u_0)},
\end{aligned}$$

where C_n is a constant depending only on n.

We used that $2\sin(\theta/2) \leq \theta$ on $[0,\pi/2]$ and that $\sin\theta \geq \theta/2$ on $[0,\pi/3]$. In addition, there were three consecutive changes of variables: $\phi = h(u_0)+h_{\max}\theta$, $\psi = \phi - h(u_0)$, and $\xi = \psi/h(u_0)$ respectively.

As u_0 was arbitrary, we conclude that

$$h_{\min} h^n_{\max}\, \mathrm{Vol}(K^\circ) \geq C(n). \qquad (3.5)$$

Recall now the Blaschke-Santaló inequality which states that for any convex body $K \subset \mathbb{R}^{n+1}$ which contains the origin in its interior the following volume product is at most that of the unit ball of $\mathbb{R}^{n+1}$: $\mathrm{Vol}(K)\cdot\mathrm{Vol}(K^\circ) \leq \omega^2_{n+1}$, [15]. Thus, we have

$$\frac{h_{\min} h^n_{\max}\, \omega^2_n}{\mathrm{Vol}(K)} \geq C_n, \qquad (3.6)$$

and, finally, using Ball's reverse isoperimetric inequality, [2], there exists a special linear transformation of $\mathbb{R}^{n+1}$, say L, and a constant C_n' depending only on the dimension n such that

$$\frac{h_{\min} h_{\max}^n \omega_n^2}{C_n' S(LK)^{\frac{n+1}{n}}} \geq C_n, \tag{3.7}$$

from which follows the claim of the proposition. □

Corollary 3.3. *Let $\{K_t\}_t \subset \mathbb{R}^{n+1}$ be a solution in $\mathcal{K}^s_{\text{reg}}$ to the flow (1.6) on the maximal time interval of existence $[0, T)$. Then there exists a sequence $\{L_t\}_t$ of $SL(n+1)$ transformations such that $L_t K_t$ collapses to a compact of dimension $(n-1)$ or lower.*

Let $K \subset \mathbb{R}^{n+1}$ be a convex body with nonzero volume and let $\partial_K = \min\{S(TK) \mid T \in SL(n+1)\}$ (which is always attained), where $S(\bar{K})$ denotes here and thereafter the surface area of the boundary of the convex body $\bar{K}$. By an abusive notation, because there is usually an additional volume normalization included in the definition, we say that K is *in isotropic position* if its surface area $S(K)$ equals ∂_K, [3]. The special transformations L_t of the above corollary are those which transform the convex bodies K_t into their corresponding isotropic position. One should mention that isotropic positions are unique up to orthogonal transformations and that many important problems in asymptotic geometric analysis are related to this notion.

Proof. The maximum of the support function of a convex body K_t evolving by the flow is bounded above by its initial value. If the minimum of the support function goes to zero before the maximum does too, then K_T will not be a point, but, by (3.3), the surface areas $S(L_t K_t)$ are forced to go to zero too. The corollary follows now either directly from the above definition or by using Proposition 3.1 in [3]. □

4 Asymptotic behavior in dimension three

We will use the previous corollary to show that in $\mathbb{R}^3$, $L_t K_t$ converges in the Hausdorff metric to a point. We believe that this method as such does not carry further in higher dimensions and one should use a different functional than surface area to study the singularity at time T. It is likely that one must consider an entropy functional in the sense encountered very often in the area of curvature flows.

Theorem 4.1. *Let $\{K_t\}_t \subset \mathbb{R}^3$ be a solution in $\mathcal{K}^s_{\text{reg}}$ to the flow* (1.6) *on the maximal time of existence* $[0, T)$. *Then there exists a sequence $\{L_t\}_t$ of $SL(n+1)$ transformations such that $L_t K_t$ shrinks to a* circular *point as t goes to T.*

Proof. There are two claims to prove. First, we need to show that $L_t K_t$ collapses to a point. To conclude, we need proof that $L_t K_t$ *approaches* a disk. We will explain the latter claim in detail in due time.

Suppose, for now, that $L_t K_t$ collapses to a segment of length $l > 0$ at time T. Let $\epsilon > 0$ be arbitrary. We may assume without any loss of generality that the segment is oriented along the x-axis or, equivalently, the $e_1 = (1, 0, 0)$ vector. Let $\{e_1, e_2, e_3\}$ be the standard orthonormal basis of $\mathbb{R}^3$ and, for each time $t \in [0, T)$, let C_t be the solid convex, centrally symmetric parallelepiped with outer normals to the faces $\pm e_i$ $(i = 1, 2, 3)$ such that, with the metric on the space of convex bodies given by the volume of the symmetric difference, $\Delta_s(L_t K_t, C_t) = \epsilon \operatorname{Vol}(K_t)^2$. Clearly, as $t \to T$, C_t converges to the segment $L_T K_T$ in both the symmetric difference metric and in the Hausdorff metric (see [16]). Since $\{C_t\}_t$ approach the segment of non-zero length, we know that, at least, a sequence of these parallelepipeds are not cubes.

Denote, for simplicity, by $\tilde{K}_t := L_t K_t$. For each t, we will now apply a special linear transformation $\tilde{L}_t$ which transforms C_t into a cube of same volume. In fact, $\Delta_s(\tilde{L}_t \tilde{K}_t, \tilde{L}_t C_t) = \Delta_s(L_t K_t, C_t) = \epsilon \operatorname{Vol}(K_t)^2$ and, as $t \to T$, we have that $\tilde{L}_t C_t$ converges to a point, in both metrics on the space of convex bodies mentioned in the previous paragraph. Note that $\tilde{L}_t C_t$ is the isotropic position of C_t, [2, Theorem 2].

Thus, for $\epsilon > 0$ sufficiently small, there exist $t \in [0, T)$ and $\delta > 0$ such that

$$\begin{aligned} |S(\tilde{K}_t) - S(C_t)| &< \delta \\ |S(\tilde{L}_t \tilde{K}_t) - S(\tilde{C}_t)| &< \delta \\ S(C_t) - S(\tilde{C}_t) &> 2\delta. \end{aligned} \tag{4.1}$$

Then

$$S(\tilde{K}_t) > S(C_t) - \delta > S(\tilde{C}_t) + \delta \quad \Rightarrow \quad S(\tilde{L}_t \tilde{K}_t) < S(\tilde{C}_t) + \delta < S(\tilde{K}_t)$$

contradicting that $\tilde{K}_t$ is in isotropic position which is known to be unique up to orthogonal transformations, [3]. This concludes the first claim that the collapsing of $L_t K_t$ is to a point and not a segment of positive length.

For the second claim, consider the evolution of the volume product functional which is known to strictly increase under the flow (1.6) unless

K_t is a centered ellipsoid in which case it will stay constant, see [17]:

$$\frac{d}{dt}\left(\ln(\mathrm{Vol}(K_t)\cdot \mathrm{Vol}(K_t)^\circ)\right) = -\frac{\Omega_p(K_t)}{\mathrm{Vol}(K_t)} + \frac{\Omega_{-((n+1)^2+2p(n+1))/p}(K_t)}{\mathrm{Vol}(K_t^\circ)}. \tag{4.2}$$

The quantity $\Omega_q(K_t) = \int_{\partial K_t} \mathcal{K}_0^{\frac{q}{n+1+q}}\, d\mu_{K_t}$, where $d\mu_{K_t} = h\, dS_{K_t}$, $S(K_t) = \int_{\partial K} dS_{K_t}$, is called the q-affine surface area and generalizes the usual affine surface area obtained for $q = 1$ (ie. the volume of ∂K_t as a hypersurface in $\mathbb{R}^{n+1}$ with respect to the Blaschke metric) in within the extended Brunn-Minkowski theory of convex bodies due to Lutwak, see [12]. Bringing in the generalized affine surface areas is another way to see that the evolution of the volume product is invariant under $SL(n+1)$ transformations. In particular, this implies that the volume product functional is strictly increasing under the flow unless $L_t K_t$ is a disk centered at the origin at some time t_0, after which all later $L_t K_t$ are smaller homothetic balls.

The volume product is scaling invariant. Thus, the Blaschke-Santaló inequality implies that $\tilde{K}_t$, normalized as to enclose constant positive volume, must approach a ball with same volume. Indeed, if not true, we will rewrite (4.2) as

$$\begin{aligned}\frac{d}{dt}&\left(\ln(\mathrm{Vol}(K_t)\cdot \mathrm{Vol}(K_t)^\circ)\right)\\ &= -\frac{\Omega_p(K_t)}{\mathrm{Vol}(K_t)}\left[1 - \frac{\Omega_{-((n+1)^2+2p(n+1))/p}(K_t)}{\mathrm{Vol}(K_t^\circ)}\,\frac{\mathrm{Vol}(K_t)}{\Omega_p(K_t)}\right],\end{aligned} \tag{4.3}$$

and, as the volume product increases, we infer that

$$\limsup_{t\to T} \frac{d}{dt}\left(\ln(\mathrm{Vol}(K_t)\cdot \mathrm{Vol}(K_t)^\circ)\right) > 0.$$

Thus, as $\frac{d}{dt}\mathrm{Vol}(K_t) = -\Omega_p(K_t)$, we have

$$\frac{d}{dt}\left(\ln(\mathrm{Vol}(K_t)\cdot \mathrm{Vol}(K_t)^\circ)\right) \geq -\varepsilon \frac{d}{dt}\ln(\mathrm{Vol}(K_t)),$$

for some positive ε and, for $0 \leq t_1 \leq t < T$,

$$\begin{aligned}\ln(\mathrm{Vol}(K_t)\cdot \mathrm{Vol}(K_t)^\circ) \geq{}& \ln(\mathrm{Vol}(K_{t_1})\cdot \mathrm{Vol}(K_{t_1})^\circ)\\ &+ \varepsilon \ln(\mathrm{Vol}(K_{t_1})) - \varepsilon \ln(\mathrm{Vol}(K_t)).\end{aligned}$$

The right hand side goes to infinity as $\mathrm{Vol}(K_t)$ goes to zero. This contradicts the Blaschke-Santaló inequality which gives an upper bound on the volume product.

Therefore, for any sequence of times converging to T, the convex bodies $\{K_t\}_t$ shrink to the origin approaching the shape of an ellipsoid. Hence, in isotropic position, the bodies $\left\{\frac{1}{\sqrt[3]{\mathrm{Vol}(K_t)}}\tilde{K}_t\right\}_t$, will approach, in the Hausdorff metric, a ball. □

Remark. The previous argument can be adapted to work for the normalization of the flow (1.6) in which one could use the standard definition of a body in isotropic position. In the normalized flow, the shapes K_t are rescaled to enclose constant volume $\mathrm{Vol}(K) = \mathrm{Vol}(K_0)$, which we can impose from the start to be 1, or the volume of $\mathbb{S}^n$. To normalize the flow (1.6) in $\mathbb{R}^3$, one changes the spatial direction by a factor of $\mathrm{Vol}(K_t)^{-1/3}$ and, by a change of variables, pushes time to infinity via $\tau = \int_0^t \left(\frac{1}{\mathrm{Vol}(K_\xi)}\right)^{\frac{2p}{p+3}} d\xi$. Hence, the *normalized* support function $\bar{h}(\tau) = \frac{h(t(\tau))}{\sqrt[3]{\mathrm{Vol}(K_{t(\tau)})}}$ satisfies the normalized flow equation:

$$\begin{cases} \dfrac{\partial \bar{h}(u,\tau)}{\partial \tau} = -\,\bar{h}(u,\tau)\,\bar{\mathcal{K}}_0^{\frac{p}{3+p}}(u,\tau) + \bar{h}(u,\tau)\dfrac{\Omega_p(\bar{K}_\tau)}{3\mathrm{Vol}(K)} \\ \bar{h}(u,0) = h_K(u). \end{cases} \tag{4.4}$$

□

References

[1] B. ANDREWS, *Contraction of convex hypersurfaces by their affine normal*, J. Diff. Geom. **43** (1996), 207–230.

[2] K. BALL, *Volume ratios and a reverse isoperimetric inequality*, J. London Math. Soc. (2) **44** (1991), 351–359.

[3] A. GIANNOPOULOS and M. PAPADIMITRAKIS, *Isotropic surface area measures*, Mathematika **46** (1999), 1–13.

[4] D. HUG, *Curvature Relations and Affine Surface Area for a General Convex Body and its Polar*, Results in Math. **29** (1996), 233–248.

[5] M. N. IVAKI, *Centro-affine curvature flows on centrally symmetric convex curves*, Trans. Amer. Math. Soc., to appear.

[6] M. N. IVAKI, *A flow approach to the L_{-2} Minkowski problem*, Adv. in Appl. Math. **50** (2013), 445–464.

[7] M. N. IVAKI and A. STANCU, *Volume preserving centro affine normal flows*, Comm. Anal. Geom. **21** (2013), 1–15.

[8] K. LEICHTWEISS, "Affine Geometry of Convex bodies", Johann Ambrosius Barth Verlag, Heidelberg, 1998.

[9] J. LU and X.-J. WANG, *Rotationally symmetric solutions to the* L_p*-Minkowski problem*, J. Differential Equations **254** (2013), 983–1005.

[10] M. LUDWIG and M. REITZNER, *A classification of* $SL(n)$ *invariant valuations*, Annals of Math. **172** (2010), 1223–1271.

[11] E. LUTWAK, *The Brunn-Minkowski-Firey theory. I: Mixed volumes and the Minkowski problem*, J. Differential Geom. **38** (1993), 131–150.

[12] E. LUTWAK, *The Brunn-Minkowski-Firey theory II: Affine and geominimal surface areas*, Adv. in Math. **118** (1996), 244–294.

[13] D. S. MITRINOVIĆ, "Analytic Inequalities", Springer-Verlag, Berlin-Heidelberg-New York, 1970.

[14] G. SAPIRO and A. TANNENBAUM, *On affine plane curve evolution*, J. Funct. Anal. **119** (1994), 79–120.

[15] R. SCHNEIDER, "Convex Bodies: The Brunn-Minkowski Theory", Cambridge Univ. Press, New York, 1993.

[16] G. C. SHEPHARD and R. J. WEBSTER, *Metrics for sets of convex bodies*, Mathematika **12** (1965), 73–88.

[17] A. STANCU, *Centro-Affine Invariants for Smooth Convex Bodies*, Int. Math. Res. Notices - IMRN (2012), 2289–2320.

[18] A. STANCU, *Some affine invariants revisited*, 2012, Asymptotic geometric analysis. Proceedings of the Fall 2010 Fields Institute Thematic Program, M. Ludwig *et al.* (eds.), Springer, 2013, 341–357.

CRM Series
Publications by the Ennio De Giorgi Mathematical Research Center Pisa

The Ennio De Giorgi Mathematical Research Center in Pisa, Italy, was established in 2001 and organizes research periods focusing on specific fields of current interest, including pure mathematics as well as applications in the natural and social sciences like physics, biology, finance and economics. The CRM series publishes volumes originating from these research periods, thus advancing particular areas of mathematics and their application to problems in the industrial and technological arena.

Published volumes

1. Matematica, cultura e società 2004 (2005). ISBN 88-7642-158-0
2. Matematica, cultura e società 2005 (2006). ISBN 88-7642-188-2
3. M. GIAQUINTA, D. MUCCI, *Maps into Manifolds and Currents: Area and $W^{1,2}$-, $W^{1/2}$-, BV-Energies*, 2006. ISBN 88-7642-200-5
4. U. ZANNIER (editor), *Diophantine Geometry*. Proceedings, 2005 (2007). ISBN 978-88-7642-206-5
5. G. MÉTIVIER, *Para-Differential Calculus and Applications to the Cauchy Problem for Nonlinear Systems*, 2008. ISBN 978-88-7642-329-1
6. F. GUERRA, N. ROBOTTI, *Ettore Majorana. Aspects of his Scientific and Academic Activity*, 2008. ISBN 978-88-7642-331-4
7. Y. CENSOR, M. JIANG, A. K. LOUISR (editors), *Mathematical Methods in Biomedical Imaging and Intensity-Modulated Radiation Therapy (IMRT)*, 2008. ISBN 978-88-7642-314-7
8. M. ERICSSON, S. MONTANGERO (editors), *Quantum Information and Many Body Quantum systems*. Proceedings, 2007 (2008). ISBN 978-88-7642-307-9
9. M. NOVAGA, G. ORLANDI (editors), *Singularities in Nonlinear Evolution Phenomena and Applications*. Proceedings, 2008 (2009). ISBN 978-88-7642-343-7

– Matematica, cultura e società 2006 (2009). ISBN 88-7642-315-4

10. H. HOSNI, F. MONTAGNA (editors), *Probability, Uncertainty and Rationality*, 2010. ISBN 978-88-7642-347-5
11. L. AMBROSIO (editor), *Optimal Transportation, Geometry and Functional Inequalities*, 2010. ISBN 978-88-7642-373-4

12*. O. COSTIN, F. FAUVET, F. MENOUS, D. SAUZIN (editors), *Asymptotics in Dynamics, Geometry and PDEs; Generalized Borel Summation*, vol. I, 2011. ISBN 978-88-7642-374-1, e-ISBN 978-88-7642-379-6

12**. O. COSTIN, F. FAUVET, F. MENOUS, D. SAUZIN (editors), *Asymptotics in Dynamics, Geometry and PDEs; Generalized Borel Summation*, vol. II, 2011. ISBN 978-88-7642-376-5, e-ISBN 978-88-7642-377-2

13. G. MINGIONE (editor), *Topics in Modern Regularity Theory*, 2011. ISBN 978-88-7642-426-7, e-ISBN 978-88-7642-427-4

– Matematica, cultura e società 2007-2008 (2012). ISBN 978-88-7642-382-6

14. A. BJORNER, F. COHEN, C. DE CONCINI, C. PROCESI, M. SALVETTI (editors), *Configuration Spaces*, Geometry, Combinatorics and Topology, 2012. ISBN 978-88-7642-430-4, e-ISBN 978-88-7642-431-1

15 A. CHAMBOLLE, M. NOVAGA E. VALDINOCI (editors), *Geometric Partial Differential Equations*, 2013. ISBN 978-88-7642-472-4, e-ISBN 978-88-7642-473-1

Volumes published earlier

Dynamical Systems. Proceedings, 2002 (2003)

Part I: *Hamiltonian Systems and Celestial Mechanics*. ISBN 978-88-7642-259-1

Part II: *Topological, Geometrical and Ergodic Properties of Dynamics*. ISBN 978-88-7642-260-1

Matematica, cultura e società 2003 (2004). ISBN 88-7642-129-7

Ricordando Franco Conti, 2004. ISBN 88-7642-137-8

N.V. KRYLOV, *Probabilistic Methods of Investigating Interior Smoothness of Harmonic Functions Associated with Degenerate Elliptic Operators*, 2004. ISBN 978-88-7642-261-1

Phase Space Analysis of Partial Differential Equations. Proceedings, vol. I, 2004 (2005). ISBN 978-88-7642-263-1

Phase Space Analysis of Partial Differential Equations. Proceedings, vol. II, 2004 (2005). ISBN 978-88-7642-263-1

Fotocomposizione "CompoMat" Loc. Braccone, 02040 Configni (RI) Italy
Finito di stampare nel mese di luglio 2013
dalla CSR, Via di Pietralata 157, 00158 Roma